BIOGAS SYSTEMS

Principles and Applications

K.M. MITAL

PUBLISHING FOR ONE WORLD

NEW AGE INTERNATIONAL (P) LIMITED, PUBLISHERS
New Delhi • Bangalore • Chennai • Cochin • Guwahati • Hyderabad
Jalandhar • Kolkata • Lucknow • Mumbai • Ranchi
Visit us at **www.newagepublishers.com**

Published by New Age International (P) Ltd., Publishers
First Edition: 1996
Reprint: 2009

Branches:

- 36, Malikarjuna Temple Street, Opp. ICWA, Basavanagudi, **Bangalore**. ✆ (080) 26677815
- 26, Damodaran Street, T. Nagar, **Chennai**. ✆ (044) 24353401
- CE–39/1016, Carrier Station Road, Near Paulson Park Hotel, Ernakulam South, **Cochin** ✆ (0484) 4051303
- Hemsen Complex, Mohd. Shah Road, Paltan Bazar, Near Starline Hotel, **Guwahati**. ✆ (0361) 2543669
- No. 105, 1st Floor, Madhiray Kaveri Tower, 3-2-19, Azam Jahi Road, Nimboliadda, **Hyderabad**. ✆ (040) 24652456
- RDB Chambers (Formerly Lotus Cinema) 106A, Ist Floor, S.N. Banerjee Road, **Kolkata**. ✆ (033) 22275247
- 16-A, Jopling Road, **Lucknow**. ✆ (0522) 2209578
- 142C, Victor House, Ground Floor, N.M. Joshi Marg, Lower Parel, **Mumbai**. ✆ (022) 24927869
- 22, Golden House, Daryaganj, **New Delhi**. ✆ (011) 23262370, 23262368

ISBN (10): 81-224-0947-4
ISBN (13): 978-81-224-0947- 5

Rs. 650.00

C-09-08-3942

Printed in India at Ramprintograph, Delhi.
Typeset at DTP System.

PUBLISHING FOR ONE WORLD
NEW AGE INTERNATIONAL (P) LIMITED, PUBLISHERS
4835/24, Ansari Road, Daryaganj, New Delhi-110002
Visit us at **www.newagepublishers.com**

Dedicated to the fond and loving memory of my illustrious father late Mr. Dwarka Prasad Mital, a leading advocate from Muzaffarnagar, a veteran freedom fighter, Congress member of the U.P. Legislative Assembly (1952-1962) and Chairman, Committee on Delegated Legislation (1952-58) who throughout his life worked for the welfare of poor and economically-deprived in true Gandhian spirit and upheld high moral values in legal and public-life.

PREFACE

Energy-deficient age in which we live today demands that new and renewable sources of energy should be fully exploited. Biologically produced energy has been identified as attractive alternative to increasingly scarce fossil-fuel supplies. Anaerobic fermentation of organic wastes leading to biogas generation is one such process which has considerable potential to supplement energy supplies at least as an interim measure. Anaerobic digestion is the treatment of animal wastes slurry, food waste and other biomass in oxygenless environment to produce valuable by-products, usable energy in the form of methane gas, organic fertiliser and protein. A biogas plant is a decentralised energy system which can lead to self-sufficiency in household heat and power needs, and at the same time reduce environmental pollution. Matching with the growing need for the development of biofuels in general and biogas technology in particular following the energy crisis that afflicted the world from 1973 onwards, growing application of scientific and engineering principles has transformed the latter into a full-fledged discipline. This book is written with a view to consolidate major developments in this field which took place in the course of last five decades. However, developments have been so rapid and vast in this field that comprehensiveness of coverage can at best be claimed only in bibliographic sense.

The book is designed to meet requirement of all those who are working for advancement of biogas technology. The book may appeal both to technical and non-technical persons alike interested in general field of energy. It is expected to be of use to energy planners, researchers, government officials, field workers, teachers, students and plant-owners. The book includes a total of thirteen chapters each of which is reasonably delinked from one another and thus appropriate for independent reading. However, while reading a chapter it is advisable that reader be aware of broad contents of the other chapters. First chapter provides detailed overview of biogas technology. Second chapter deals with microbiology and biochemistry of anaerobic fermentation. Properties of biogas and methods for its purification are given in chapter three. A compendium of biogas plant designs commonly used world over is given in chapter four. Design, construction, operation and maintenance of biogas plants is covered in chapter five. Review of factors affecting biogas yield is given in chapter six. Comparative biogas yield from different organic wastes are given in chapters seven to ten. Applications of biogas and utilisation of biogas plant effluent as enriched fertiliser are given in chapters eleven and twelve, respectively. The concluding chapter thirteen deals with identification of areas for R&D efforts and aspects relating to promotion and implementation of biogas programme. While some highly technical terminology might have inadvertently intruded here and there, it has on the whole been endeavoured to keep the language of the book easily comprehensible for a general reader.

The book is deliberately selective and subjective in character reflecting author's prioritisation of topics covered. The author has drawn a number of sources all of which may not be easily accessible. Each chapter was written on the basis of information from select sources in keeping with priority accorded. Developments in India have been more comprehensively covered though national boundaries had no bearing in general. This was both inevitable and desirable, inevitable following the author's greater familiarity with developments in this country and desirable because biogas technology forms part of select group of disciplines which have been developed and utilised to a far greater degree in countries like India, China, Nepal and Philippines than western world.

Biogas is not an exact fuel, its composition and characteristics vary from source to source. Variation in feedstock composition from place to place poses problems in standardising biogas yield. For example, in case of agricultural crop residues, composition depends on type of plant, soil conditions, fertiliser used and in case of animal faeces it depends on animal breed, grazing conditions, feed consumed and state of animal health. In view of inconsistencies, yield values and their system of measurement have been given as reported in the original work without changing to one common uniform system throughout the book. A set of conversion factors are given in the end to help the reader in changing one unit to another.

The author is grateful to Dr. R.S. Venkataraman, Chairman and Managing Director, Engineers India Limited, New Delhi, for providing a progressive and learning environment in the organisation which is conducive for innovation and development. The author would like to convey his sincere gratitude to Mr. B.N. Jha, Director (Personnel), EIL, for his moral encouragement and support in writing this book. A very sincere gratitude is conveyed to Mr. R. Srinivasan, Ex-Member (Personnel), ONGC, for his guidance and support; and Dr. S.P. Agarwal, a former ILO expert and Gita Scholar who has been a path-finder and source of inspiration all along.

Grateful thanks are is conveyed to Mr. Anil Kumar, IAS, Vice-Chairman, Delhi Development Authority; Mr. Vishnu Bhagwan, IAS, Financial Commissioner and Secretary (Irrigation and Power), Haryana Government; Mr. V.K. Sardana, Chief Executive Officer, Aga Khan Foundation; and Mr. A.K. Gupta, Executive Director, Railway Board, New Delhi for their invaluable suggestions.

The author would also like to thank Dr. Prem Vrat, Professor of Industrial Engineering and Dean, Post-Graduate Studies & Research, IIT, Delhi; Dr. Nishkam S. Agarwal, an eminent economist from U.S. Environment Protection Agency (Washington); Dr. A.K. Jain, Reader and Head, Department of Botany, S.D. College, Muzaffarnagar; and Dr. Sushil, Professor, Department of Management Studies, IIT, Delhi for their valuable ideas in improving the manuscript. Gratitude is also conveyed to Mrs. Twinkle Gupta, entrepreneur and social activist (Secunderabad); Mrs. Poonam Goel, Sub-Editor, The Hindustan Times; and Mr. Deepak Garg, Senior Human Resource Officer, EIL, New Delhi for varied assistance and cooperation.

Gratitude is conveyed to concerned staff of Tata Energy Research Institute (TERI) for making their library and documentation facilities available which proved to be very fruitful. In particular, grateful thanks are conveyed to Mr. Jitendra Bakshi, Information Analyst, TERI, New Delhi for his varied support. Gratitude is also conveyed to Ms Meena Bilgi, Programme Executive, Aga Khan Rural Support Programme (Ahmedabad) for fruitful interaction and cooperation. Very hearty and grateful thanks are conveyed to Mr. Suresh Chandwani of EIL for his varied help and cooperation.

Words fail the author to express gratitude to his mother Mrs Sarswati Mital and father-in-law Mr. T.P. Gupta for their moral and financial support. Gratitude is also conveyed to Mr Man Mohan Mital, author's elder brother and a leading advocate from Muzaffarnagar, for his all-round help and and support. A very hearty gratitude is conveyed to my brother-in-law and family veteran, late Lt. Col. I.C. Gupta who inspired and kept my spirits high all throughout the period of book-writing and was looking forward for its release but suddenly died on March 26, 1996.

In a book-writing project the family of an author has to invariably bear the burden. A book often - indeed always - interferes with family-life. Throughout my career, my wife, Renu, has amazed me with her love and cooperation. My son, Vivek, had to subsist on less than the rightful share of my attention during the long and arduous period the book-writing was in progress. I wonder if I would ever be able to make up their lost weekends and evenings.

New Delhi K.M. Mital

August 15, 1996

CONTENTS

Chapter 1

AN OVERVIEW OF BIOGAS TECHNOLOGY

1.1 Introduction

Biogas can be produced by fermenting organic materials in absence of air (or oxygen) with the help of bacteria (micro-organisms) to break down materials to intermediates such as alcohols and fatty acids and finally to methane, carbon dioxide and water. This process is called *anaerobic fermentation* and was known to exist from quite long time back. Biogas has also been known as the swamp gas, sewer gas, fuel gas, marsh gas, wet gas, and in India more commonly as '*gobar*' gas. Natural gas is also produced by the action of anaerobic bacteria on plants that grew thousands of years ago. Biogas and natural gas are therefore very much akin to one another. The main fuel component of both is methane gas. However, over the years, pressure and temperature of underground rocks have converted part of the methane in natural gas to other gases such as ethane, propane, butane and the condensate. In contrast, biogas is produced in a digester by anaerobic fermentation. A period of 15 days or so enables anaerobic bacteria to convert organic matter to biogas which however is too short for the conversion of methane to other gases like ethane, propane, butane [35]. The enormous potential of the smallest living organisms such as bacteria, yeasts and fungi to transform organic wastes into valuable source of fuel and enriched fertiliser through a simple process of anaerobic fermentation has been widely recognised. Anaerobic fermentation is a simple and low cost process which can be economically carried out in rural areas where organic wastes are generated aplenty which otherwise pollute environment and pose health hazards [15]. Animal and human wastes are excellent feedstock for biomethanation which are available plentifully all over. Wastes in large quantities on renewable basis are also available from agricultural crops and residues, fruit and vegetable plants and municipal refuse. The potential for generating gaseous fuel and enriched fertiliser through biomethanation is enormous which can bring economic and environmental gain to a vast population [12].

Anaerobic fermentation as described earlier is a biological process which takes place in the absence of air. A number of stages are involved in this process; initially organic material is hydrolysed by enzymes into simple sugars, alcohol peptides and amino acids. These are then converted to volatile fatty acids, hydrogen, carbon dioxide, water and some amount of methane. Methane forming bacteria then convert the fatty acids into methane, carbon dioxide, and water. The process is temperature-dependent and slows down considerably below 30°C , the optimum being 35-38°C which is known as the *mesophyllic* range. Above the optimum temperature, the process again slows down between 40 and 45°C and then increases to a peak between 55 and 60°C, known as the *thermophyllic* range. The pH of digester contents also affects for fermentation as methane bacteria is less active below pH of 6 while the ideal pH is 7. Fermentation breaks down wastes at 35°C under controlled conditions that produces a mixture of methane and carbon dioxide in proportion of about 70 to 30 per cent. The biogas produced can be used directly for heating purposes or used in an engine-driven generator to generate electricity [21]. Anaerobic fermentation produces biogas on renewable basis, eliminates foul smell and reduces harmful bacteria of organic wastes, improves nitrogen and phosphate content of resulting sludge to yield highly enriched fertiliser, and reduces biological oxygen demand (BOD) and chemical oxygen demand (COD).

1.2 Historical Background of Biomethanation

Perhaps the first person to observe the phenomenon of biomethanation was Alessandro Volta of Italy. Way back in 1776 he wrote to a friend that 'combustible air' was being produced continuously in lakes and ponds in the vicinity of Como in northern Italy. Volta observed that whenever he disturbed the sediments of the lake, bubbles of gas rose to the surface. He also noticed that if the sediment contained more plant material, more bubbles came up. In 1806 William Henery showed that Volta's gas was identical with methane gas. Humphrey Davy in the early 1800's observed that methane was present in farmyard manure piles [11]. Davy conducted the first laboratory experiment to produce methane by anaerobic fermentation of wastes. In earlier period anaerobic fermentation was carried out mainly as a municipal waste treatment process and energy recovery was not the primary concern. In 1895, biogas from a waste treatment plant in Exter in England was collected and used to light nearby streets [21]. Interest in biogas as a fuel received fillip during second World War. French scientists took particular interest in advancing biogas technology in forties and installed large number of plants in French colonies in Africa. During this period fuel-starved French and Germans used biogas as a fuel for vehicles and farm tractors. Following the war, several nations such as England, USA, Canada, Russia, Japan, China, Kenya, Uganda, South Africa, New Zealand and India showed interest in biomethanation but later it waned following cheap availability of fossil fuels for over next three decades. However, series of energy shocks which rocked the world from 1973 onwards along with concern for environment protection revived interest in biomethanation. In view of its vast potential, large number of community and family-size plants have been set up in recent years in countries like China, India, Philippines and Nepal [29].

1.3 An Overview of Biomethanation in India

A number of estimates have been made from time to time about potential of biogas generation in India. With livestock population of over 300 million, India can set up over 30 million individual biogas plants and 560,000 community plants at a simple norm of minimum one in each village of the country. With a modest norm of 10 biogas plants per village, it is feasible to set up some 5.5 million plants in country's 0.56 million villages. If this becomes reality, it would provide at 1981-82 price level fuel equivalent to 5341 million litres of kerosene valued at Rs. 5484 million and conserve organic manure to the extent of 89 million tonnes valued at Rs. 4464 million otherwise burnt as dung cake [32, 33]. Effective utilisation of organic wastes can yield gas equivalent of 285.5 million tonnes of coal, almost equal to the amount of energy consumed in the country during 1978-79 [20]. A small biogas plant of 4 cubic meter gas per day capacity set up at a cost of Rs. 6000 including Rs. 2200 availed as Central Government subsidy, operating for 300 days a years can save 4.16 tonnes of fuelwood equivalent to 744 litres of kerosene, and produces manure worth Rs. 1656 per year at the then subsidised market price of fertilisers [4, 7].

As per some broad estimates, over 30 per cent of 980 million tonnes of cattle dung generated annually on renewable basis is directly burnt as fuel without harnessing its full potential as gas and rich fertiliser following fermentation [9]. Assessed in quantitative terms, it is equivalent to a third of India's chemical fertiliser used. If the entire quantity of 980 million tonnes of cattle dung is first used for biomethanation, it can meet domestic fuel needs of 437 million people, raise manure production by 350 million tonnes and substantially improve hygienic conditions in rural areas [10]. If the entire quantity of dung produced is available for biomethanation, it can produce about 6375×10^7 cubic metre of gas having 3825×10^7 cubic metre of methane with net heat value as 312×10^{12} kcal [31]. In addition, resulting plant effluent can generate one million tonnes of nitrogen, one million tonnes of phosphate, and half million tonnes of potash that can be used as fertiliser.

Based on several designs initially developed by scientists like N.V. Joshi, S.V. Desai, S.C. Biswas, Y.N. Kotwal and C.N. Acharya, the noted agricultural scientist J.J. Patel while at the Agricultural College in Pune developed a biogas plant which later on came to be known as the '*Gramlakshmi*' Gas Plant. Distinctive features of this plant covered integration of a gasholder and digester in one unit, underground positioning of the gasholder, daily feeding of slurry, facility of automatic overflow of the digested slurry, and gasholder rotation to break scum formation. Based on pioneering studies carried out by S.C. Dasgupta, Swami Vishwakarmanand and C.N. Acharya in early fifties, original design of Gramlakshmi Gas Plant underwent further changes and finally the year in 1954, a much modified version emerged. This improved design had two chambers, one to act as primary and the other as secondary digester with facility for gasholder to have rotary as well as vertical movements. Its digester made of brickwork was kept underground and fitted with two pipes one for feeding and the other for removing digested slurry. The gasholder was designed in such a way as to maintain uniformity of pressure at 3 to 4 inch of water column but without using counterweights as was used in earlier design. This modified design of the Gramlakshmi Gas Plant became the major landmark in the programme for biomethanation in the country. It was adopted by the Khadi and Village Industries Commission (KVIC) for its countrywide popularisation during sixties and seventies [27]. Apart from the KVIC, a Lucknow-based agency called Planning Research and Action Division (PRAD) also adopted it for promotion. The main pioneer at the PRAD in promoting biomethanation was Mr Rambux Singh who has now become a legend in this field. The KVIC and PRAD also tried to promote plants of the IARI (The Indian Agricultural Research Institute) design. However, plants made to the IARI design with pulley and counterweights created alignment problems often causing tilting of the gasholder sideways which resulted in uneven and interrupted gas supply. The KVIC, therefore, preferred to popularise plants which had no attached pulley and counterweights. Gas pressure was instead sought to be built-up with the weight of gasholders. KVIC also initiated R&D activities during sixties and seventies. Based on Chinese design and field trials carried out at the Gobar Gas Research Station, Ajitmal, Etawah in U.P., PRAD developed a fixed dome type plant which has come to be known as the *Janata* Biogas Plant (JBP).

From 1962 onwards, the KVIC launched a well-coordimated biogas programme in the country under which over a lakh biogas plants were set up till the beginning of eighties when the programme was finally taken over by the DNES (Department of Non-conventional Energy Sources) [25, 26]. During the Sixth Plan (1980-85), the government set up the National Biogas Development Board (NBDB) with a view to advance grants, subsidies and loans to organisations associated with the implementation of the biogas programme. It launched the National Project for Biogas Development (NPBD) to give a renewed thrust to biomethanation the country. DNES under the Ministry of Energy was in September, 1982 assigned the task to pursue the NPBD [8]. The DNES also initiated several measures for the success of NPBD which covered upward revision of the Central Government subsidy for plant construction, availability of more effective post-installation follow-up services, creation of large cadre of trained manpower in the field, and supply of raw materials [22, 23]. During the Seventh Plan (1985-90), it was planned to set up 15 lakh biogas plants with their break-up for successive years as 2.5, 2.75, 3, 3.25 and 3.5 lakh respectively [10, 13]. Under the NPBD, 18.52 lakh family type and 1009 community, institutional and night-soil based biogas plants had been set up by Dec. 31, 1993. According to the annual MNES sources, 23.05 lakh plants had been installed till Feb. 29, 1996 which are estimated to generate gas equivalent of over 72 lakh tonnes of fuelwood approximately valued at Rs. 360 crores per annum besides producing 345 lakh tonnes of enriched fertiliser. Over two lakh biogas plants were installed during 1993-94 against a target of 1.75 lakh plants. Matching with the average cattle holding size of rural families, emphasis in recent years has shifted towards installing small-sized plants of 2 m^3 capacity. Considering that 23.05 lakh biogas plants constructed upto 1995-96 represent just over 1 per cent of the total rural households in the country, there is vast untapped potential for installing many times the existing number of plants in the country. With an assumed norm that adult bovine generates 10 kg of animal wastes per day, 75 per cent of which is collected, 262 million adult bovines can generate nearly 2 million tonnes of animal wastes per day in the country. With the established norm that 25 kg of animal wastes can produce 1 m^3 of biogas, it is possible to feed some 40 million biogas plants of 2 m^3 capacity daily which are 25 times the plants (16 lakhs) installed upto 1991-92. The biogas produced by these plants can meet most of the fuel needs of 120 million rural households in the country. By the end of Eighth plan (1992-97), biogas plants tally is expected to cross 26 lakh mark in the country. MNES is planning to draw ambitious targets for the Ninth Plan (1997-2002).

1.4 Basics of Anaerobic Fermentation

Anaerobic fermentation process can be broadly divided into two phases, acid forming and methanogenesis. In the first phase, acid forming bacteria break-down and liquefy volatile solids of feed into fatty acids. In the second phase, the methane-forming bacteria convert these volatile acids into methane and carbon dioxide which are the main constituents of biogas [5]. These bacteria are sensitive to changes in their environment. Rapid digestion and efficient biogas generation occur within narrow ranges of temperature and are influenced by the composition of raw material [24]. Consistent temperature, pH and use of fresh organic matter as feed help to maximise the gas yield. Temperature is usually maintained at approximately 95°F (35°C) but different temperatures can also be used provided no variation is allowed. For every 20°F (11°C) decrease, gas production is nearly halved. Hence a constant temperature is essential for higher yield. Temperature variations of as little as 5°F (2.7°C) can inhibit methane forming bacteria causing acid accumulation and possible digester failure [20].

Biogas plant requires no special skill or scientific knowledge to operate. However, acidity of digester contents needs to be occasionally checked. This can be accomplished by dipping a strip of litmus paper in the effluent and noting its colour. Increasing acidity indicates an imbalance in various bacteria which can be corrected by adding burnt lime until the slurry becomes neutral. Generally maximum biogas production requires daily loading of the digester. When the loading is done once a week or even less, use of warm water for slurry preparation can compensate provided biogas demand is low. When the demand rises, output can be quickly increased by resorting to daily loading [19].

A simple apparatus or plant is enough to produce biogas. Desired capacity and life of a plant affects cost and complexity of a biogas system. A simple batch loaded digester requires an oxygen free container, relatively constant temperature, a means for collecting gas and some mixing device. A comparison of different types of commonly used digester systems is given in Table 1.1. As biogas is explosive gas, adequate safety precautions need to be observed while handling it [20].

Size of biogas plant digester is controlled by a number of factors such as number, size and type of animals providing feed, slurry dilution and retention time. The most easily controllable factor (that can be changed with tank size) is retention time. Ten days is minimum retention time but a longer period can be used. The longer the retention time, the larger is the size of digester.

Loading rate gives an indication of weight of volatile solids fed to a digester per day. Volatile solids are that portion of total solids which are organic in content. Biological organisms utilise a portion of this material as a substrate to produce gas. The degree of utilisation of the material affects gas production rate. Not all volatile solids present in feed are biodegradable and normally around 50 per cent of volatile solids are destroyed depending upon feed composition. Volatile solids concentration in the digester determines the rate of gas production. For instance, a digester loaded with four units by weight of volatile solids produces twice the volume gas as the same digester loaded with only two units. Gas production rate is quite often expressed in terms of volume of the gas produced per unit weight of the volatile solids added or destroyed [24, 28].

Proper mixing of digester contents provides intimate contact between micro-organisms and the substrate and maintains uniform temperature, distribution of bacteria and volatile solids throughout the slurry. It also minimises sludge formation and prevents a crust or scum from being formed on the surface of the slurry which interferes with release of biogas. Mixing also prevents accumulation of undigested feed in digester. Scum is a collection of light weight, inert material that collects on liquid surface in digester. This accumulation, unless broken up or removed, tends to markedly reduce the effective volume of digester and obstruct release of gas from digester contents. Methods of mixing or agitation include slurry recirculation pumps and mechanical paddles. However mechanical components exposed to slurry are susceptible to corrosion which makes them difficult to repair without disrupting digestion by opening the digester. Mixing can also be accomplished by recycling gas through diffusers at the bottom of digester tank. An electrically driven vacuum pump draws biogas from storage under roof and injects it at the bottom of both stages of digester. Recirculation of biogas helps in agitation of slurry [24, 28].

Amount of water required in slurry preparation is an important factor in digestion process. Considering that optimum total solids (TS) concentration in an anaerobic digester is normally 7 to 9 per cent, and since animal faeces may contain 10 to 25 per cent solids in general, considerable dilution is quite often

Table 1.1 Comparison of Different Types of Digester Systems

Digester type	*Suitable wastes*	*Volumes solids content*	*Typical Retention Periods (days)*	*Degree of mixing*	*Operating temperature (°C)*	*Gas Yield*	*Degree of control required*	*Remarks*
1. Batch	Agricultural irregular, or seasonal fibrous or difficult to digest	Low volumes up to 25% solid	60 or more	Little needed	Usually 30-35	Irregular and discontinuous	Little once started	Messy and time-consuming to start
2. Plug-flow horizontal vertical	Agricultural continuous or regular flows, less fibre content	Larger volume 5-15 per cent solids	30-60	Occasional	30-35	Continuous	Simple	Loading and scum removal can be messy
3. Conventional Sewage Works	Continuous sewage sludge	Less than 5 per cent solids	30-60	Occasional	30-35 or unheated	Continuous	Simple	Not very effective
4. High Rate Sewage Digester								
4.1 Primary	Sewage sludge	4-10 per cent solids	10-30	Regular	30-35	Continuous	More sophisticated	Can be automated
4.2 Secondary	From primary	4-10 per cent solids	20-60	None	Unheated	None collected	simple	
5. High rate	Agricultural industrial	4-15 per cent solids	5-20	Continuous	30-35	Continuous	More sophisticated	Can be automated
6. Anaerobic contact	Industrial Agricultural	Low solids	0.5-5[1]	Continuous	30-35	-do-	Sophisticated	Automatic
7. Anaerobic filter	Industrial	Low solids (Low organic contact)	0.5-5[1]	None needed	30-35	-do-	-do-	-do-

Note: 1. Liquid Retention time.

required. Furthermore, as far as possible, total solids in digester content should be maintained fairly uniform, allowing only gradual changes whenever it becomes necessary [28].

Very little reduction of feedstock occurs in a plant by volume. In fact, there occurs an increase in the volume of effluent on account of slurry preparation with water. Digested slurry after anaerobic fermentation has far less odour. As it still contains most of the original nitrogen, phosphorus and potassium and is still highly polluted, affluent cannot enter a sewer stream after it leaves the digester. Lagoons are commonly used to hold the waste in most European countries until it can be disposed of by either hauling or pumping onto the agricultural land. As a result of fermentation, there occurs no increase in the amount of nitrogen, phosphorus or potassium although it ensures their larger availability through concentration.

Biogas plants can be fed with slurry continuously or in batches. A batch loaded digester is filled to capacity sealed until it has produced all it can, emptied and filled again. Gas production is uneven because bacterial digestion starts slowly, peaks and then tapers off with growing consumption of volatile solids. This difficulty can be overcome by connecting the batch loaded digesters in series loaded at different times so that adequate biogas is available daily. This method uses feed efficiently but is less efficient in terms of digester space. In a continuous type biogas plant fresh slurry is added daily. Gas production in this method is consistent because the bacteria always have a fresh supply of volatile solids to digest. The continuously loaded digester uses expensive digester space efficiently although it may not produce quite as much gas per kg of feed as a batch type may do [24].

Incoming slurry displaces an equal amount of processed slurry from digester each time it is fed since maximum allowable volume of slurry in a digester is fixed for a particular plant. In view of digester volume being constant, fraction of digester's liquid volume replaced each day determines retention period. For example, if slurry equal to one-tenth of digester's liquid volume is added daily, digester slurry has an average retention time of 10 days. Considering that retention period depends on digester's volume, its selection to suit a particular feedstock and digester determines size or capacity of a digester. Other factors that have bearing on plant capacity include number, size and type of animals providing feed and optimum amount of dilution-water required for slurry preparation. A very brief retention time does not allow bacteria enough time to digest and hence results in low gas production. Long retention time allows complete decomposition of wastes and hence increased gas production rate follows. However, if retention time is made very long, it does not furnish enough fresh slurry to promote bacterial growth, and hence again a low gas production rate results. In most biogas plants, fifteen days is usually chosen retention period [20, 24, 28].

Considering that peaks of gas generation and usage of biogas do not always match, some gas storage either within gasholder or outside is necessary. Low pressure storage systems usually suffice but if the gas is to be used for such application as vehicle fuel, high pressure storage is required. At low pressures of 100 to 150 mm of water column, polystyrene or butyl rubber bags can be used for storage. However, for storing several days gas output, volume needs become excessive. For example, to store biogas equivalent of 200 litres of petrol, a low pressure storage system requires a volume of about 300 cubic metres which is broadly equivalent to the volume of average size house. High pressure systems use a three or four stage compressor to compress biogas to pressures of around 3500 psi which can be stored only in a high pressure cylinder [35]. Even at such high pressures which are expensive to attain, biogas has only one-sixth of energy density of liquid fuels.

Methane which is the principal constituent of biogas cannot be liquefied by any pressure at ambient temperature. The critical pressure of methane is 4710 kPA at $-82.3°C$. In contrast, propane has a critical pressure of 1290.3 kPA at 38°C and can thus be liquefied at ordinary temperatures (Critical temperature of a gas is that temperature above which it is not possible to liquefy it under any pressure. Critical pressure is the pressure needed to liquefy a gas at its critical temperature). Liquefaction poses further problems following presence of large quantities of carbon dioxide and other impurities. In case of natural gas, it is possible to store it by adsorption on propane or adsorption on activated charcoal. If this method is tried on biogas, it may pose several difficulties such as high initial cost of adsorbent, large storage needs, and also release of gas from adsorbent. Propane tanks, have been successfully used for biogas storage at pressures upto 1400 kPA (200 psi) on farm digester installations in Canada. The Biomass Energy Institute at Manitoba (Canada) demonstrated that this pressure is easily achievable by using a conventional cold storage compressor [28, 30].

1.5 Movable Drum Versus Fixed Dome Type Plant

The conventional movable drum type model comprises a masonry digester with an inlet on one side for feeding slurry and an outlet on the other side for removing digested slurry. The gas collects in a steel gasholder which is inverted over the slurry and moves up and down depending upon accumulation and discharge of gas guided by a central guide pipe (Fig. 1.1). The main drawback of this design is that metal construction cost are very high. The movable gasholder made of mild steel alone accounts for about 40 per cent of the total cost of plant. The maintenance costs of this type of plant are also high considering the fact that vital components like gasholder are corrosion-prone which need to be given anti-corrosive painting at least once a year. Moreover, several construction materials are not readily and locally available at plant sites [6]. *Janata* Biogas Plant (JBP) which is a fixed dome plant dispenses with movable steel drum type gasholder, and its gasholder is built with bricks and cement as immovable and fixed dome cover of the well-like digester itself (Fig. 1.2). When biogas is formed inside the digester, it rises towards the top of the

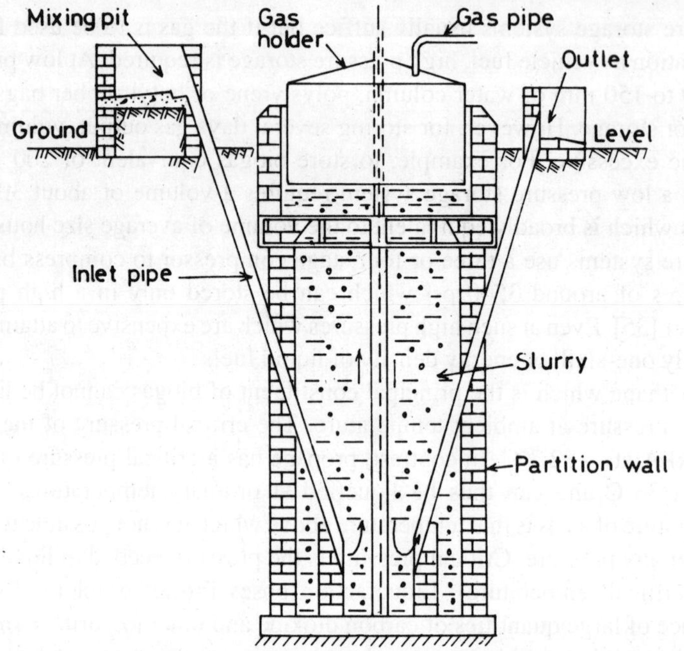

Fig. 1.1 Cross-sectional view of floating drum type biogas plant promoted and popularised by KVIC. [Source: UN, Guide-book on Biogas Development, New York, 1980, p. 13]

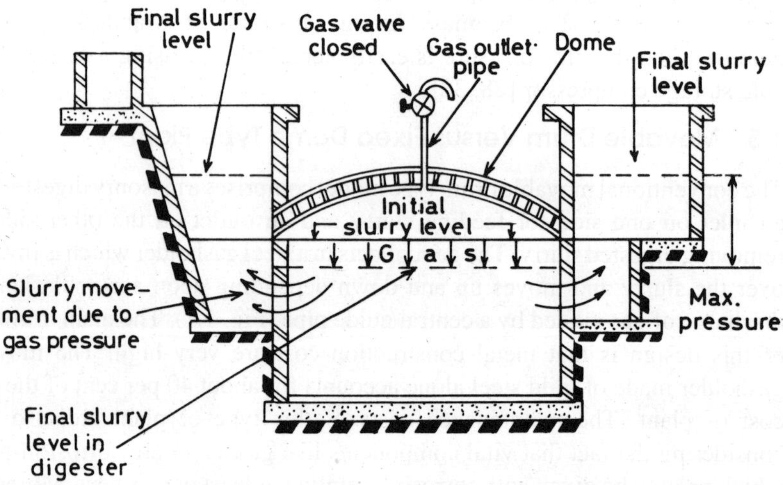

Fig. 1.2 Cross-sectional view of fixed dome Janata biogas plant showing working principle, initially developed by the PRAD, Lucknow and later widely popularised in India. [Source: Rajput. R.K., A Treatise on Janata Model Biogas Plant, Changing Village, 6(4), 1984, p. 265]

dome and pushes slurry down. Displaced level of slurry provides the necessary pressure for release of the gas up to the point of its use. The first pilot drumless JBP was developed by the Gobar Gas Research Station, Ajitmal in early part of 1977. The depth and diameter of digester were 2 m and 3 m, respectively with the digester so designed that it could hold about 50 days feed.

The JBP's have no moving parts and hence the wear and tear and consequent maintenance problems are relatively few. Accordingly these plants have longer working life. Service life of a biogas plant can be expected to range anywhere between 15 to 25 years. In a movable drum type plant, digester is generally expected to last for 40 years, gasholder for 10 years and pipes and accessories for about 20 years. Gas yield of JBPs is considered to be higher (0.04 m^3) per kg of cattle dung at 24-27°C, than those of other plants. However, gas yield per unit volume of the digester is more for a movable drum type than for a fixed dome type plant. Operating costs per cubic meter of biogas is generally more for a movable drum type than a fixed dome type biogas plant. Effluent obtained from JBPs has the same carbon and nitrogen contents and it is richer in phosphorus and potassium contents by 2.5 to 6.5 per cent and 7.2 to 18.5 per cent, respectively as compared to those from other plants. On account of JBP's being underground, rate of gas yield from them shows relatively uniform pattern during days and nights due to their being less susceptible to temperature variations. One side advantage of JBP's is that the space available above the plant can be gainfully used for a variety of applications such as providing a cattleshed. A comparison between fixed dome type (JBP) and movable drum type biogas plant is summarised in Table 1.2 [5, 18].

Table 1.2 A Comparison Between Fixed Dome Versus Movable Drum Type Biogas Plant

Comparison Factor	*Fixed Dome Type Plant*	*Movable Drum Type Plant*
0. System	Regular filling, & irregular discharge.	Regular filling and regular discharge through over-flow.
1. Material	Concrete blocks, bricks, sand, quarry-stones, lime, cement & steel.	Concrete blocks, bricks, sand, quarry-stones, lime, cement & steel.
2. Main cost factor	Cement and bricks.	Steel, cement and bricks
3. Construction aspects	Suitable to construct locally but considerable masonry skills required.	Possible to construct locally but gasholder is normally required to be constructed in a workshop.
4. Thermal insulation aspects	Since constructed underground, it ensures uniformity of temperature and heat insulation.	Considerable digester heat is lost through steel gasholder which is difficult to insulate; it is, therefore, less suitable for use in colder regions.

contd...

5. Gas-tightness	Gas storage dome is required to be given special treatment for gas tightness and painted periodically.	No such problems.
6. Feedstock	Agricultural waste, other organic matter (including fibrous material), animal & human faeces.	Animal and human faeces, chopped agricultural wastes as additive.
7. Biogas yield	Digestion period normally between 30-60 days, gas yield 0.15-0.35 m^3 per m^3 of digester volume and per day which can go upto 0.4-0.6 m^3 if plant is absolutely air-tight.	Digestion period normally between 40-60 days, gas yield 0.4-0.6 m^3 per m^3 of digester volume per day.
8. Maintenance aspect	If fibrous material is used as a feedstock, plant is normally required to be cleared once or twice a year for maintenance purposes.	Regardless of feed, gasholder is required to be painted every year and replaced after every 10 year.
9. Cost consideration	Since construction is more labour-intensive, hence it is relatively inexpensive.	Due to greater use of steel, it is relatively more expensive.
10. Scope for possible improvements.	Scope exists for providing facilities for dome-tightness, slurry-mixing and digester-heating.	Scope exists for improving thermal insulation facilities, and in improving rust-proofness in gasholder.

Source: 1. BORDA, Biogas: Manual for the Realisation of Biogas Programmes, Breamen Overseas Research and Development Association, German Association for Technical Cooperative (GTZ), Gmb H, Eschoborn, Germany (erstwhile FRG), 1980, p. 61-62.

2. GATE, Biogas Plants: Building Instructions, German Appropriate Technology Exchange, German Agency for Technical Cooperation, P.O. Box 5180, 5236, Eschoborn 1, (erstwhile FRG) Germany, 1980, p. 15.

1.6 Factors Affecting Biogas Yield

Regardless of specific feed used, biogas generation proceeds most efficiently when raw materials fed to the digester have a desired pH and carbon-to-nitrogen ratio (C : N). Bacteria thrive in a slurry with a pH value of around 7 and consequently if the input slurry has a pH close to 7, fermentation proceeds more smoothly. Under normal conditions, fermentation process balances excess acidity or alkalinity on its own. Carbon is the major chemical element in organic wastes which bacteria digest releasing finally methane and carbon dioxide in the process. However, microorganisms require certain amount of nitrogen in the feed to perform their function. Ratio of carbon-to-nitrogen is thus crucial for anaerobic fermentation. A high C : N ratio means that the nitrogen will be exhausted before carbon is digested. Conversely, a low C : N ratio means too much nitrogen in relation to carbon. It results in high ammonium concentrations which tend to become toxic to anaerobic bacteria. It is, possible to adjust the C : N ratio in a biogas plant by adding another material to supplement or counter material already in the plant. For instance, saw dust which has a high C : N ratio could be added to poultry droppings which have a low C : N ratio. Dairy cow manure has a C : N ratio just slightly below that required by bacteria [14].

Total biogas production rate depends on organic material being digested, digester loading rate and environmental conditions. Under ideal condition (95°F or 35°C temperature and proper pH, it is possible to produce about 45 ft^3 (1.4 m^3) of biogas at atmospheric pressure from one day's faeces of a 100-pound (54 kg) cow. Not all of the biogas produced is available for utilisation as part of it goes waste [20]. In very aggregate and general terms, 1 m^3 of slurry fed to biogas plant produces an average of 0.15 to 0.2 m^3 of biogas daily; one cubic metre of biogas produces about 5500 kcal of heat; and each member of a family requires 1.5 to 2 m^3 of digester space to meet his mean requirement of lighting and cooking [18]; and 32 kg of cattle dung per day, 20 kg of pig faeces per day, and 12 kg of poultry droppings per day produce 1 m^3 (34 ft^3) of biogas every day under normal conditions [36]

1.7 Commonly Used Feed Materials for Biomethanation

Common feed materials include animal wastes, crop-residues and urban wastes including night-soil. These are broad categories which cover variety of materials. Some of the waste materials are directly used as fuel or fertiliser. Utilisation of these materials for biogas generation harnesses additional benefit while retaining their existing benefits which are anyway tapped. The list of materials which fall under the above three broad categories can be further expanded to include dairy cow wastes, swine wastes, poultry litter, sheep and goat droppings, crop-residues of various plants, sugar-cane trash, bagasse, rice-husk, rice-bran, tobacco wastes and seeds, cotton dust from textile mills, tea waste, fruit and vegetable wastes, marine algae, seaweeds, water – hyacinth, press-mud from sugar plants, twigs, bark, tree branches, plant leaves, municipal wastes and night-soil etc. [30].

Another development in the area of anaerobic digestion is application of industrial wastes which constitute an important category of feedstocks. Wastes available from food and drink industry are particularly useful for this purpose. Vegetables and fruit-processing and canning units produce large quantities of carbohydrates-rich-wastes which are easily amenable to anaerobic fermentation. Drink manufacturers use enormous quantities of biomass in the form of barley, apples, grapes etc. and residues from these plants can be anaerobically fermented. Methane gas production from vegetable wastes has in past been particularly popular in countries like Australia and South Africa. Wastes from sugar distilleries which have high carbohydrate contents are also a valuable feedstock. Several sugar factories world over have established fermentation units. One manufacturer has successfully completed trials for a digestion plant utilising wastes from an olive-oil factory in Spain which besides generating usable gas also solved problems of pollution-control.

1.8 Properties and Characteristics of Biogas

Biogas which consists of 60-65 per cent methane, 35-40 per cent carbon dioxide and traces of hydrogen sulphide, ammonia and other impurities is an explosive and toxic gas. Methane which is the principal constituent of biogas, is colourless, odourless and tasteless. Methane is also combustible in a concentration of 5 to 15 per cent in air and, therefore, potentially hazardous. Considering that biogas is warm when it leaves digester, it contains water vapour which condenses when exposed to colder temperatures. Condensation traps devised in gas pipelines are necessary to prevent water from blocking low points in pipelines. Biogas after condensation of water vapour present therein or even without condensation, can be used as a fuel for cooking, lighting and several other applications. However, in stoves gas-burner-nozzle needs to be enlarged to compensate low heat value of biogas. Tests using biogas as exclusive fuel in a petrol engine showed that energy present in 200 ft^3 (5.7 m^3) of biogas is equal to energy present in 1 gallon of petrol.

Calorific value of biogas ranges between 500 to 700 BTUs per cubic foot (0.03 cubic metre). In comparison, natural gas has a heat value of 850 BTUs per cubic foot (0.03 cubic metre), petrol contains nearly 120,000 BTUs per gallon (3.8 litres), diesel oil comprises 133,000 BTUs per gallon and propane contains 92,000 BTUs per gallon. Biogas is difficult to store, compress or liquefy. Biogas unlike Liquified Petroleum Gas (LPG) cannot be converted to liquid state under normal temperatures. A temperature of – 260°F (– 162°C) is needed to liquefy methane at atmospheric pressure. Liquefaction of methane requires pressures of nearly 5000 psi (LPG liquefies at 160 psi) at a temperature of – 117°F (– 83°C) and is thus not practical. If biogas is compressed to 1000 psi, it needs about 1320 BTUs of energy to put 6350 BTU into a storage container. As biogas cannot be easily liquefied economically, it is best suited for such applications as cooking, lighting, grain-drying or operating stationary internal combustion (IC) engines. Biogas is normally produced on continuous basis. Most applications are intermittent and hence some kind of storage facilities are required. Amount of gas storage is dependent on storage time and pressure. High demand applications such as grain drying are normally impractical following excessive storage capacity requirement. Although biogas can be used for variety of applications including crop-drying, refrigeration and running of diesel engines, so far it is mainly used as cooking and lighting fuel in most developing countries [20, 24].

Methane content in biogas is affected by digester conditions. Biogas contains traces of hydrogen sulphide which is highly corrosive. This can be removed by passing the gas through steel wool or iron filings. Following condensation of water vapour present in biogas, it can be used as fuel for any standard application. Biogas contains nearly forty per cent carbon dioxide which can be removed by bubbling it through lime water. Lime water is a mixture of 1.8 kg burnt lime in 1000 litres of water. 1000 litres of lime water removes 560 litres of carbon dioxide. For removing 35 per cent carbon dioxide in biogas, 625 litres of lime water is needed to purify 1 m^3 of biogas. Lime water becomes milky following absorption of carbon dioxide [36].

1.9 Potential of Biogas Plant Sludge as Enriched Fertiliser

Effluent or sludge resulting from a biogas plant is a resource as it contains almost all of nitrogen present in raw feed material prior to digestion. The effluent is an excellent fertiliser because nitrogen in the effluent is more readily absorbed by plants than nitrogen in raw manure [1]. The sludge that collects at the bottom of a digester should be regularly removed because accumulated sludge tends to reduce active digester space. The end result of applying anaerobically digested sludge to soils has the same beneficial effect as that obtained by applying any other organic matter. The humus materials formed in sludge tend to improve physical properties of soil: for example, aeration, moisture-holding and water-infiltration capacity of soils are improved. Furthermore, sludge serves as a source of energy and nutrient for development of microbial populations which directly or indirectly improve the solubility and thus availability of essential chemical nutrients contained in soil minerals [24, 30].

The organic fraction of sludges from an anaerobic digester operating on plant and animal waste may contain 30 to 40 per cent lignin, undigested cellulose and lipid materials on dry weight basis depending upon the type of raw material used. The remainder consists of substances originally present in the raw material but protected from bacterial decomposition by lignin, newly synthesized bacterial cellular substances and relatively small amount of volatile fatty acids [24, 30]. The amount of bacterial cell mass is small as 10 – 20 per cent of substrate is converted to cells. Thus there is less risk of creating odour and insect-breeding problems when anaerobically digested sludges are stored or spread on land than when untreated or partially treated organic waste materials are similarly handled, indiscriminately disposed off or stored. The sludge is an almost odourless, homogenised liquid which does not attract rodents or flies.

1.10 Biomethanation as an Aid for Environment Improvement

Biogas plants help to bring improvements in ecology and environment. Biomethanation provides a non-traditional way for disposal of sewage, animal and human faeces in urban and rural areas. In biogas plants as these wastes are no longer exposed to open air, menace of mosquitoes, flies and hookworms is sufficiently controlled. Following check on hookworms, Biomethanation also helps to control diseases like Schistosomiasis, hepatitis and intestinal infections. It improves household hygiene as it prevents occurrence of polluting smoke that emanates for traditional fuelwood stoves (*chulha's*). Faeces contains countless parasitic ova and pathogenic organisms which often cause ailments like typhoid, dysentery, hepatitis and poliomyelitis (Table 1.3). Chances of health hazards posed, basically depend upon three factors, namely, proportion of viable pathogenic organisms in faeces, survival rate of these organisms in sludge, and length of time the sludge is stored prior to use.

Table 1.3 A Set of Diseases caused by Improper Handling of Night-soil and their Causative Micro-organisms

Disease Category	*Disease Description*	*Causative Micro-organisms*
Viral	Infectious hepatitis	
	Gastroenteritis	
	Respiratory illness	*Adenovirus*
		Reovirus
	Poliomyelitis	*Entenovirus (poliovirus)*
Bacterial	Typhoid fever	*Salmonella typhosa*
	Salmonellosis	*Salmonella spp.*
	Bacillary dysentery	*Shigella spp.*
	(Shigellosis)	*Vibrio cholerae*
	Cholera Tuberculosis	*Mycobacterium tuberculosis*
Protozoan	Amebrasis (Amebic dysentery)	*Entemoeba histopytica*
Helminthic	Roundworm	*Ascaris lumbricoides*
	Pinworm	*Oxy aris vermicularis*
	Whipworm	*Trichurus trichiura*
	Tapeworm	*Taenia saginate*
	Hookworm	*Ankylostoma duodenale*
		Necator americanus.

Source: NAS, Methane Generation from Human, Animal and Agricultural Wastes, Report of an Ad Hoc Panel of the Advisory Committee on Technology Innovation, National Academy of Sciences, Washington, D.C., 1977.

Survival rate of pathogenic micro-organisms varies from situation to situation and there are no fixed rates available for this purpose (Table 1.4). Most of these bacteria do not survive beyond a period of 14 days. Anaerobic fermentation process is considered to be one of the most effective ways of killing enteric micro-organisms which is highly beneficial to public health. The survival of enteric micro-organisms has been the subject of many investigations, the outcome which depends on numerous factors such as the type of organism, ambient temperature, type of soil, moisture and pH of digester contents, and as a result survival is found to vary from few hours to several months. Free ammonia which is released during anaerobic fermentation is another factor that influences survival and growth of micro-organisms. The

Table 1.4 Kill-rates of Enteric Micro-organism During Anaerobic Fermentation

Micro-organisms	*Temperature °C*	*Retention Time (Days)*	*Kill Rate of Pathogens (Per cent)*	*Reference*
Poliovirus	35	2	98.5	1,2
Salmonella ssp.	22-37	6-20	82-96	3
Salmonella typhosa	22-37	6	99	3
Mycobacterium tuberculosis	30	Not reported	100	4
Ascaris	29	15	90	3
Parasite cyst	30	10	100	5

Source:
1. NAS, *Methane Generation from Human, Animal and Agricultural Wastes*, Report of an Ad. Hoc Panel of the Advisory Committee on Technology Innovation, National Academy of Sciences, Washington, D.C. 1977.
2. Bertucci, J., et al., Inactivation of Viruses During Anaerobic Sludge Digestion, *Journal Water Pollution Control Federation*, 49(7), 1642-1651 (1977).
3. Foster, D.H. and Engelbrecht, R.S., Microbial Hazards in Dispersing of Wastewater on Soil, in 'W.F. Sopper and L.T. Kardos (eds.), *Recycling Treated Municipal Wastewater and Sludge through Forest and Cropland.* Pennsylvania State University Press, PA, 1973.'
4. Briscoe, J., *Public Health in Rural India: The Case of Excreta Disposal,* Ph.D thesis, Harvard University, Boston, MA, 1976.
5. Muller, G., Tubercle Bacteria present in Sludge Produced in Mechanical and Biological Sewage Treatment Plants. Abstracted by W.J. Mulles., *Sewage and Industrial Wastes*, 32(9), 1030 (1960).

usual proportion of ammonia in digester contents is of the order of 0.07 per cent. If free ammonia level is around 0.2 per cent, the ova of *Schistosoma* can hardly survive beyond six days and hookworm and *Ascaris* only marginally longer. Free ammonia reduces the survival period of micro-organisms as it tends to permeate egg-shell and cell membrane and in the process kills ova and bacteria. Addition of certain plant materials like algae to sludge checks the growth of some bacteria. For instance, algae when added to effluent retards the activity of *Anopheles* larvae following change in pH level and in the process makes the bacteria ineffective [30]. In spite of the fact that certain pathogens and parasites do survive even after anaerobic fermentation, and some remain active much longer, no cases of outbreak of diseases have been reported from any quarter. This can be explained by the fact that most of the rural-folk are anyway vulnerable to higher health risks, and biomethanation reduces the health hazards significantly, if not completely.

1.11 Applications of Biogas

Biogas provides a clean fuel for cooking and lighting, biogas burner (Fig. 1.3) and lamp (Fig. 1.4) are structurally similar with their common items being a nozzle, an air inlet and a mixing chamber. The uncommon items are mantle in case of lamp and fire-stove plate in case of stove. A schematic plan for using biogas as a cooking and lighting fuel is given in Fig. 1.5. Biogas lamp works on a simple principle that when gas is burnt, mantle of the lamp glows and emits light. Mantle in a biogas lamp is similar to one used in a 'coleman' or propane lamp. Brightness of biogas lamp mainly depends on factors like gas pressure, gas/air ratio and extent of mixing. Proper nozzle adjustment is necessary to achieve requisite light intensity. Low gas pressure causes poor light intensity but higher pressure though it improves brightness, lowers mantle life. Biogas lamps can be either suspended (hanging type) or table (standing) type. As biogas is almost half as light as air and as hot air rises upwards, the brightness of a standing biogas lamp is somewhat greater than that of a hanging type. Biogas lamps are generally designed to produce 100 candle power and consume 0.11 to 0.15 m^3 (4 to 5.5 ft^3) biogas per hour. Several designs of biogas lamps are popular but most common ones are what have come to be known as the Indian, Chinese and Pakistani types.

Fig. 1.3 Biogas burner

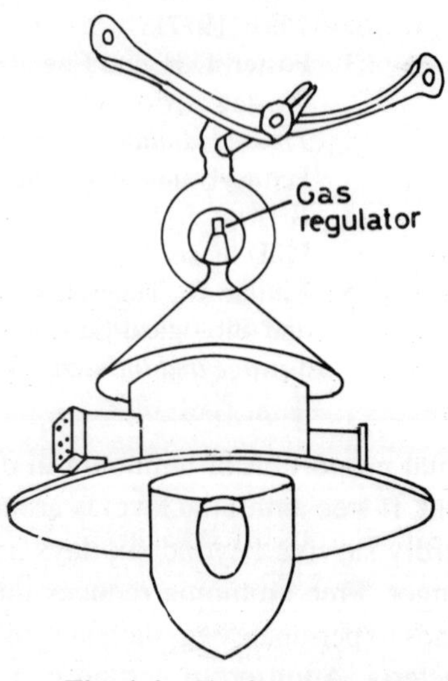

Fig. 1.4 Single mantle biogas lamp

Source: BORDA, Survey of Biogas Plants, Bremen, Germany, 1978 (For both the figures)

Biogas stoves normally operate at gas pressure of 75 to 90 mm. (3 – 3.5 inch) water column. As biogas passes through the nozzle, air is allowed to be drawn in mixing chamber. Nozzle adjustment is generally necessary to achieve desired flame temperature. Brightness and combustibility of gas is regulated by con-

Fig. 1.5 A schematic plan for using biogas as a cooking, lighting and engine fuel
Source: Buren, A.V. (ed.) A Chinese Biogas Manual, Inter. Tech. Pub. Ltd., London, 1979.

trolling gas pressure and air fuel ratio which is generally maintained as 10 : 1. Combustibility of gas is maximum when the flame appears to be slightly yellow, a bit bright and burnished. When the flame turns blue and smokeless after air jet adjustment, it burns at a temperature of about 800°C. For meeting cooking needs of an individual 0.28 to 0.42 m^3 (10 to 15 ft^3) of biogas is needed daily. Cooking appliances in wide range are available both in single and double burner category with each burner consuming anywhere between 0.25 to 1.25 m^3 of gas per hour. Biogas cannot be burnt on stoves designed for LPG/natural gas as it tends to lift off the stoves on account of its flame-speed factor being lower than these gases. This problem can be somewhat overcome by reducing gas pressure but this causes higher gas consumption and reduces burning efficiently.

Biogas provides reasonably efficient fuel for both spark-ignition (petrol) and compression – ignition (diesel) engines. Whereas diesel engines can run on biogas as a dual-fuel comprising both oil and gas, petrol engines can be made to run on cent per cent biogas. However in the latter case, it is initially necessary to use little petrol for start-up. While running IC engine on biogas, some modification in the fuel injection system is required. In a partially converted system, a gas tap is fitted onto the inlet manifold of the engine. The engine is intially started on liquid fuel and a little gas supply, and later the proportion is reversed when engine develops desired speed. Opening of gas tap and adjustment of air supply is carried out simultaneously. In steady engine running, dual fuel helps to achieve about 80 per cent saving in diesel consumption. (Fig. 1.6).

Biogas can also be utilised for electricity generation. As per some studies carried out in the past, 1 kwh of electricity can be generated from 0.7 m^3 of gas which can light about 15 electric bulbs of 60 watt rating for one hour. In contrast if 0.7 m^3 of gas is directly burnt to light gas lamps, only 5 gas lamps of 100 candle power (CP) rating equivalent to 60 watt electric bulb can be lighted for one hour duration. This shows that power-route is more efficient for lighting

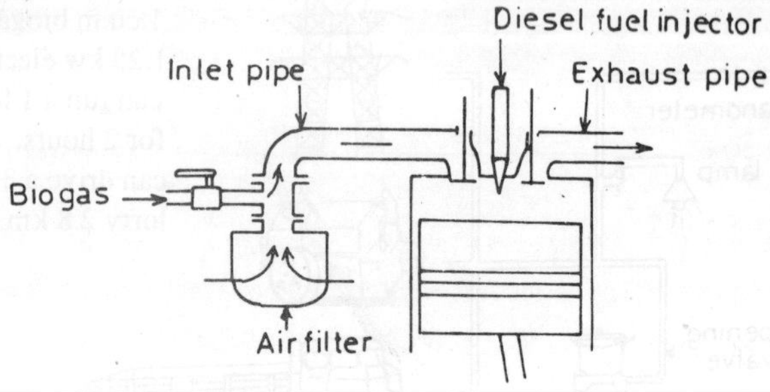

Fig. 1.6 Diesel engine running on dual-fuel
Source: UN, Guidebook on Biogas Development, New York, 1980, p. 84.

than direct gas burning. However, electricity generation from biogas is economical only from large sized plants and it requires high initial capital investment. Asynchronous generators of 3, 5.5 and 7.5 kw rating are generally coupled with IC engines running on biogas.

Biogas can also be used as a vehicle fuel. It has been successfully used on a commercial scale in countries like Italy and New Zealand. Conversion kit suitable for biogas is virtually the same as used for compressed natural gas (CNG) which is marketed as an automotive fuel in many parts of the world [34]. For use as a vehicle fuel, biogas is first compressed in a three or four stage compressor upto a pressure of 2500–3500 psi and stored in a high pressure gas storage cascade. Cascade helps to facilitate quick refuelling of storage cylinders. The number of cylinders in a cascade depends on the size of digester and the number of cylinders fitted to each vehicle. Being gaseous fuel, biogas provides excellent cold starting for vehicles because unlike petrol it does not first have to be vapourised [35]. Once started, the engine performs well immediately with no sluggish warm-up. This is because unlike petrol, methane does not condense on cold intake manifolds, cylinder heads and pistons. Biogas is carried in compressed state in cylinders which are fitted on to the vehicles. Each cylinder is about 1 m long, 0.3 m in diameter and weighs close to 50 kg. Two cylinders can be easily mounted on a biogas-driven car and four on a truck.

1.12 Community Versus Family Size Plants

Biogas plants can be constructed either as individual family units or community plants. Whereas individual plants are appropriate for families, community plants are set up to meet fuel and fertiliser needs of groups, institutions and village as a whole. Community plants are particularly appropriate where individuals do not have adequate manure or funds to set up family units. Community plants can be operated by a group of families, cooperatives or members of a locality. Utility of a community plant is further enhanced if some small scale industry is made to run on biogas as a fuel which may also create some employment opportunities for rural youth. In case of community plants, common utilisation of gas in the direct neighbourhood of the plant, such as for running a community kitchen is generally desirable. Many energy planners argue that individual plants should be constructed only when idea of setting up community plant found is infeasible. It is often seen that a farmer who is reluctant to invest a large sum of money for his own private plant may not find it difficult to make a small deposit and pay monthly charges as his share in the running of the plant. Community plants provide economies of scale which arise from savings in initial capital cost, land utilised and labour charges needed in running such plants. Community plants make large-scale composting of organic waste feasible by mixing them with plant sludge.

Community biogas plants are socially more egalitarian. Out of some 100 odd million households in India, only about 20 million own three to four heads of cattle needed to feed a family-size biogas plant. According to the report of the Planning Commission's Working Group on Energy Policy (1979), even if all those households which can set-up family-size plants were to do so, still enough cattle would be available to feed 560,000 community plants at a simple norm of one in each village of the country. In view of their vast potential, community plants received major thrust from the Sixth Plan (1980-85) onwards when an ambitious target of 100 plants was fixed [8]. Plant economics can be further improved if night-soil from community latrines is also used as valuable feedstock. Animal wastes often lying on village roads and grazing lands which otherwise go waste can also be gainfully utilised in the community plants. Community plants ensure higher gas yield per kg of feed on account of economic viability of providing facilities such as warming up of slurry during winter season, which is not cost-effective in family units. Higher biogas supply from community plants besides meeting fuel requirement of households, can be used for generating power, fuelling rural industrial units, operating irrigation pump sets, and lighting village streets.

Community plants are considered to be more cost-effective. The capital cost needed to set up a 500-ft^3/day (CFD) capacity plant is nearly half of that required to install 50 plants of 10 CFD each. However, running of these plants is not without problems. For instance, due to countrywide social and economic disparities task of dung collection and gas distribution poses practical difficulties. Similarly, evolving a pricing policy for the gas and fertiliser produced by a plant which meets the requirement of both rich and poor farmers alike is difficult to work out.

1.13 Integrated Farming System

Integrated farming system may comprise a biogas plant, energy farming, wind-mills; solar heating and photovoltaic systems; food, fodder and algal cultivation and pisciculture (Fig. 1.7). Effluent from the biogas plant provides a nutritious protein-rich feed for animals, particularly fish. Solids are recovered from sludge by allowing it to settle in tanks or by draining and drying. These solids along with other additives after being meshed and detoxified can be used

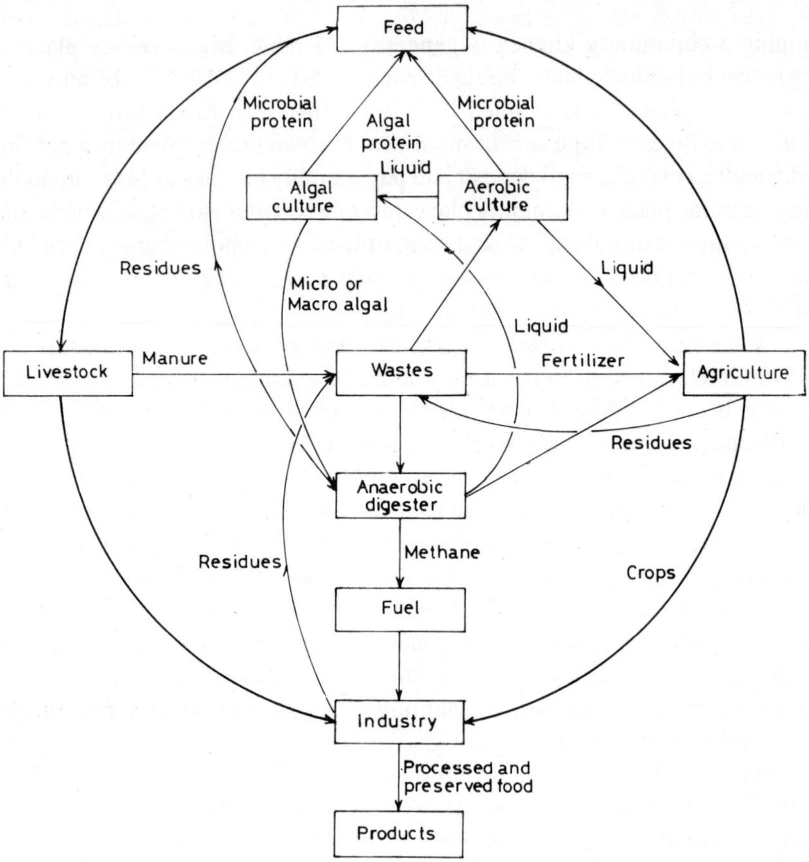

Fig. 1.7 Elements of Integrated Farming System

Source: Huang, C.J., et al., Design and Operational Model for the Integrated Swine Waste Management System in Hawaii, Trans. ASAE Pac. Reg. Ann. Meeting, Hawaii, March 18 – 20, 1980.

as animal feed. After separation of solids from sludge, remaining liquid can be utilised as a nutrient for algal growth [37]. *Chlorella*, a single-celled high protein, can be harvested with the liquid separated from sludge in a shallow pond. The pond is usually lined with concrete, metal or plastic material which makes harvesting easier and minimises contamination. Chlorella can be used in proportions upto 10 per cent of animal feed to replace soyabean used as a meal for protein supplementation. With the availability of fish meal from sludge, it is advisable to have a fish-pond as part of the integrated farming system [5]. Algae produced can be utilised in two ways. Part of it can be used as feedstock for producing biogas and part as fish meal along with other additives. A large portion of biogas plant sludge is used as fertiliser for growing food, fodder and energy crops. Biogas plant sludge contains a variety of products including nitrogen which is mostly lost during aerobic composting. Nitrogen is one of the most important nutrients for plant growth. Nitrogen content can be increased by adding animal urine to influent slurry of a biogas plant. For maximising fertilising effect, slurry should be ploughed into the soil about one week before sowing is planned. In integrated farming system, sludge is stored till the time of its spread. In a farm, it can be stored in two ways, as a liquid fertiliser in a covered pit and as solid manure in a compost pit [30]. As per rough norm, one cubic metre of slurry can fertilise abo ut 100 sq. m of land annually.

1.14 Concluding Remarks

Biogas systems are appropriate for promotion is small scale decentralised energy systems which are very relevant for the country's rural and agriculture economy. Considering that organic wastes are anyway generated on renewable basis in any nation, it is always advantageous to view biogas as supplementary fuel which can go a long way in meeting part of the energy needs of the nation [15]. It should be realised that production of biogas from various types of wastes will not completely solve the energy problems of a nation even if all wastes are available for biomethanation. With a high rate of deforestation which must be prevented for ecological reasons, potential for utilising renewable sources of energy in rural areas is sufficiently high. Success in biomethanation on a large scale is linked with success in afforestation. Biogas programme cannot make significant headway unless biomass availability is also increased simultaneously so that it can supplement cattle dung which is also directly used as cooking fuel [2]. An integrated energy system based on bio-energy, recycling of wastes and biomass agro-forestry for meeting rural energy needs can also help to prevent soil-erosion, improve rural health, environment and ecology [3, 25]. Reliance of biogas as a fuel requires neither any import nor dependence on diminishing supply of indigenous fossil fuel resources. Whereas fuelwood, charcoal, kerosene, petroleum and coal resources may not last for ever, biogas shall be available as long as cattle plays a major role in a country's agriculture and milk production [14]. In view of these factors, all-out efforts be made to popularise biogas programme and enrich related technology.

REFERENCES

1. Balasubramanian, P.R. and Kasturibai, K., Biogas Plant Effluent in Aquaculture. *The Hindu*, Visakhapatnam, Sep. 4, 1991, p. 19.
2. Bembakar, N.L., Energy Crisis : Indian Response, *Indian Express* (*Science Express*), Visakhapatnam, Feb. 26, 1991, III.
3. Bhushan, B., Biogas Plan Under Attack : Devaluing Decentralised Energy, *Times of India*, New Delhi, July 31, 1986, p. 8.
4. Biswas, T.D., *Biogas Technology and Rural Development*, IARI, New Delhi, 1979.
5. BORDA, *Biogas : Manual for the Realisation of Biogas Programmes,* Bremen Overseas Research and Development Association, Bremen, Germany (erstwhile FRG), 1980.

6. Brown, N., Biogas Systems in Development, *Appropriate Technology,* **14(3)**, 57 1987.
7. Chawla, O.P., *Advances in Biogas Technology*, Information Division, ICAR, New Delhi, 1986, p. 17-18.
8. Chenogappa, R., National Biogas Project : Raising a Stink, *India Today*, July 31, 1986, p. 138.
9. Directorate of Economics and Statistics, *Indian Agriculture in Brief*, Ministry of Agriculture, New Delhi, 1974.
10. DNES, *A Personnel Communication*, Department of Non-Conventional Energy Sources, New Delhi, Feb. 16, 1989.
11. Dodson, C.E. and Meynall, P.J., *Anaerobic Treatment of Organic Wastes*, General Information and Biomass Advancement Company Ltd., Marlborough, Wiltz, SM8 2HZ, 1981.
12. *Economic Times*, Untapped Potential (Editorial), New Delhi Saturday, May 4, 1985, p. 5.
13. *Economic Times*, Biogas Technology : India's Achievement Lauded, New Delhi, Nov 10, 1985, p. 5.
14. *Economic Times*, Time-bound Programme for Non-conventional Energy Sources Urged, New Delhi, Jan. 29, 1986, p. 6.
15. *Economic Times*, Energy Minister Reiterates Commitment to Alternative Energy Sources, New Delhi, Jan. 13, 1990, p. 6.
16. *Economic Times*, Economic Survey : 1990-91, New Delhi, July 21, 1991, p. 14.
17. *Financial Express*, Wider Use of Solar Heating Systems Favoured, Madras, Aug. 11, 1991, p. 13.
18. GATE. *Survey of Biogas Plants*, German Appropriate Technology Exchange, Postfach 5180, D-6236 Eschborn 1, Germany (erstwhile FRG), 1978.
19. Golueke, C.G. and Mc-Gauhey, P.H., Waste Materials, in '*Annual Review of Energy*, Vol 1, Palo Alto, California, 1976,' p. 257-277.
20. Hansen, R., *Methane Generation from Livestock Wastes*, Circular No 5002, Colorado State University Extension Service, Fort Collins, Colorado April 15, 1977.
21. HEL, *Why Anaerobic Digestion of Farm Wastes?* Hamworthy Engineering Ltd, Fleets Corner, Poole, Dorset BH 17 7LA, UK.
22. *The Hindu*, Role of Renewable Sources of Energy Stressed, Visakhapatnam, Dec. 27, 1990.
23. *The Hindu*, Biomass : A Potential Energy Source, Visakhapatnam, May 14, 1991, p. 17.
24. Homan, E., *et al.*, *Biogas from Manure*, Special Circular 260, College of Agriculture Extension Service, Pennsylvania State University, University Park, Pennsylvania, 1980.
25. Khandelwal, K.C., Biogas for Better Life, *Changing Villages*, **6(4)**, 274-283 (1984).
26. Khandelwal, K.C. and Mahdi, S.S., *Biogas Technology: A Practical Handbook*, Tata McGraw-Hill Publishing Co. Ltd., New Delhi, 1986.
27. KVIC, *Gobar Gas on the March*, A Report, Mumbai, March, 1979.
28. Lapp, H.M., *et al.*, Methane Production from Animal Wastes – 1: Fundamental Considerations, *Canadian Agricultural Engineering*, **17(2)**, 97-102 1975.
29. Leach, G., Household Energy in South Asia, *Biomass*, **12(3),** 155–184 1987.

30. NAS, *Methane Generation from Human, Animal and Agricultural Wastes*, Report of an Ad. Hoc Panel of the Advisory Committee on Technology Innovation, National Academy of Sciences, Washington, DC, 1977.
31. Neelakantan, S., Assessment of the Availability of Cowdung for Potential Gobar Gas Production in India, *Agricultural Situation in India*, **30(9)**, 633-636 1975.
32. NSS, *National Sample Survey, 17th Round*, 1961-62, New Delhi.
33. Parikh, K.S., *Energy, Second India Statistics*, Macmillan Co of India Ltd., Delhi, 1976, p. 66-98.
34. Ramarao, B.V., Energy-Related Options in Transport Sector, *The Hindu*, Visakhapatnam, April 11, 1991.
35. Stewart, D.J., *Energy: Biogas Production*, Media Services, FPP 375, MAF, Box 2298, Wellington, New Zealand.
36. United Nations, *Guidebook on Biogas Development*, Energy Resources Development Series No. 21, New York, 1980.
37. Hung, C.J., *et al.*, *Design and Operational Model for the Integrated Swine Waste Management System in Hawaii*, paper presented at the ASAE Pacific Region Annual Meeting, Hilo Lagoon Hotel, Hilo, Hawaii, March 18-20, 1980.
38. TERI, *Biogas Technology : A Promise for the Future*, A monograph sponsored by the MNES, Tata Energy Research Institute, New Delhi, 1994, p. 30.

Chapter 2

MICROBIOLOGY OF ANAEROBIC FERMENTATION

2.1 Introduction

In biomethanation process, micro-organisms decompose feedstock in oxygen-less environment releasing methane and carbon dioxide. Acid-forming bacteria break down or liquefy volatile solids changing them into simple fatty acids. The methane-forming bacteria then convert these volatile acids to methane and carbon dioxide which are sensitive to changes in their environment. Rapid digestion and efficient biogas production occur within narrow ranges of temperature and are influenced by the composition of raw material [1, 6]. Methane is also formed from some alcohols and other organic compounds, anaerobic oxidation of hydrogen and carbon mono-oxide and reduction of carbon dioxide to methane. Acid formation phase is considered to be a bit faster. Acid-forming bacteria are somewhat less sensitive to environment. If acids formed during acid-phase are not used up in methane formation, unused acids tend to lower the gas yield.

Biomethanation involves conversion of cellulose into simple carbohydrates, carbohydrates into fatty acids by non-methanogenic bacteria and finally conversion of fatty acids into methane and carbon dioxide [1, 9]. Carbohydrates cover cellulose, semi-cellulose, lignin, sugars, starch, plant alkaloid, gum resin, pectin, plastics, pigments etc. Being the main component of organic wastes, fermentability of cellulose governs rate of anaerobic fermentation.

2.2 Biodegradation and Biodegradability of Substrate

Biodegradation involves destruction of a chemical compound by the action of living organism such as microbe. Degradation of biomass by micro-organisms under controlled conditions leads to the production of biogas. Whereas technology of biogas is sufficiently advanced, various physio-chemical processes leading to the biogas generation are still obscure and need to be studied further.

Biodegradation can be of different varieties. Biodegradation which completely changes the characteristic properties of a compound is called primary biodegradation in which case biodegraded compound no longer responds to analytical procedures prescribed for its detection. Environmentally acceptable biodegradation refers to the minimum alteration of the original compound by micro-organisms necessary to remove its undesirable properties. Ultimate biodegradation refers to the complete conversion of a compound into inorganic end products and products associated with the microbes' normal metabolic processes.

Degree of biodegradation of a chemical compound is called biodegradability [15]. Biodegradability can be measured, among others, by following a procedure which involves incubating a solution of the compound under test with micro-organisms at a suitable pH and temperature. This is followed by analysing the samples at specified times, and such experiments continue over a period of three to four months. For illustration purposes and understanding the concepts, biodegradability of four compounds *A*, *B*, *C* and *D* on the basis of experimental observations can follow similar pattern (Fig. 2.1). Curves *A* and *B* indicate rapid biodegradation whereas *C* refers to a slow process. However, either of the curves *A* and *B* do not establish that ultimate biodegradation has taken place. Thus further tests on the products derived from *A* and *B* are necessary to establish the occurrence of ultimate biodegradation. Neither it can be assumed that they will be biodegraded in all conditions. As for compound *D*, it cannot be concluded with definiteness that *D* is non-biodegradable. Flatness of curve *D* may be due to the absence of any microbes to biodegrade it or it could be due to unfavourable environmental conditions. Curve *C* indicates that after slow start, compound *C* is around 60 per cent biodegradable. It is just possible that *C* is not a pure compound and that it is impurity which is responsible for its low biodegradation. This can be checked by adding more quantity of *C* and if biodegradation does not improve, it could be due to the presence of impurities or due to unfavourable environmental factors [2].

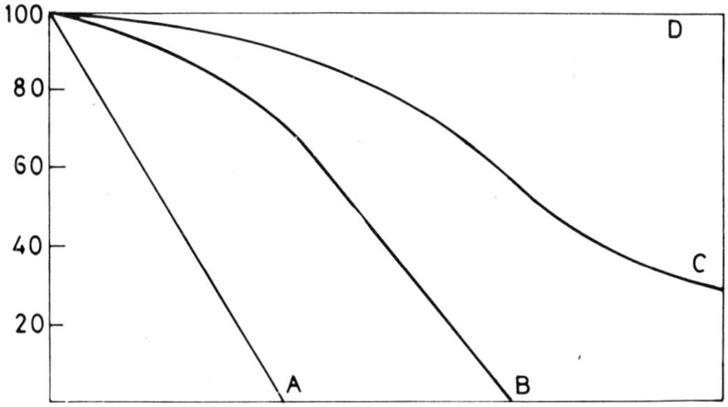

Fig. 2.1 Biodegradation of four compounds *A*, *B*, *C* and *D* following fermentation with the help of bacteria

Biodegradability does not solely depend on physical and chemical composition of a compound or a substrate. It is just possible that though a substrate may be prone to biodegradation, yet following retarded growth of microbes, biogas yield may be low. Addition of certain stimulating materials including animal urine can accelerate biodegradation. Reduced growth of methane bacteria causes accumulation of fatty acids, drop in pH and low gas yield. Methane bacteria have a generation time of over ten days. Studies for evaluating suitability of substrates for biogas yield should normally be continued for a period of five to six times the retention time. Several terms are used to describe different aspects of biodegradation of substrates. A stable process can be defined as one which lasts five times the retention period. A process which is not stable is called unstable process. An unstable process can be improved by certain course-corrections. Addition of certain salts and alkaline matter is known to improve its biodegradability. Different growth patterns of various kinds of methane bacteria in relation to the substrate and other bacteria populations make it difficult to apply a general solution. Thus every organic substrate of unknown composition needs a long run test to establish its suitability for biodegradation.

Fermentation can be either wet or dry. In wet fermentation digester contents are mixed with excessive water so that proportion of dry matter is within 10 per cent. Animal wastes are particularly amenable for wet fermentation. On the other hand agricultural plant material following problems of mixing and consequent scum formation pose problems in wet fermentation. For preventing scum formation on account of floating of plant materials, amount of water in digester is kept to a minimum which is just sufficient to keep the raw materials wet for active fermentation [5]. This process is called dry fermentation wherein total solids range between 25 to 30 per cent with no free water.

2.3 Biochemical Oxygen Demand and Oxidative Reduction Potential

Oxygen in slurry is consumed by a quantity what is called biochemical oxygen demand (BOD) which is defined as the quantity of oxygen needed by the bacteria to oxidise an organic matter. The BOD of organic waste is determined by a standard laboratory procedure. Rate of reaction is proportional to balance oxygen left in organic matter. The rate of reaction and BOD at time t (L_t) can be described by the following mathematical relationship:

$$\frac{dL}{dt} = -\ k_1 \cdot L$$

and,

$$L_t = L_0 \cdot e^{-k_1 \cdot t}$$

where,

L_0 = Ultimate BOD

k_1 = Coefficient representing reaction velocity (value for a typical sewage is 0.4 per day and for industrial wastes between 0.15 to 0.75 per day at 20°C)

Methane forming bacteria are highly sensitive to oxygen. Presence of oxygen prevents them from functioning. Scientists in China often use a term Oxidative Reduction Potential (ORP) to describe the activity of aerobic and anaerobic bacteria which they call aerobiosis and anaerobiosis respectively [11]. Initially when a biogas plant is charged with feed, environment within digester is aerobic following presence of air. Oxygen present in the beginning activates aerobic bacteria which in turn exhausts digester oxygen creating anaerobic environment. Besides aerobic bacteria, facultative bacteria which can function both in aerobic and anaerobic environment also help to create anaerobic environment within the digester by reacting with oxygen. Efficiency of aerobic and facultative bacteria in reducing oxygen level is measured by the ORP. Biogas yield, ORP, volatile acids and ammoniacal nitrogen are related to one another. For instance when ammoniacal nitrogen increases gas yield rises accompanied by corresponding decrease in volatile acid concentration and ORP. Initially the ORP is usually high but starts declining as fermentation progresses and stabilises at a value around tenth day of fermentation which roughly corresponds to the period of changeover from acid forming to methane forming phase.

Table 2.1 Types of Methane Bacteria Helpful in Biomethanation

Microorganism	*Substrates-Sole Electron Donor for Methanogenesis*
Methanobacterium arbophilicum	Hydrogen
Methanobacterium formicicum	Hydrogen or formate
Methanobacterium ruminantium	Hydrogen or formate
Methanobacterium mobile	Hydrogen or formate
Methanobacterium thermoautotrophicum	Hydrogen
Methanococcus vannieli	Hydrogen or formate
Methanosarcina barkeri	Hydrogen or methanol
Methanospirillum hangatii	Hydrogen or formate
Methanobacillus	Hydrogen or formate.

Source: Stafford, D.A. and Hughes, D.E., Microbial Production of Fuels, in 'D. Merrick and H. Marshall, Energy: Present and Future Options, Vol I, John Wiley and Sons Ltd, 1981,' p. 71.

2.4 Biochemistry of Anaerobic Fermentation

Anaerobic fermentation basically involves three phases, namely, hydrolysis, an acid phase and the methane phase. Hydrolysis phase covers break-down of large molecules into smaller ones by enzymes that are decomposable by bacteria. During the acid phase, complicated molecules such as proteins, fats and carbohydrates are decomposed by acid-forming bacteria into organic acids, carbon dioxide, hydrogen, ammonia and some impurities. Organic acids are mainly short chain fatty acids [10]. It should be noted that bacteria can utilise only hydrogen, carbon dioxide and formic acid (H·COOH). Volatile fatty acids formed during fermentation include formic acid (H·COOH), acetic acid (CH_3·COOH), propionic acid (CH_3CH_2·COOH). butyric acid (CH_3·CH_2·CH_2 ·COOH), valeric acid (CH_3·CH_2·CH_2·COOH), isovaleric acid (CH_3·CH_2·CH_2·COOH) and cacproic acid (CH_3·CH_2·CH_2·CH_2·CH_2·COOH). Proportion of different volatile acids formed is shown in Fig. 2.2. During methanogenic phase, methane-forming bacteria convert fatty acids into methane. Broad details of biochemistry of anaerobic fermentation process are shown in Fig. 2.3. In simple terms the anaerobic fermentation process can be described by a set of following chemical equations [7].

$$H{\cdot}COOH \longrightarrow CO_2 + H_2$$

$$4\,H_2 + CO_2 \longrightarrow 2\,H_2O + CH_4$$

$$CH_3{\cdot}COOH \longrightarrow CH_4 + CO_2$$

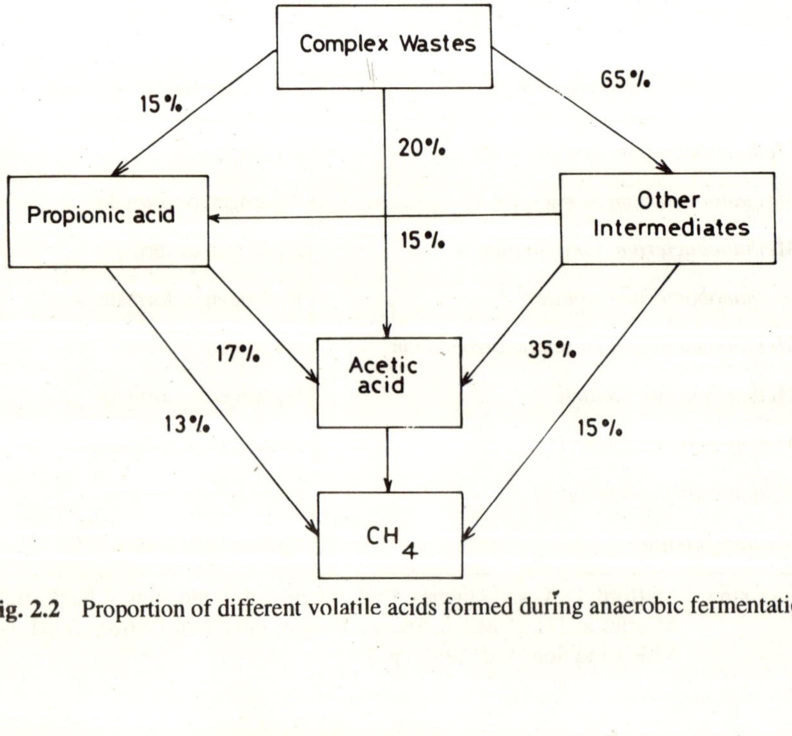

Fig. 2.2 Proportion of different volatile acids formed during anaerobic fermentation

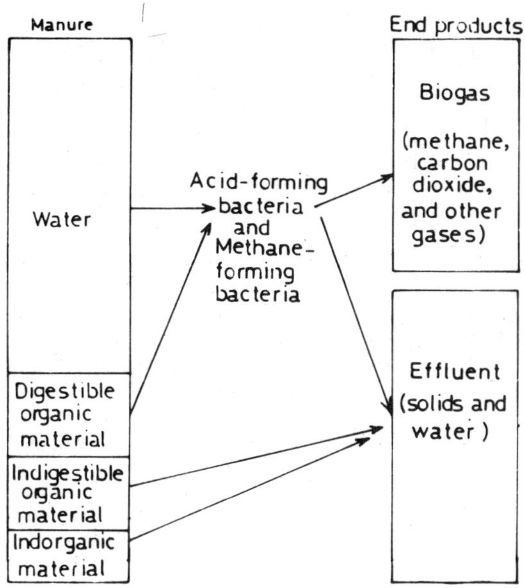

Fig 2.3 Illustration depicting biochemistry of anaerobic fermentation process [Source: Homan, E., Biogas from Manure, Special Circular 260, Pennsylvania State University, Pennsylvania, 1986]

Methane is also formed from some alcohols and other organic compounds, anaerobic oxidation of hydrogen and carbon dioxide and reduction of carbon dioxide to methane. Normally in a biogas plant all three phases occur in balanced fashion and any imbalance tends to lower gas yield [3]. Complete details of biochemistry of anaerobic fermentation process are shown in Fig. 2.4. If acids produced during acid phase are excessive and not fully utilised in other chemical reactions, it retards methane formation [9]. For optimum results, pH of digester contents needs to be maintained between 6.8 and 7.2.

2.5 Microbiology of Anaerobic Fermentation

Methane bacteria can be divided into four broad categories covering rod, coccus, sarcina and spiral form. Linked with these four categories there are four broad morphological group of methane bacteria which includes rod-shaped sporulating cells such as *methanobacillus;* rod-shaped non-sporulating cells such as *methanobacterium,* spherical cells in sarcina arrangement such as *methanosarcina*; and spherical cells not in sarcina arrangement such as *methanocoocus* [12]. Types of methane bacteria helpful for anaerobic fermentation are given in Table 2.1. These bacterias help to oxidise hydrogen by using available carbon dioxide. Reproductive cycle time of methane bacteria is found to vary from 4 to 6 days. There are some species whose reproductive cycle time is shorter than this period. These include *methanobacterium, thermoautotrophicum, M. arbophicum* and *Hungattii.* Among them reproductive cycle time of *M. thermoautotrophicum* is even shorter than three hours and

Cellulose + Nutrients

Extracellulor Enzymes
(Step 1 Hydrolysis

i.e.

$$(C_6\,H_{10}\,O_5)_n + n\,H_2O \xrightarrow[\text{Enzymes}]{\text{Hydrolysis}} n\,(C_6\,H_{12}\,O_6)$$

↓

Soluble Glucose + Nutrients

Acid Producing Bacteria
(Step 2) Acids

e.g.

$C_{16}\,H_{12}\,O_6$ = $2\,CH_3{\cdot}CHOH{\cdot}COOH$
Glucose — Lactic Acid'

$C_{16}\,H_{12}\,O_6$ = $CH_3{\cdot}CH_2{\cdot}CH_2{\cdot}COOH + 2\,CO_2 + 2\,H_2$
Glucose — Butyric Acid

$C_{16}\,H_{12}\,O_6$ = $2\,CH_3{\cdot}CH_2{\cdot}OH + 2\,CO_2$
Glucose — Ethyl Alcohol

↓

$$\begin{bmatrix}\text{Bacteria}\\ \text{Cells}\end{bmatrix} + \begin{bmatrix}\text{Volatile}\\ \text{Fatty Acids}\end{bmatrix} + CO_2 + H_2 + \begin{bmatrix}\text{Other products, alcohols,}\\ \text{lactate etc.}\end{bmatrix}$$

Methane Producing
Bacteria
(Step 3) Methane
Fermentation

e.g.

$4\,H_2 + CO_2 = CH_4 + 2\,H_2O$

$2\,CH_3{\cdot}CH_2{\cdot}OH + CO_2 = 2\,CH_3{\cdot}COOH + CH_4$

$CH_3{\cdot}COOH = CH_4 + CO_2$

$2\,CH_3\,(CH_2)_2\,COOH + 2\,H_2O + CO_2$
$= 4\,CH_3{\cdot}COOH + CH_4$

↓

Bacteria Cells + CH_4 + CO_2

Fig. 2.4 Details of biochemistry of Anaerobic Fermentation process

hence this is particularly appropriate for breeding methane bacteria. Methane bacteria utilise hydrogen electron donor to support their growth and breeding.

Most of the above species of methane bacteria are sensitive to temperature and are most effective at 35°C (308°K) and they become ineffective at temperatures below 10°C (283°K). It should be noted that whereas acid-forming bacteria are not very sensitive to oxygen, reproduce and multiply fast, the methane-forming bacteria are quite sensitive to oxygen and are slow to reproduce and multiply in its presence. As methane-forming bacteria cannot use nitrogen or carbon directly available from sources such as carbohydrates and proteins, intermediate acid-forming phase is desirable in between in order to maintain correct balance between the two bacterial populations. This in turn depends on temperature, level of acidity, and feedstock composition. As bacteria use carbon atoms nearly thirty times faster than the nitrogen, desirable carbon nitrogen ratio in feedstock is having more carbon than optimum, anaerobic fermentation decreases as all nitrogen is used; similarly if too much nitrogen is present digestion stops when the carbon is exhausted and excess nitrogen is lost as ammonia. During the acid phase, pH of digester contents normally drops to a value marginally below 6 in two weeks. After two weeks as digestion continues further and ammonia is formed, pH again rises to around 7 with methane formation and then further upto 8.2. At this stage mixture is regarded to be fairly well buffered and fresh feedstock can be added without any harmful effect. The bacteria are most efficient when they have intimate contact with feedstock particles and it is in this context that stirring of digester contents is conducive for efficient fermentation [11].

Hydrolysis phase. During hydrolysis phase, large molecules are broken down into smaller sizes by enzymes which in turn are decomposable by bacteria. During enzymatic hydrolysis, polymers are converted into soluble monomers that act as substrates for microorganisms during acid phase in which soluble organic compounds are converted into organic acids. Sewage sludge on dry weight basis broadly comprises hemicellulose 6 per cent, cellulose 34.5 per cent, lipids 14 per cent etc. Thus cellulose forms the largest fraction of carbohydrates in organic matter. Not only sewage but other wastes like animal manure and crop residues mainly contain carbohydrates alongwith lipid, proteins, and inorganic material. In organic wastes carbohydrates are mostly in the form of cellulose and other components of plant fiber such as hemicellulose and lignin. Thus, for breaking these components during anaerobic fermentation, bacteria need to possess what is known as cellulolytic, lipolytic and proteolytic activity.

Cellulolytic activity is helpful in reducing complex raw materials into simple and soluble organic components. Cellulose comprises polymerised glucose units as chains of indefinite length and complex branching patterns. Cellulolytic bacteria help to reduce chains and branches to dimeric and then to monomeric sugar molecules which are finally converted to organic acids. Cellulolytic bacteria is of two types *mesophilic* and *thermophilic*. Mesophilic bacteria are most active when digestion ranges from 30°C to 40°C whereas thermophilic bacteria are most active between 50°C and 60°C. For efficient working of both these bacteria, pH value of digester content should be between 6 and 7. As acid formation begins, pH tends to drop which can also be controlled by adding lime. When activities of acid forming and methane forming bacteria are in balance, pH of digester contents stabilises around 7. Rate of hydrolysis is somewhat slower than subsequent rate of acid and methane formation. This indicates that conversion of cellulose and other complex raw materials to simple monomers is a kind of rate limiting operation. Rate of hydrolysis depends on nature of substrate, bacterial concentration, pH and temperature of digester contents.

Carbohydrates form an important part of raw material for anaerobic fermentation [14]. Semi-cellulose is mostly in the form of polycondense pentose. It can be hydrolysed into pentose and small traces of hexose, xylose, arabinose, mannitol and glactose. Pectin which is present in traces is another kind of polycondense pentose. Pectin can be hydrolysed into galactose, gum sugar, xylose, galacturonic acid, methyl alcohol and acetic acid. Starch can be hydrolysed to form glucose, lignin is difficult to biodegrade and various celluloses have different biodegradability. With the formation of butyric and propionic acids as intermediate products, cellulose finally gets transformed into methane and carbon dioxide which are the main constituents of biogas [11, 13].

Acid phase. The monomeric components released during hydrolysis phase act as substrate for acid-forming phase. Acid phase can be viewed as periodic during which complex organic materials are converted into simpler acids such as volatile fatty acids. Acetic acid is one common by-product of digestion of fats, starch and protein. Methane bacteria are strictly anaerobic and can produce methane either by fermenting acetic acid to form methane and carbon dioxide, or by reducing carbon dioxide to methane using hydrogen gas or formate produced by other bacteria. Acetic acid is by far the single most acid prone to react with methanogenic bacteria resulting in methane formation. Acetic acid accounts for 70 per cent of methane produced.

Methane-forming phase. Methanogenic bacteria are selective in their reaction with substrate components. Methanogenic bacteria react with acetic acid, methanol, carbon dioxide and hydrogen to produce methane. Presence of oxygen retards the activity of methanogenic bacteria. Not only oxygen but even products like nitrites and nitrates which contain oxygen inhibit the activity of methanogenic bacteria. Methanogenic bacteria are sensitive to the pH of digester contents. The optimum pH lies between 7 and 7.2 though gas production is fairly stable for pH between 6.6 and 7.6. When pH drops below 6.6 there occurs a significant inhibition of methanogenic bacteria, and when it drops to a value of 6.2, digester contents become toxic to these bacteria. Acid-forming bacteria continue to be active even when pH level drops below 6.2 and in fact acidogenic bacteria produce acids until the pH drops to 4.5-5.0. When anaerobic fermentation proceeds in balanced state, pH automatically comes within desired range due to various biochemical reactions. It should be noted that volatile organic acids produced during acid-forming phase tend to lower the pH, a phenomenon which is countered by destruction of volatile acids and reformation of bicarbonate buffer during the methane-forming phase. In the event of imbalance, acid formers become more active than methane formers resulting in volatile fatty acid accumulation. This problem can be alleviated by adding lime which tries to arrest drop in pH. Chemicals such as ammonium hydroxide can also be used as buffering agents but it should be added in requisite quantity only otherwise it may release excessive ammonia and ammonium ions which are toxic to methanogenic bacteria. Common toxins include ammonia (> 1500–3000 mg/l of total ammonia nitrogen at pH > 7.4); ammonium ion (> 3000 mg/l of total ammonia nitrogen at any pH); soluble sulphides (> 50 – 100 mg/l, possibly > 200 mg/l); and soluble salts of metal such as copper, zinc and nickel.

Alkali an alkaline-earth metal salts of sodium, potassium, calcium or magnesium may be either stimulatory or inhibitory for methane bacteria depending upon their concentration. Toxic effect of harmful material can often be appreciably reduced by permitting only gradual build up in their concentration which helps bacteria to acclimatize. It is possible to minimise the adverse impact of soluble sulphides and ammonia on bacterial activity by allowing their concentration to grow gradually. Concentrations of soluble sulphides varying from 50 to 100 mg/l can be tolerated with little or no acclimatisation, but concentrations above 200 mg/l are considerably toxic. Heavy metal ions such as copper, zinc and nickel are toxic at concentrations as low as few ppm. However, when these concentrations are upto 100 mg/l (nearly 100 ppm) or more, no toxic effect is reported because at such concentrations they react with soluble sulphides and thus contain toxicity. There are some organic materials as well which act toxic to anaerobic fermentation. Their harmful effect depends on their concentration and rate of generation [9, 11]. For example, alcohols act toxic at high concentrations but when introduced gradually at low concentrations, they are easily degraded, and if added in controlled manner their addition is non-toxic.

Biodegradation of protein or urea in digester leads to the production of ammonia. Ammonia provides nitrogen to bacteria and its presence upto a certain level is desirable as it acts stimulatory to the bacterial activity. If ammonia concentration exceeds desired limit, it becomes toxic or inhibitory to bacterial activity. It is to be underscored that ammonium ion exists in equilibrium with dissolved ammonia gas $NH_4^+ \Leftrightarrow NH_3 + H^+$ and the latter is inhibitory at a much lower concentration than ammonium ion. Degree of impact of ammonia on bacterial activity depends on ammonia concentration and pH of the digester contents. When pH is high, there occurs increased concentration of dissolved ammonia gas as compared to ammonium ions which again causes inhibition. It has been found that if ammonia concentration exceeds 3000 mg/l or ranges between 1500 and 3000 mg/l it acts inhibitory to bacterial activity.

Process temperatures have considerable bearing on microbial activity. During mesophilic fermentations, as temperature falls below the optimum range of 33-38°C, microbial activity declines following increase in minimum solids-retention time for process stability. Same phenomenon takes place when temperature of digester contents during thermophilic fermentation falls below the desired range. Effect of sudden temperature changes is more deeper during thermophilic than mesophilic fermentation.

During conversion of carbohydrates to carbon dioxide and methane, almost equal volumes of each gas are produced. As carbon dioxide is soluble in water, not all of the carbon dioxide produced during fermentation is released as part of it gets dissolved in water. Carbon dioxide is also known to react with hydroxyl ion to form bicarbonates whose formation depends on alkalinity, temperature and composition of digester contents. Hydroxyl ion is mainly produced following reaction of ammonia with water to form ammonium hydroxide during deamination of biodegradable protein. Protein content of substrate thus affects amount of carbon dioxide formed. Amount of carbon dioxide associated with liquid stream as bicarbonate declines with rise in temperature and decrease in pH. Retention time of slurry in digester also affects proportion of carbon dioxide produced. Based on experimental studies for a given substrate, anaerobic fermentation at shorter retention times produces more carbon dioxide during acid phase and a gas with higher methane content during methane forming phase [4].

2.6 Kinetics of Anaerobic Fermentation

Several kinetic models have been developed to describe the anaerobic fermentation process which were developed initially in the context of anaerobic digestion of sewage sludge. J. Monod [8] developed a hyperbolic relationship between the exponential microbial growth rate and substrate concentration. In this model kinetic parameters, namely, microorganism growth rate and half velocity constants are deterministic in nature which help to predict condition when maximum biological activity occurs and when activity ceases. This model can be used to determine substrate utilisation (r_S):

$$r_S = \frac{q_{max} \cdot S_x}{K + S}$$

where,

S = limiting substrate concentration
K = half constant
x = concentration of bacterial cells
q_{max} = maximum substrate utilisation rate

For low substrate concentrations,

$$r_s \longrightarrow \frac{q_{max} \cdot S_x}{K + S}$$

and for high concentrations,

$$r_S \longrightarrow q_{max} \cdot x$$

The Monod model suffers from the drawback that one set of kinetic parameters cannot describe biological process at short and long retention times, and that kinetic parameters cannot be obtained for some complex substrates. To alleviate limitations of the Monod model while retaining its advantages,

A.G. Hashimoto, *et al.* [6] developed an alternative equation which attempts to describe kinetics of methane fermentation in terms of several parameters. According to this equation given below, for a given loading rate S_0/θ daily volume of methane per volume of digester depends on the biodegradability of the material (B_0) and kinetic parameters μ_m and K.

$$r_V = \frac{B_0 \cdot S_0}{\theta} \cdot \left[1 - \frac{K}{\theta\,\mu_m - 1 + K}\right]$$

where,

r = volumetric methane production rate, LCH_4/L. fermentor. day

S_0 = influent total volatile solids (VS) concentration, *g/L*

B_0 = ultimate CH_4 yield, LCH_4/g.VS added as $\theta \to \infty$

θ = hydraulic retention time/day

μ_m = maximum specific growth of microorganism per day

K = kinetic parameter, dimensionless

P.Y. Yang [17] modelled the initial biodegradable substrate concentration by using total exponential biogas production rate, the main relationship of which is as follows:

$$\beta = \frac{\beta_m \cdot S_0}{K_S + S_0}$$

whence,

$$\frac{1}{\beta} = \frac{1}{\beta_m} + \frac{K_S}{\beta_m} \cdot \frac{1}{S_0}$$

where,

β = exponential gas production rate

β_m = maximum gas production rate

K_S = saturation constant or the substrate concentration,

S_0 = initial biodegradable substrate concentration, TVS or total COD removed while the waste is stabilized.

Continuous vs. batch process. Anaerobic fermentation can be carried out as continuous or batch process. Continuous digestion in turn can be carried out either in a plug flow reactor (PFR) or continuous stirred tank reactor (CSTR). In the plug system each particle has the same residence time in the reactor (digester) while in CSTR there is a broad distribution of residence times. Traditionally CSTR digesters have been used for sewage treatment plants based on continuous anaerobic fermentation. Plug flow reactors though sufficiently developed are still much less in number [16].

Assuming that digestion process follc ws first order kinetics, plug flow or continuous stirred tank digestion can be modelled by utilising chemical reaction theory. For first order kinetics,

$$\frac{dC_B}{dt} = -k \cdot C_B$$

where C_B is the concentration of biodegradable volatile solids, k is the first order reaction rate constant and t is time. A material balance equation for the digester can be written as follows:

Material out = Material in – Material digested

Exit concentration of biodegradable volatile solids for the two processes in relation to the initial concentration can be suitably expressed through a set of mathematical relationships [16].

For continuously stirred tank reaction, the relationship becomes

$$C_{B_e} = \frac{C_{B_0}}{(1 + k \cdot T)}$$

and for the plug flow system the corresponding relationship is

$$C_{B_e} = C_{B_0} \cdot \exp(-k \cdot T)$$

where,

C_{B_0} = inlet concentration of biodegradable volatile solids

C_{B_e} = exit concentration of biodegradable volatile solids

T = hydraulic retention time defined as rate reactor volume to inlet volumetric flow rate

A comparison of above equations shows that for the same retention time, plug flow process achieves higher substrate utilisation or equivalently lower exit concentration.

The gas production rate for the two processes can be represented by the following equations:

$$G_{\text{CSTR}} = A\,(C_{B_0} - C_{B_e})$$

$$= \frac{A \cdot C_{B_0} \cdot k \cdot T}{(1 - k \cdot T)}, \text{ and}$$

$$G_{\text{Plug}} = A \cdot C_{B_0} \cdot [1 - \exp(-k \cdot T)]$$

where A is the volume of gas produced per mass converted. The rate of gas yield can be determined by their respective derivatives with respect to T. Hence (dG/dT) represents the volume of gas produced per reactor volume per day (VVD). Thus,

$$\frac{dG_{\mathrm{CSTR}}}{dT} = \frac{A \cdot C_{B_0} \cdot k}{(1 - k \cdot T)}$$

$$\frac{dG_{\mathrm{Plug}}}{dT} = A \cdot C_{B_0} \cdot k \exp(-k \cdot T)$$

C_{B_0} and k can be calculated with the help of the experimental data.

For a batch process, the rate of substrate removal and rate of gas production is the same as for the plug flow system. Hence, for a batch process, the substrate concentration after time t is given by

$$C_{B_e} = C_{B_0} \cdot \exp(-k \cdot T)$$

and the total gas yield is given by

$$G = A \cdot C_{B_0} [1 - \exp(-k \cdot T)]$$

The rate of gas yield (VVD) decreases with time and is proportional to e^{-kT}. Experimental results indicate that there occurs first an increase in gas yield with time followed by a decrease. This difference can be attributed to the assumption of first order kinetics which does not hold completely at all times. To elaborate further, this assumes that the microorganism level, volatile acid concentration, pH, nutrients, and other digestion conditions are at appropriate levels for first order kinetics to be applicable which in reality is not true.

2.7 Hydrogen Production by Hydrogen Producing Bacteria of Digester Sludge

Scientists at the CIB (Chengdu Institute of Biology) in China in past also isolated 14 strains which comprised 63.6 per cent of total hydrogen producing bacteria obtained from enriched digester sludge. This included 4 strains of *Serratia marcescens* (18.2 per cent), 1 strain of *Escherichia coli* (4.5 per cent), 1 strain of *Citrobacter freundii* (4.5 per cent) and 2 strains of *Clostridium acetobutylicum* (9.1 per cent). While carrying out experiments for analysing the effect of hydrogen producing bacteria on anaerobic fermentation the above strains were mixed with enriched cultures of methane producing bacteria. The results are summarised in Table 2.2 from which it can be seen that with the mixed cultures, methane contents in biogas increased to appreciably high value of 90.3 per cent with carbon dioxide reduced to insignificant proportions. This is a very welcome improvement in biogas composition and needs to be studied further to make a commercial success.

During hydrolysis and acid-forming phase, a group of bacteria produce hydrogen. Further research needs to be carried out to study kind, number, nutrition, metabolism and behavioural pattern of micro-organisms. Only 20 per cent of hydrogen producing bacteria produce hydrogen. Considerable work has been recently done at the CIB, Academia Sinica in China to study the role of dynamics of hydrogen producing bacteria. Scientists (China) obtained

Table 2.2 Effect of mixing Hydrogen-producing Bacteria with Enriched Culture of Methane-producing Bacteria

Fermentation Period (days)	*Biogas Constituents*	*Per cent Proportion*	
		With Hydrogen Producing Bacteria	*Without Hydrogen Producing Bacteria*
2	CH_4	77.0	77.5
	CO_2	8.5	11.6
5	CH_4	69.5	66.2
	CO_2	17.3	16.4
7	CH_4	69.2	72.0
	CO_2	1.1	18.0
9	CH_4	90.3	64.5
	CO_2		23.8
12	CH_4	94.6	66.3
	CO_2		23.8
16	CH_4	90.5	58.5
	CO_2		34.0
19	CH_4		67.0
	CO_2		11.7

hydrogen producing algae by isolating them from biogas plant sludge. These bacteria were later on enriched which gave hydrogen yield of more than 50 per cent. A total of 51 bacteria strains were isolated from enriched cultures containing 24 (47 per cent) hydrogen producing bacteria. As expected hydrogen yield was found to be different from different types of bacteria. Summary of hydrogen yield of different hydrogen producing bacteria is given in Table 2.3. Hydrogen-forming bacteria belong to *Escherichiaccae* and *Bacillaceae* families. The former category comprises five species, namely *Escherichia coli, Enterbacter, Cloacae, Hafnia alvei, Citrobacter freundii* and *Serratia marcescens* whereas the latter comprises only one specie, *Clostridium acetobutylicum.*

2.8 Concluding Remarks

Bacterial growth occurs during all phases of anaerobic fermentation. Anaerobic bacteria utilise substrate to somewhat lower level as compared to aerobic bacteria for supporting their bacterial growth. Process temperatures have considerable bearing on microbial activity. Methane bacteria are considered to be very sensitive to sudden temperature changes [4]. For optimum plant performance, it is necessary to have controlled temperature within the desired range which is quite narrow. In anaerobic fermentation process there are specific micro-organisms which play important role. These, among others, include *methanobacterium, methanobacillus, methanococcus* and *methanosarcina.* Several difficulties are faced while carrying out experimental studies for anaerobic fermentation process. For instance, it is difficult to obtain active bacteria from digester and keeping them intact while delineating their species.

Table 2.3 Summary of hydrogen yield of various hydrogen producing bacteria

Hydrogen Producing	*Bacteria Strains*	*Average Hydrogen Yield (ml/day)*	*Per cent Hydrogen Content in Biogas*	
			Average content	*Range*
6-1	25	270	57.6	52.5-63.9
H-1	R	366	55.6	51.5-69.9
H-1	W	386	46.8	22.1-69.9
6-2	28	250	41.8	11.0-66.1
S-2	27	180	51.8	43.6-65.0
6-2	34	413	53.7	46.6-65.0
6-2	33	321	49.8	36.0-63.8
6-2	32	317	51.5	42.7-63.8
6-2	30	273	47.0	18.3-63.8
1-2	30	(trace)	23.5	0-63.0
T3-2	100	94	36.7	10-63.0
6-1	50	190	27.7	4.5-61.8
6-2	54	142	51.9	34.5-63.0
6-2	27	240	44.4	38.9-53.9
6-2	29	113	47.9	42.9-53.9
1-2	32	108	49.7	45.4-52.5
6-2	55	330	39.7	30.4-52.0
6-2	49	103	46.6	43.8-52.4
6-2	53	50	30.9	0-47.5
6-1	31	147	46.5	43.3-51.5
1-1	4	(trace)	16.3	0-49.5
6-2	48	353	46.3	44-47.9
1-2	1	(trace)	15.4	0-25.1
6-1	20	378	52.9	48.7-75.5

Counting of bacteria is also not a simple affair and many times study results differ because of different counting techniques used.

REFERENCES

1. BEI (Biomass Energy Institute), *Biogas Production from Animal Manure,* Winnipeg, Manitoba, 1978.
2. Calley, A.G., *et al., Treatment of Industrial Effluents,* Hodder and Stoughton, London, 133 1977.
3. Chan, D.B. and Pearson, E.A., *Comprehensive Studies of Solid Wastes Management: Hydrolysis Rate of Cellulose in Anaerobic Fermentation,* Soil Report No. 70-3, Sanitary Engineering Research Laboratory, Univ. of California, Berkeley, 1970.
4. Chhonkar, P.K., *Microbiological Aspects of Biogas Production,* Proc. Nat. Symp. on Biogas Technology and Usage, IARI, New Delhi, Nov. 29-30, 1977.
5. Goswami, K.P., Appropriate Technology for Rural India to Produce Biogas from Vegetable Wastes, *Energy Sources,* **11**, 59-67 1989.

6. Hashimoto, A.G., *et al., Anaerobic Fermentation of Beef Cattle Manure,* Report No. TR-98372-1, SERI, Colorado, 1981, p. 9.
7. Homan, E., *Biogas from Manure*, Special Circular No. 260, Pennsylvania State University, Pennsylvania, 1980.
8. Monod, J., The Growth of Bacterial Cultures, *Ann. Rev. Microbiology,* **3**, 371-394 1941.
9. NAS, *Methane Generation from Human, Animal and Agricultural Wastes,* Report of an Ad Hoc Panel of the Advisory Committee on Technology Innovation, National Academy of Sciences, Washington, DC, 1977.
10. Schroader, E.D., *Water and Waste Water Treatment,* McGraw-Hill Book Co, NY, 1977.
11. SPOBD, *Biogas Technology and Utilisation,* Proc. Chengdu Seminar, Sichuan Provincial Office of Biogas Development, P.R. China, 1979.
12. Stafford, D.A. and Hughes, D.E., Microbial Production of Fuels, in 'D. Merrick and H. Marshall, *Energy: Present and Future Options, Vol I,* John Wiley and Sons Ltd, 1981', p. 71.
13. Tumwasron, S., *Biogas Technology and its Application at Different Farm Levels in Central Thai Village,* A Report, Animal Science Department, Kasetsart University, Thailand, 1981.
14. United Nations, *Guidebook on Biogas Development,* U.N., New York, 1980.
15. Waart, J.D., *Limitation of Biodegradability of Substrate for Biogas Production,* paper presented at the European Economical Society Conf. on Bioenergy, Dyon, Oct. 29-30, 1980.
16. West, C.E. and Ashare, E., *Feasibility Study for Anaerobic Digestion of Agricultural Crop Residues – 1980 Update,* Report TR-98175-1, SERI, Colorado, March, 1981.
17. Yang, P.Y., *et al., Kinetic Aspects of the Anaerobic Fermentation of Swine Waste With and Without Algal Biomass, A Report,* Agricultural Engg. Dept., Univ. of Hawaii, Honolulu, Hawaii, 1980.

Chapter 3

PROPERTIES OF BIOGAS AND METHODS FOR ITS PURIFICATION

3.1 Introduction

Biogas consists of about 60 to 70 per cent methane, 30 to 40 per cent carbon dioxide and traces of other impurities like ammonia, hydrogen sulfide, mercaptans and other noxious gases [8]. Biogas is about 20 per cent lighter than air and its ignition temperature ranges between 650 to 750°C. Biogas because of its unique characteristics is beneficial in many ways. Large-scale promotion of biogas plants helps to generate substantial employment for entrepreneurs, masons, village technicians and unskilled workers in rural area. Apart from its use as cooking, lighting and IC engine fuel, biogas saves women and children from hardships and drudgery of collecting and headloading of heavy bundles of fuelwood, exposure to smoke in the kitchen and time-consuming cooking. Biomethanation improves village sanitation and upgrades environmental conditions as the forest cover is protected due to saving of fuelwood [18]. Carbon dioxide which is its second major component being non-combustible tends to lower combustible characteristics of biogas. Hydrogen sulfide which is one of its main impurities has an odour similar to one from rotten eggs and can be detected in concentrations as low as 0.0001. Due to its peculiar smell, hydrogen sulfide is helpful in easy leakage detection of biogas whenever it takes place. Presence of hydrogen sulfide is also responsible for the corrosive nature of biogas since it tends to form an acid in reaction with water in slurry which corrodes metallic parts including gas pipelines. This process also contributes to water-vapour formation which is a commonly overlooked contaminant in biogas that makes it corrosive. As biogas leaves the digester it gets saturated with water vapour making it wet. It is because of this property that it is sometimes called wet gas. Furthermore, whenever ambient temperatures fall appreciably water vapour gets condensed causing blockages in pipelines. Water vapour corrodes pipelines, damages internal combustion engines using biogas as a fuel and causes failures in compressors. Several methods have been developed for purification of biogas which are quite effective but with the limitation that these add to cost of production. Process of removing impurities from biogas is called scrubbing [4, 5].

3.2 Important Properties

In view of the fact that biogas is rarely available in its purest form, its composition and characteristics vary from source to source depending upon feed materials used. Accordingly, calorific value of biogas is also not fixed and varies within 500 to 700 BTU per cubic foot (0.03 cubic metre). As compared to this, natural gas has a value of 850 BTU per cubic foot (0.03 cubic metre) and petrol 120,000 BTU per gallon (3.8 litres). In metric units, calorific value of biogas ranges between 4700 to 6000 kcal/m^3. Biogas burns with a blue flame. It can be directly used in gas burning appliances used for heating, cooking, lighting and running internal combustion engines with a compression ratio of 8 : 1 or even greater [17]. Amount of air needed for combustion depends upon methane content of biogas. For example, for combustion of 1 m^3 of biogas with 60 per cent methane, 8 m^3 of air is needed. Flame speed factor of biogas is 11.1 which affects the manner in which the flame tends to lift off burners. Critical pressure of methane is 4710 KPA at – 82.3°C which makes its liquefaction easier [11]. The high critical pressure of methane is the main factor which acts as hurdle in bottling of biogas and consequent portability like LPG. General properties of biogas and its constituent gases are given in Tables 3.1 to 3.7.

Table 3.1 Pressure and Temperature Needed to Liquify Biogas

Pressure (kg/cm^2)	*Temperature (°C)*	*Volume of Cylinder (m^3)*
80	– 70	0.10
100	– 50	0.14
130	– 25	0.14
160	0	0.15
160	25	0.20

Source: Singhal, O.P., Development of Biogas Technology in India, Steam and Fuel Users Journal, 33 (3-4), 53 (1984)

Table 3.2. Select of Properties of Biogas after Purification (Scrubbing)

Properties	*Value along with unit*
1. Molecular weight	16 g $mole^{-1}$
2. Density	0.72 g L^{-1} (air = 1.29 gl^{-1})
3. Energy content	37.75 MJ/m^3
4. Octane rating	130
5. Ignition temperature	651°C
6. Air to CH_4 ratio needed for complete combustion	10 to 1 by volume
7. Explosion limits in air	5 to 14%
8. Wobbe No.	732
9. Flame speed factor	11.1

Source: BEI, Biogas Production from Animal Manure, Biomass Energy Institute Inc, 304-870, Cambridge Street, Winnipeg, Manitoba, Canada, 1978, p. 9.

Table 3.3 Important General and Combustion Properties of Methane which is the Main Constituent of Biogas

Properties	*Value along with unit*
General	
1. Chemical formula	CH_4
2. Characteristics	Colourless, odourless
3. Molecular weight	16.04
4. Density	0.5576 g/cc (air = 1)
5. Heat value	1012 BTU/cft at 14.4 psi, 60°F
6. Critical pressure	684 psi
7. Critical temperature	– 116°F
8. Heat capacity	8.4072 cal. deg^{-1}. $mole^{-1}$ at 300°K
9. Heat of combustion	210.8 kg-cal/g. mole at 1 atm, 68°F
10. Heat of formation	20.34 kg-cal/g. mole at 1 atm, 68°F
Related to combustion	
11. Octane rating	130
12. Ignition temperature	1202°F
13. Limits of flammability	5-15% by weight
14. O_2/CH_4 for complete combustion	3.98 by weight, 2.0 by volume
15. Products of combustion	$CH_4 + 2O_2 \longrightarrow CO_2 + 2\,H_2O$
16. CO_2/CH_4 for complete combustion	2.74 by weight; 1.0 by volume.

Source: 1. Lapp, H.M., *et al.*, Methane Production From Animal Wastes: Fundamental Considerations, Canadian Agricultural Engineering, 17(2), 1975, p. 101.

2. Johnson, A.J. and Auth, G.H., Fuels and Combustion Handbook, Mc-Graw Hill, NY, 1951.
3. Weast, R.C. (ed)., Handbook of Chemistry and Physics, 51st edition, Chemical Rubber Co., Cleveland, Ohio, 1971.

Table 3.4 Physiological Properties of Constituent Gases of Biogas

Gas	*Specific*[1] *Gravity*	*Odour*	*Colour*	*Min %*	*Max %*	*MIO*[2] *(ppm)*	*MAC*[3] *(ppm)*	*Concentration*[4] *(ppm)*	*Exposure*[5] *Period (min)*	*Physiological*[6] *Effects*
Ammonia (NH_3)	0.6	sharp pungent	none	16	–	53	100		–	Irritant
								400	–	Irritation of throat
								700	–	Irritation of eyes
								1700	–	Coughing & frothing
								3000	30	Asphyxiating. Could be fatal
Carbon Diodixe	1.5	none	none	–	–	–		5500		Asphyxiant
								20000		Safe
								30000		Increased breathing
								40000		Drowsiness, headaches.
								60000	30	Heavy, asphyxiating breathing
								300000	30	Could be fatal
Hydrogen Sulfide (H_2S)	1.2	rotten smell	none	4	46	0.7	20			Poisonous
		nauseating						100	hours	
								200	60	
								500	30	
								1000	–	
Methane (CH_4)	0.5	none	none	5	15	–	–	500000	–	Asphyxiant
										Headache, nontoxic

Note: 1. Specific gravity is the ratio of the weight of pure gas to standard atmospheric air. If number is less than one, the gas is lighter than air otherwise it is heavier.

2. MIO in expanded form is the Minimum Identifiable Odour which is threshold odour, i.e. the lowest concentration (highest dilution) level that enables detection.

3. MAC in expanded form is the Maximum Allowable Concentration which is the concentration prescribed by the health authorities as the maximum allowable in atmosphere in which people can work during 8 to 10 hour period. These levels need be lower in confinement units as animals stay in such environment continuously for 24 hours.
4. Concentration in parts of the pure gas in million parts of atmospheric air. For changing the concentration to per cent by volume, the listed number is to be divided by 10000.
5. Exposure period is the time during which effects of noxious gas are felt by an adult human being or an animal (especially pig) of about 150 lb weight.
6. Physiological effects are those that occur in adult human beings; similar effects would be felt by animals weighing 150 lb, lighter animals will be affected sooner and at lower levels, heavier animals at later times and higher concentrations.

Source: Hills, D.J., and Firbank, W.C., Methane Generation from Agricultural Wastes, A Report, Department of Agricultural Engineering, University of California, Davis, Jan, 1979.

Table 3.5 Comparative Properties of Different Fuels Commonly Used in India

Biogas	*Unit*	*Calorific Value (kcal)/ vol.*	*Mode of Burning*	*Thermal Efficiency*	*Effective Heat Available (Kcal)/vol.*
1. Biogas	m^3	4713	In standard burner	60	2828
2. Kerosene	litre	9122	Pressure stove	50	4561
3. Cowdung	kg	4708	In open chulha	17	814
4. Cowdung cakes	kg	2092	–do–	11	230
5. Charcoal	kg	6930	–do–	28	1940
6. Soft coke	kg	6292	–do–	28	1762
7. Butane*	kg	10882	In standard burners	60	6529
8. Furnace oil	litre	9041	In water tube boiler	75	6781
9. Coal gas	m^3	4004	In standard burners	60	2402
10. Electricity	kwh	860	Hot plate	70	602

*Butane: Bottled domestic cooking gas sold as Indane, Burshane, Esso etc.

Source: KVIC, Gobar gas: Why and How, Khadi and Village Industries Commission, Mumbai, 1979.

Table 3.6 Some Important Characteristics of Biogas and its Constituent Gases

Characteristic	*Unit*	*Characteristic of Constitutent Gases*				*Biogas (60% CH_4 & 40% CO_2)*
		CH_4	CO_2	H_2	H_2S	
1. Volume share	%	55-70	27-44	1	3	100
2. Calorific value	MJ/m^3	35.80	–	10.8	22.8	21.50
3. Limit of inflammability (explosive range) in air	Vol%	5.15	–	4.80	4.45	6.12
4. Ignition temperature	°C	650.750	–	585	–	650.750
5. Density (°C, 760 mm)	g/cc	0.74	1.98	0.98	1.54	1.20
6. Specific gravity (relative to air)	–	0.55	1.5	0.07	1.2	0.83

Source: BORDA, Biogas: Manual for the Realisation of Biogas Programmes. Bremen Overseas Research and Development Association, Bremen, Germany (erstwhile FRG), 1980, p. 81.

3.3 Safety Aspects for Handling Biogas

Biogas is explosive gas and accordingly it needs to be handled with care. The most critical time to observe safety precautions is during periods of digester and gas pipeline repairs. Biogas in a concentration of 6 to 15 per cent with air forms an explosive mixture. While carrying out periodic repairs, it is necessary that all gas in gasholder be allowed to escape first to avoid presence of any air gas mixture inside. In case of fixed dome biogas plant it can be achieved by allowing sufficient time for gas to escape. In a movable drum type plant this objective can be met by making gasholder to settle down to maximum possible depth over slurry. Soapy water may be applied on gas taps, gas cocks, joints and fittings for checking for possible leakages as indicated by bubble formation. Small leakages at joints can be repaired by applying bitumen coating. After coating, affected joints can be rechecked for residual leakage. Fire-prone appliances should be prevented from coming in contact with leaking gas. Gas pipelines leading to flame in a burner or lamp be fitted with fire-arrester to prevent back-travelling of flame into plant. Flame traps in the form of wire mesh can be installed in gas delivery lines near gas-appliances. In kitchen, vent can be provided near roof-top to enable the accumulated gas to move out which due to its being lighter than air tends to accumulate near ceiling. Plants should be periodically inspected for possible leakages to minimise involved risks and hazards [9]. A primitive method of bringing fire close to the point of suspected leakage is highly risky and dangerous which is to be avoided. Inlet and outlet of biogas plant should be covered with strong lids to avoid falling of objects into them. Areas where biogas may accumulate following leakage should have

Table 3.7 Replacement Values of Different Fuels Commonly Used in India

Fuel	*Unit*	*Biogas m^3*	*Kerosene 1 litre*	*Fuelwood 1 kg*	*Cowdung cakes 1 kg*	*Charcoal 1 kg*	*Soft coke 1 kg*	*Butane 1 kg*	*Furnace Oil 1 litre*	*Coal gas 1 m^3*	*Electricity (1 kwh)*
1. Biogas	m^3	1.000	1.613	0.288	0.081	0.686	0.623	0.309	2.398	0.849	0.213
2. Kerosene	litre	0.620	1.000	0.178	0.050	0.425	0.386	1.431	1.487	0.527	0.132
3. Fuelwood	kg	3.474	5.603	1.000	0.283	2.383	2.165	8.210	8.330	2.951	0.741
4. Cowdung cakes	kg	12.296	19.830	3.539	1.000	8.435	7.640	28.387	29.483	10.443	2.617
5. Charcoal	kg	1.458	2.351	0.420	0.119	1.000	0.908	3.365	3.495	1.238	0.310
6. Soft coke	kg	1.605	2.589	0.462	0.130	1.101	1.000	3.705	3.848	1.363	0.342
7. Butane*	kg	0.433	0.699	0.125	0.035	0.297	0.270	1.000	1.039	0.368	0.092
8. Furnace Oil	litre	0.417	0.673	0.120	0.034	0.286	0.260	0.963	1.000	0.354	0.089
9. Coal Gas	m^3	1.177	1.899	0.339	0.096	0.808	0.734	2.788	2.823	1.000	0.251
10 Electricity	kwh	4.698	7.576	1.352	0.383	3.223	2.927	10.846	11.264	3.990	1.000

*Butane: Bottled domestic cooking gas sold as Indane, Burshane, Esso etc.
Source: KVIC, Gobar Gas: Why and How; Khadi and Village Industries Commission, Mumbai, 1979.

alarm facilities. Biogas should be prevented from discharging into air, closed and confined areas. This requires use of gas-tight lines, fittings and pressure relief valves vented to exterior of buildings and confined spaces. It is also advisable to have fire-fighting facilities available in areas where large number of plants exist in clusters. Necessary safety instructions prescribed by implementing agencies should be affixed near biogas plants [12].

3.4 Removal of Carbon-Dioxide from Biogas

Carbon dioxide being noncombustible constituent of biogas lowers its heat value. Removal of carbon dioxide is not necessary when gas is to be used for cooking or lighting purposes only. A number of methods have been developed for CO_2 removal (scrubbing) which depending upon the technique involved are called water-scrubbing, caustic-scrubbing, solid absorption, liquid absorption and pressure separation. A brief description of these methods is as follows [11].

Water scrubbing. In this method gas is made to pass through water which absorbs part of CO_2. The inherent limitation of this method is that it requires large quantity of water. Based on studies carried out by H.M. Lapp [11], 7 ft^3 (0.2 m^3) of biogas at 68°F (20°C) and 1 atmospheric pressure (1.03 kg/cm^2) requires 2.7 gallons (12.3 litres) of water for CO_2 removal. CO_2 is highly soluble in water (Table 3.8). Spent water following absorption of CO_2 becomes acidic and hence unsuitable for several applications as it corrodes metallic surface it comes in contact with.

Table 3.8 Approximate Solubility of Carbon Dioxide in Water

Pressure (Atmospheres)	*Solubility Expressible in lb of Carbon Dioxide per lb of Water at Different Temperatures*				
	32°F	*50°F*	*68°F*	*86°F*	*104°C*
1	0.40	0.25	0.15	0.10	0.10
10	3.15	2.15	1.30	0.90	0.75
50	7.70	6.95	6.00	4.80	3.90
100	8.00	7.20	6.60	6.00	5.40
200	–	7.95	7.20	6.55	6.05

Source: Nonhebel, G., Gas Purification Processes. George Newness Ltd., London, 1964.

Caustic scrubbing. This method works on the principle that when caustic solutions are made to react with CO_2 bearing gas streams, an irreversible carbonate-forming reaction followed by a reversible bicarbonate forming reaction take place as per the following equations. This process involves use of hydroxides of sodium, potassium and calcium.

$$2\,KOH + CO_2 \longrightarrow K_2CO_3 + H_2O$$

$$K_2CO_3 + CO_2 + H_2O \Leftrightarrow 2\,KHCO_3$$

In most industrial applications, no attempt is made to regenerate spent bicarbonate solution due to high steam requirement for this process. Carbon dioxide absorption into alkaline solution is adversely affected by slow conversion of dissolved CO_2 molecules into more reactive ionic species. Mixing of liquid during absorption helps to achieve diffusion of gas molecules in liquid and prolongs their contact time which adds to former's absorptivity. Normality of caustic solution also affects rate of absorptivity. With sodium hydroxide solution (NaOH), for instance, it was found that rate of reaction is more rapid if normality lies between 2.5 to 3. Potassium hydroxide (KOH) is more commonly used in industrial scrubbing but it suffers from limitation that it is not readily available in rural areas where biogas plants are normally located. Calcium hydroxide [$Ca(OH)_2$] on the other hand is generally preferred for biogas scrubbing as this chemical is more readily available and cost of operating a lime-water scrubber is also relatively less. The main limitation of lime-water scrubbing are difficulties faced in controlling solution strength, and removal of large amounts of precipitate from mixing tank and scrubber. In most cases, sediment and suspended particulate matter need be removed in order to avoid clogging in pumps, high pressure spray nozzles, packing and bubbling apparatus. Sodium hydroxide has major advantage of being available in easily handled pellet forms that enables rapid and simple recharging of scrubber. However with NaOH solutions problems of suspended particulate matter are not totally eliminated.

Method Developed by the IARI, New Delhi. T.D. Biswas, *et al.* [15] developed another method for CO_2 scrubbing. It was found that biogas can be removed by bubbling it through 10 per cent aqueous solution of mono-ethanolamine (MEA). By single bubbling through a plain column of 6 cm height, carbon dioxide content in biogas was reduced to 0.5-1 per cent by volume from initial value of 40 per cent. Scrubbing column was made of an inexpensive plastic bubbler of 5 cm diameter and 15 cm height with only one orifice. Maximum removal of carbon dioxide was observed when bubbles moved out individually without colliding one another to form continuous stream. Optimum gas flow rate to the regulator was estimated as 100 ml per minute which gave 60 ml of purified gas per minute in the reservoir. Decrease in this rate of flow was not found to cause any further scrubbing. Initial pressure of the gas introduced into the bubbler was 10 cm of water column and drop in pressure head was about 5 cm of water column. Both caustic potash and monoethanolamine solution were effective in reducing the carbon dioxide content to 0.5 to 1 per cent. Whereas spent caustic potash solution cannot be regenerated, MEA solution can be completely regenerated by boiling for five minutes and thus can be used again and again. Furthermore, MEA solution is far less caustic than other solutions used and therefore pose much less hazards for skin. This method is thus very practical and economic for biogas scrubbing.

Pressure separation. This method works on the principle of compressing biogas beyond the limit of critical partial pressure of impurities (CO_2) with temperature greater than critical temperature of methane but below those of impurities. For instance carbon dioxide liquifies when gas temperature falls below 89.6°F (32°C) after compressing beyond 1106 psi (77.76 kg/cm^2). The method can be used to separate hydrogen sulfide also but it suffers from drawback that it tends to corrode compressor. It is particularly appropriate when methane is desired at pressures greater than 1100 psi (77.34 kg/cm^2). For necessary reference while using this process, critical temperature and pressure of methane and primary impurities in biogas are included in Table 3.9 [11, 16].

Table 3.9 Critical Temperatures and Pressure of Constituent Gases of Biogas

Constituents	*Critical Temperature (°F)*	*Critical Pressure (psi)*
Methane	– 116	683.6
Carbon dioxide	88	1106.5
Hydrogen sulfide	213	1328.9
Ammonia	270	1626.6

Source: West, R.C. (ed.), Handbook of Chemistry and Physics, 51st edition, Chemical Rubber Co, Cleveland, 1971. Quoted in 'Lapp, H.M., Methane Production from Animal Wastes-1: Fundamental Considerations, Canadian Agricultural Engineering, 17 (2), 1975, p 100'.

3.5 Removal of Hydrogen Sulphide from Biogas

Hydrogen sulphide which is usually present in trace quantities in biogas is poisonous and harmful for health, and also imparts corrosive property to it. Process of removing hydrogen sulfide from gas is called desulfurisation. Like carbon dioxide, it can also be removed by water and caustic scrubbing. A brief description of some of the commonly used desulfurisation methods are as follows [11].

Water scrubbing. According to this method gas is passed through water which absorbs part of hydrogen sulfide. However, water-scrubbing for desulfurisation is not that effective as it is for CO_2 removal. Low partial pressure of H_2S leads to low mole fraction absorption of hydrogen sulfide in water which makes it relatively less effective for desulfurisation.

Caustic scrubbing. This method works on the principle that when caustic solution are made to react with CO_2 bearing gas streams, an irreversible carbonate-forming reaction followed by another irreversible potassium hydrosulfide forming reaction takes place which appears as a yellow precipitate. Following chemical equations represent the two chemical reactions:

$$2\,KOH + CO_2 \longrightarrow K_2CO_3 + H_2O$$

$$H_2S + K_2CO_3 \longrightarrow KHS + KHCO_3$$

It is possible to remove hydrogen sulfide and carbon dioxide by using other chemicals like ammonical solutions, trisodium phosphate, sodium phenolate and other alkacid processes but these suffer from limitation of high costs of purification. In caustic scrubbing hydroxides of potassium, sodium and calcium [KOH, NaOH and Ca $(OH)_2$] are used for purification.

Solid chemical absorption. This is one of the simple and practical methods for desulfurisation which is especially appropriate for situations where other impurities are not necessary to be removed. In this method 'iron sponge' consisting of ferric oxide and wood shavings and kept in dry scrubber is made to react with gas to be purified. According to studies carried by H.M. Lapp *et al.* [11], one bushel (0.0532 m^3) of iron sponge can remove 8.2 lbs (3.7 kg) of sulfur. For removing upto 0.2 per cent of hydrogen sulfide which is the normal range, 1.1 ft^3 (0.03 m^3) of iron sponge is needed for a plant running on wastes from 1000 hogs with each hog contributing 7 ft^3 (0.2 m^3) of biogas per day. It is possible to convert (regenerate) used sponge back to free sulfur and ferric oxide which can be re-used for de-sulfurisation. Following chemical reactions take place in this process.

$$2\,Fe_2O_3 + 6\,H_2S \xrightarrow[\text{Scrubbing}]{H_2O} 2Fe_2S_3 + 6\,H_2O$$

$$2\,Fe_2S_3 + 3O_2 \xrightarrow[\text{Regeneration}]{H_2O} 2\,Fe_2O_3 + 3\,S_2$$

The above method is quite economic and easy to operate and maintain. In place of ferric oxide, zinc oxide can also be used for desulfurisation which has added advantage of removing organic sulfur compounds such as carbonyl sulfide and mercaptans. Use of zinc oxide is however less economic as compared to ferric oxide which is easily available from iron filings and scrap.

Some other commonly used absorbents are activated alumina, activated carbon, silica gel and molecular sieves. Molecular sieves are preferred as they help to remove both carbon dioxide and hydrogen sulfide. Molecular sieves are also used to remove contaminants from industrial gases. One commonly used class of absorbents are alumino-silicates which are found to be quite effective in absorbing hydrogen sulfide and water vapour. These absorbents produce clean dry gas without use of high pressure components. As an added advantage molecular sieves do not absorb methane unless bed temperature falls below –20°F (–28.8°C) which barring high altitude hilly-regions is never the case in this country. It is to be kept in mind that molecular sieves are not inexpensive and quickly get saturated with contaminants. As reported by H.M. Lapp. *et al.* [11], based on results from industry, 2.7 lb of molecular sieve is needed to remove 1 ft^3 of CO_2 from biogas, or equivalently 42.3 kg of molecular sieve is needed to remove 1 m^3 of CO_2.

3.6 Microbiological Methods of Purification

P.G Anglo, *et al.* [1] carried out studies at the National Institute of Science and Technology in Manila (Philippines) on production, and desulfurisation of biogas from swine wastes using microbiological methods. A purple photosynthetic bacterium, *Chromatium sp.* was isolated from the methane fermentor and was found as an active agent for removing hydrogen sulfide from biogas. Separate 7-day old cultures of the two purple bacteria, *Chromatium sp.* and *Rhodopseudomonas spheroides* briefly called Rs-1 and Rs-2 respectively, were inoculated in meat boullion agar in 300 ml effluent medium (1 : 3) and incubated at 35°C in lighted anaerobic conditions. Sodium sulfide solution (24 per cent, 0.5 ml) was introduced into the liquid cultures and air was replaced with nitrogen gas. For comparison, the two organisms were also inoculated separately in 300 ml meat boullion containing the same amount. The residual hydrogen sulfide was noted daily until all the gas was consumed. Further experiments were conducted to study the desulfurisation activity of Rs-1. Duplicate flasks of meat boullion with sodium sulfide as above were used throughout. In the temperature experiment, incubation at 18°C, 25°C and 35°C was tested under lighted conditions. In the light and dark experiments, one set of flasks was covered with tinfoil and another set was exposed to light bulbs at 35°C. To test the desulfurisation efficiency of isolates Rs-1 and Rs-2, two-weak old separate flask cultures of the bacteria in meat boullion were connected to the 6-litre biogas jar fermentors. Percentage removal of hydrogen sulfide was determined every day. The study established capability of certain photosynthetic bacteria like *Chromatium sp.* for use in desulfurisation of biogas. According to P.G. Angle *et al.* [1], desulfurisation was most effective at 35°C under lighting conditions of 1000 to 1200 lux.

T.S. Chou, *et al.* [6] studied the effectiveness of hydrogen sulfide removal with silica gel. Hydrogen sulfide can be effectively removed till silica gel is saturated as indicated by the change in colour from blue to black. It is possible to reuse silica gel is following regeneration. Flow rate of biogas, regeneration temperature, degree of saturation of silica gel and its water content help to evaluate scrubbing efficiency of silica gel. A mathematical model for evaluating adsorber efficiency was also developed.

D.R. Pande [14] made a comparative study of some of the common and efficient scrubbing methods for biogas purification. He discussed adsorption method using molecular sieves made of natural zeolites which is claimed to be economic, efficient and simple.

3.7 Concluding Remarks

It has been attempted to review properties of biogas culled out from various sources. Biogas which consists of 60-65 per cent methane, 35-40 per cent carbon dioxide, traces of hydrogen sulfide, ammonia and other impurities is explosive and toxic in nature. Methane which is the main component of biogas is colourless, odourless, tasteless, and is combustible in nature. Being its main constituent several properties of biogas are sufficiently close to methane. As methane is not easily liquifiable at normal temperatures it is thus also sometimes referred to as permanent gas. The high critical pressure of methane creates difficulties in attaining concentrated storage, and consequent portability like LPG. Carbon dioxide, second main constituent of biogas and being non-combustible tends to lower inflammability of biogas. Presence of impurities like hydrogen sulfide and water vapour imparts corrosive characteristics to biogas. Water vapour corrodes gas pipelines, causes damage in internal combustion engines using biogas and creates problems in compressors. Several properties of biogas and methane are substitutable without much loss of accuracy due to commonality of methane. Being a toxic and explosive gas, biogas should be handled with requisite caution and care.

REFERENCES

1. Anglo, P.G., Tanaka, H. and Haga, K., Production and Desulfurisation of Biogas from Swine Waste, Kalikasan, *Philippines Journal of Biology*, **7(3)**, 247-258 1978.
2. BEI, *Biogas Production from Animal Manure*, Biomass Energy Institute Inc., Winnipeg, Manitola, Canada, 1979.
3. BORDA, Biogas: *Manual for the Realisation of Biogas Programmes*, Bermen Overseas Research and Development Association, Bermen, Germany, 1980.
4. Chawla, O.P., Physical and Chemical Properties of Biogas, *Khadi Gramodyog,* April, 259-298 1980.
5. Chawla, O.P., *Advances in Biogas Technology*, Publications and Information Division, Indian Council of Agricultural Research, New Delhi, 96-107 1986.
6. Chou, T.S., *et al.*, Selective Removal of Hydrogen Sulfide from Biogas by a Packed Silica Gel Adsorber Tower, *Biotechnology Progress*, **2(4)**, 203-209 1986.
7. Hills, D.J. and Firbank, W.C., *Methane Generation from Agricultural Wastes,* A Report, Department of Agricultural Engineering, University of California, Davis, Jan, 1979.
8. Johnson, A.J. and Auth, G.H., *Fuels from Combustion Handbook*, McGraw-Hill Book Co, New York, 1951.
9. Khandelwal, K.C. and Mahdi, S.S., *Biogas Technology: A Practical Handbook,* Tata McGraw-Hill Publishing Company Limited, New Delhi, 1986, p. 51.
10. KVIC, *Gobar Gas: Why and How*? Khadi and Village Industries Commission, Mumbai, 1979.
11. Lapp, H.M., *et al.*, Methane Production from Animal Wastes: Fundamental Considerations, *Canadian Agricultural Engineering,* **17(2)**, 101 1975.

12. NAS, *Methane Generation from Human, Animal and Agricultural Wastes*, Report of an Ad. Hoc. Panel of the Advisory Committee on Technology Innovation, National Academy of Sciences, Washington, DC, 1977.
13. Nonhebel, G., *Gas Purification Processes*, George Newness Ltd., London, 1964.
14. Pande, D.R., *Some Methods of Biogas Purification*, Proc. of National Seminar on Solar Energy and Rural Development held at Kolhapur, May 29-31, 1987, Shivaji Univ., 89-101 1988.
15. Singhal, O.P., Development of Biogas Technology in India, *Steam and Fuel Users Journal*, **33(3-4)**, 53 1984..
16. Weast, R.C. (ed.), *Handbook of Chemistry and Physics*, 51st edition, Chemical Rubber Co, Cleveland, Ohio, 1971.
17. Mital, K.M., *A Survey of Properties of Biogas*, Proc. 6th Annual Conf. of the Indian Soc. of Mech. Eng., Delhi College of Engineering, New Delhi, Dec 22-23, 1985.
18. MNES, *Biogas: A Rural Energy Source*, 14 C.G.O. Complex, Lodhi Road, New Delhi 110 003, Oct., 1995.

Chapter 4

A COMPENDIUM OF BIOGAS PLANT DESIGNS

4.1 Introduction

With the advent of biogas technology a number of plant designs have been developed over the years. Biogas plants can be classified in several ways depending upon the plant design and mode of working. One common way to classify them is 'movable drum type' and 'fixed dome type'. These two models in India are popularly known as the KVIC or the conventional type and the Janata or Deenbandhu model, respectively. Both these types have their own merits and demerits though the fixed dome plants based on Chinese design are expanding fast in numbers. Plants can also be classified as batch-fed and continuous. Batch-fed plants are appropriate where daily supplies of raw waste materials are difficult to obtain. A batch loaded digester is filled to capacity, sealed until it has produced all it can, emptied and filled again. Gas production is uneven because bacterial digestion starts slowly, peaks, and then tapers off with growing consumption of volatile solids. This difficulty can be overcome by having minimum two digesters so that least one is always in operation. This problem can also be alleviated by connecting the batch loaded digesters in series fed at different times so that adequate biogas is available for use daily.

Several connected digesters though advantageous, however, occupy more space. Digesters can also be vertical and horizontal displacement type (plug flow) depending on the orientation of digester (Fig. 4.1). Horizontal displacement types face less problems of scum formation than vertical types. Digesters can have one or more partitions and accordingly they can be categorised as single chamber and multi-chamber type. Compartments in single or multiple digesters in series help to achieve added waste stabilisation and higher gas yield. Digestion of waste materials in a single chamber or digester is called single stage process, in two chambers or digesters is called double-stage process, and several chambers or digesters is called multi-stage process. In double-stage process, acidogenic and methanogenic stage are physically separated into two chambers. Thus the first stage of acid production is carried out in a separate chamber and only the diluted acids are fed into the second chamber where

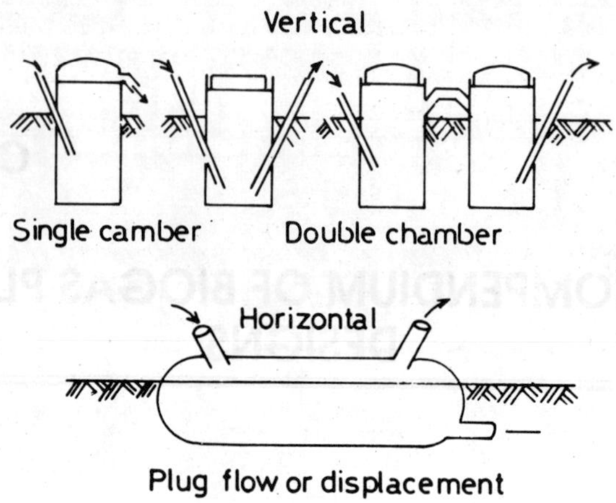

Fig. 4.1 Designs of continuous-fed digesters. [Source: Stoner, Producing Your Own Power, p. 163]

biomethanation takes place. In this chapter it has been attempted to give broad features of several plant designs commonly used in India and world at large.

4.2 Movable Drum Type Plants

India can be rightly claimed as birth place of movable drum type biogas plants, where number of designs have been developed [17, 18]. It basically comprises an underground brick masonary digester connected with an inlet and outlet and covered by a movable steel gasholder for gas collection. Gasholder moves up and down guided by a central guide pipe depending upon accumulation and discharge of gas. Movable gasholder made of mild steel alone accounts for some 40 per cent of the total plant cost and accordingly these plants are much more expensive than fixed dome type. Maintenance costs of these plants are also high in view of much the need to paint gasholder every year to prevent corrosion [21]. This plant helps achieve more consistent gas pressure which can be adjusted by regulating weight. Based on series of chronological developments in India, IARI and KVIC developed two plant designs which have come to be known as the IARI and the KVIC models (Figs. 4.2 and 4.3). Cross-sectional view of floating drum type plant developed and widely used in Nepal is shown in Fig. 4.4 one distinctive feature of which is a central guide-pipe [25]. Tables 4.1, 4.2 and 4.3 enlist quantities of materials needed for their construction. Based on average conditions with pig manure as feed it is possible to get yield of 0.4 to 0.55 m^3 gas/day/m^3 of digester volume of a movable drum type plant whereas with cattle dung as feed it is possible to obtain 0.3 to 0.4 m^3 of gas yield, in the same units of measurement.

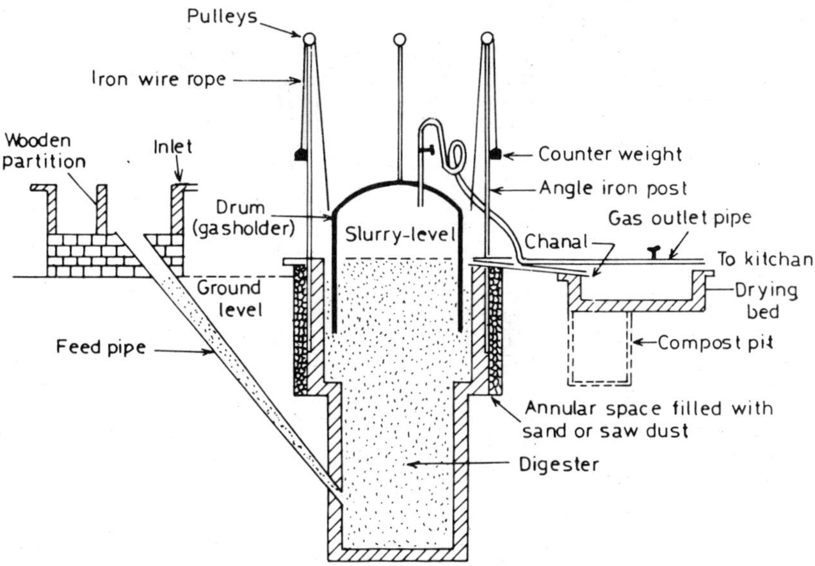

Fig. 4.2 A schematic view of an IARI plant

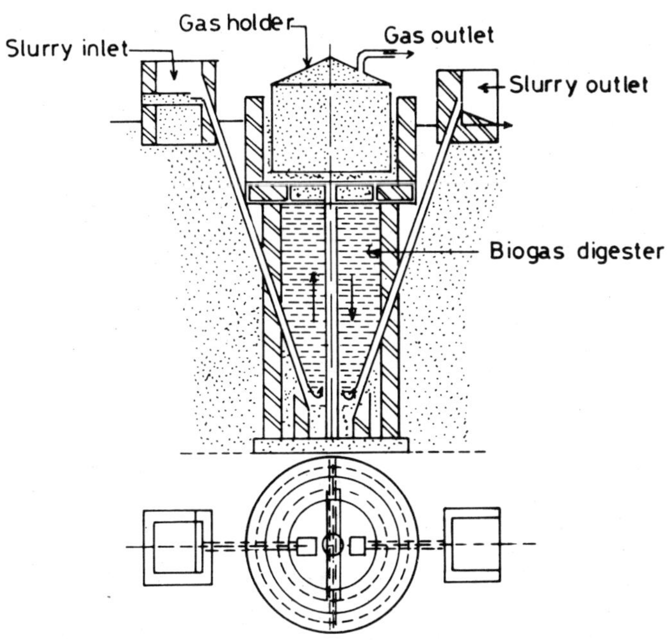

Fig. 4.3 A schematic view of a KVIC biogas plant [Source: Bansal, N.K., A Techno-economic Assessment of Solar-Assisted Biogas Systems, Energy Sources, Vol. 10, 1988, p. 216.]

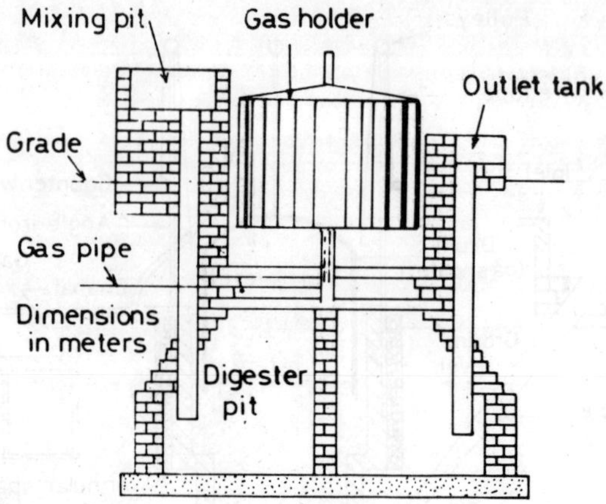

Fig. 4.4 Cross-sectional view of floating drum type biogas plant developed and commonly used in Nepal [Source: TERI, Biogas Technology: An information Package, Oct., 1985, p. 29]

Table 4.1 Bill of Materials Needed for Constructing KVIC Plants

List of Materials	*Unit*	*Quantity Needed for Different Capacity Biogas Plants*		
		4.36 m³ (1.0-1.8 m³ gas/day)	*10.8 m³ (4.0-6.9 m³ gas/day)*	*20.3 m³ (8.0-12.6 m³ gas/day)*
Digester inlet and outlet				
1. Bricks (10 × 15 × 23)	cm^3	6450	8000	9600
2. Sand	m^3	2.2	3.6	5
3. Gravel	m^3	0.6	0.95	1.4
4. Cement	Sacks	14 (700 kg each)	33 (1150 kg each)	33 (1650 kg each)
5. Pipes (diameter 100 mm)	m	5	9.5	10.5
6. Gasholder central guide				
7. Angle iron	m	11	12.7	15
8. Pipe (diameter 50 mm)	m	1.85	1.85	2.15
9. Sheet metal (2 mm gauge)	m^2	0.5	0.5	0.5
10. Angle iron	m	17	21	36
11. Pipe (diameter 65 mm)	m	1.15	1.25	1.45
12. Iron strip	m	4.2	43	4.05
13. Sheet metal (2 mm gauge)	m^2	6	8.25	11.25
14. Other sheet metal	m^2	0.5	0.50	0.5
15. Corrosion resistant paint	kg	20	25	30

Table 4.2 Bill of Materials Needed for Constructing a 3 m^3 (100 ft^3) Capacity Biogas Plant as per the IARI Design

List of Material	*Quantity*
1. *For the gasholder*	
(a) Mild steel sheet (16 SWG), 250 cm × 125 cm	Three
(b) Iron rod, 1.3 cm (1/2 in.) diameter	6 m
(c) G.I. nipple, 1.3 cm (1/2 in.) diameter, 6 cm long	One
(d) Socket, 3.8 cm (1½ in.) diameter, fitted with screw cap	One
2. *For the supporting pillars*	
(a) Black iron pipe, 3.8 cm (1½ in.) diameter, 240 cm (8 ft) long	Four
(b) Angle iron, 3.8 cm × 3.8 cm × 30 cm or iron flats, 5 mm × 30 cm × 8 mm	Sixteen
(c) Iron pulley, 10 cm diameter	Eight
(d) Mild steel sheet (SWG), 10 cm × 10 cm	Four
(e) Nut and bolt, 8 cm × 12 mm	Sixteen
3. (a) Slotted counter weights (approx. 5 kg each)	Sixteen
(b) Cast iron weight hanger for above	Four
4. Wire rope, 6 mm (1/2 in.) thick (can be substituted by 8 mm nylon rope)	10 m
5. Conduit hose pipe made of cement, 180 cm (6 ft) long × 10 cm diameter	Two
6. Flexible rubber or polyethylene hose pipe, 1.9 cm (3/4 in.) diameter	2 m
7. G.I. pipe, 1.3 cm (1/2 in.) diameter, for main gas line leading to kitchen	As required
8. G.I. elbows, sockets, bends and tees for 1.3 cm pipe	Two
9. Wheelcock, 1.3 cm (1/2 in.) diameter	Two
10. Water tap, single turn 1.3 cm (1/2 in.) diameter	One
11. Anti-corrosive paint (black)	5 litres
12. Bricks	2,500
13. Sand	500 kg (19 bags)
14. Cement	100 kg (2 bags)

Source: IARI, Cow Dung Gas Plant, Indian Agricultural Research Institute, New Delhi, 1979.

Table 4.3 Bill of Materials Needed for Constructing a 9 m^3 (300 ft^3) Capacity Biogas Plant as per IARI Design

List of Material	*Quantity*
1. *For the gas holder*	
(a) Mild steel sheet (14 SWG), 250 cm × 125 cm	Five
(b) Iron rod, 1.8 cm (3/4 in.) diameter	2 m
(c) G.I. nipple, 2.5 cm (1 in.) diameter, 6 cm long	One
(d) Socket, 5 cm (2 in.) diameter, fitted with screw cap	One
(e) Angle iron, 2.5 cm × 2.5 cm (1 in. × 1 in.)	20 m
2. *For the supporting pillars*	
(a) Black iron pipe, 3.8 cm (1½ in.) diameter, 240 cm (8 ft.) long	Four
(b) Angle iron, 3.8 cm × 3.8 cm × 30 cm or iron flats (5 cm × 30 cm × 8 mm)	Sixteen
(c) Iron pulley, 15 cm diameter	Eight
(d) Mild steel sheet (8 CWG), 10 cm × 10 cm	Four
(e) Nut and bolt, 8 cm × 12 mm	Sixteen
3. (a) Slotted counter weights (approx. 10 kg each)	Sixteen
(b) Cast iron weight hanger for above	Four
4. Wire rope, 8 mm (3/8 in.)	10 m
5. Conduit huge pipe made of cement, 180 cm (6 ft.) long × 10 cm (4 in.) diameter	Two
6. Flexible rubber or polyethylene hose pipe, 2.5 cm (1 in.) diameter	2 m
7. G.I. pipe, 2.5 cm (1 in.) diameter, for main gas line leading to kitchen	As required
8. G.I. elbows, sockets, bends and tees for 2.5 cm pipe	Two
9. Wheelcock, 2.5 cm (1 in.) diameter	Two
10. Water tap, single turn, 1.3 cm (1/2 in.) diameter	One
11. Anti-corrosive paint (black)	7 litres
12. Bricks	4,000
13. Sand	1,500 kg (30 bags)
14. Cement	300 kg (6 bags)

Source: IARI, Cow Dung Plant, Indian Agricultural Research Institute, New Delhi, 1979, p. 3.

Complete details of movable drum type biogas plant with facility of counterweights to build adequate gas pressure approved vide IS : 9478-1980 are given in Fig. 4.7. Sketch of gasholder and recommended method of joining it to the pipeline leading to the biogas appliances are shown in Fig. 4.5 and 4.6 respectively. The digester is to be made of RCC foundation with its minimum thickness as 150 mm for plants with capacities upto and including 10 m^3 and 230 mm for plants with capacities more than 10 m^3. The digester wall is made of brick with top edge having a slope to a depth of 150 mm on the outerside (Fig. 4.7). The digester wall includes an annular projecting support for gasholder. Stone, iron or RCC beam is used for fixing a detachable central guide pipe above the partition wall. Two pulleys with hanging counterweights are provided on both sides of the gasholder to build adequate gas pressure. Gasholder is made of GI sheet or welded steel construction (Table 4.4). Both inlet and outlet tanks need be made of bricks which can have any shape. The bottom of the inlet should always be above the level of the digester. The inlet and outlet tanks should be connected to the digester with an asbestos cement pipe conforming to IS : 6908-1975. For plant capacities up to and including 15 m^3 recommended diameter of pipeline for connecting digester to inlet and outlet is 100 and 150 mm respectively as per IS: 9478-1980 (Tables 4.5 and 4.6). Gas outlet pipe leading to the biogas appliances is taken from the gasholder top.

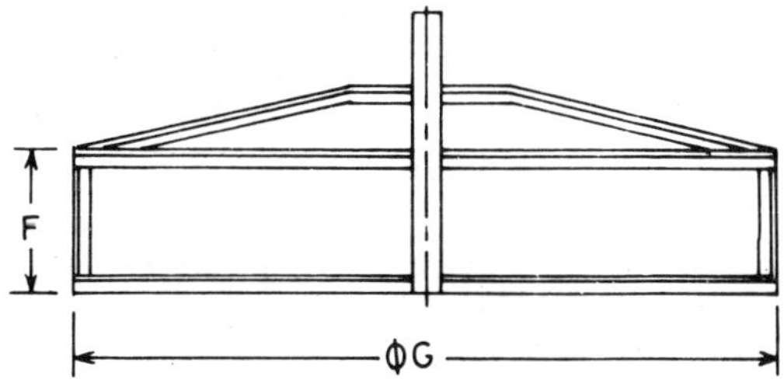

Fig. 4.5 Broad dimension of gasholder (Table 4.4)

Table 4.4 Standardised Specification of Gasholder

Plant Capacity (m^3)	*External Diameter (G) (mm)*	*Minimum Length of Straight Flange (F) (mm)*
3	1400 ± 50	1000
4	1600 ± 50	1000
6	2000 ± 50	1000
8	2200 ± 50	1250
10	2250 ± 50	1250
15	3000 ± 50	1250

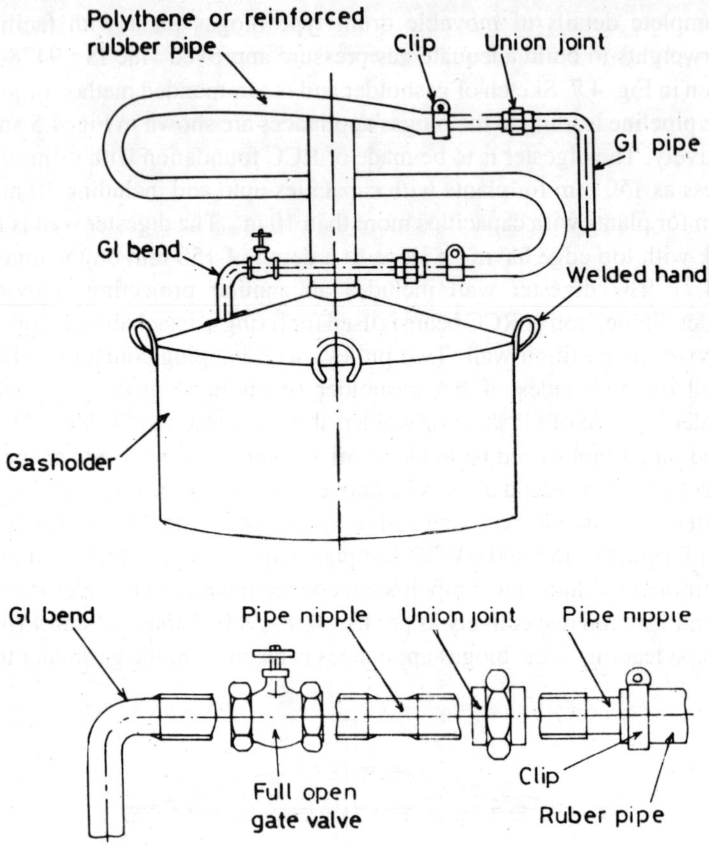

Fig. 4.6 Recommended method of joining gasholder to main gas-pipe (IS: 9478-1980)

Table 4.5 Standardised Specification of Plant Digester

Plant Capacity (m^3)	*Inner Diameter (A) (mm)*	*Minimum Wall Thickness (B) (mm)*	*Approximate Depth (C) (mm)*	*Inner Edge (D) (mm)*	*Maximum Projected Support (E) (mm)*
3	1600 + 50	230	3770	75	1000
4	1800 + 50	230	4270	75	1000
6	2200 + 50	230	4270	75	1000
8	2400 + 50	230	4770	75	1250
10	2750 + 50	230	4770	125	1250
15	3200 + 50	230	4770	125	1250

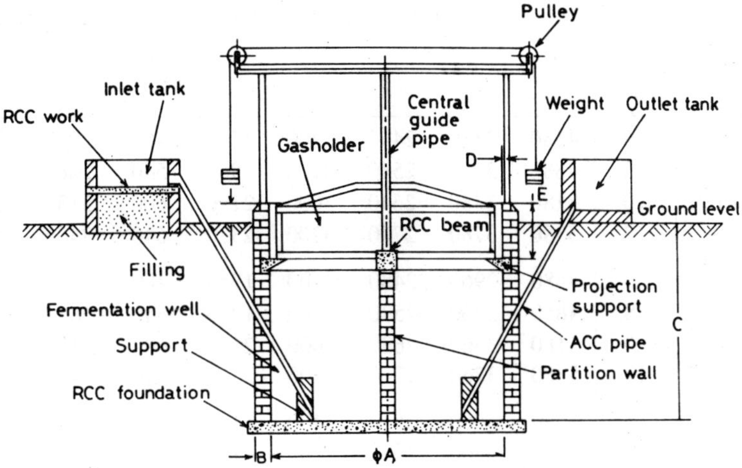

Fig. 4.7 A detailed cross-sectional view of Gobar Gas Plant Adopted by the Bureau of Indian Standard vide IS : 9478-1980

Table 4.6 Standardised Specifications of Plant Inlet and Outlet

Plant Capacity (mm^3)	*Size of Inlet*		*Size of Outlet*	
	Sides of Square (mm)	*Depth (mm)*	*Sides of Square (mm)*	*Depth (mm)*
3	750	450	450	450
4	750	450	450	450
6	900	450	450	450
8	900	525	450	450
10	900	600	450	450
15	900	600	450	450

Table 4.7 Important Dimensions of Chinese Fixed Dome type Biogas Plants Constructed from Brick (All dimensions in mm)

Size (m^3)	*L*	*B*	*H*	*D*	r_1	f_1	r_2	f_2
6	4700	3080	2550	2600	1410	870	2760	330
8	4900	3280	2630	2800	1510	930	2980	350
10	5200	3580	2760	3100	1680	1030	3300	380
12	5400	3780	2870	3300	1790	1100	3510	410

Table 4.8 Dimensions of Chinese Fixed Dome type Biogas Plants Constructed from Concrete, Lime-concrete and Lime-clay
(All dimensions in millimetres)

Material	*Size (m^3)*	*L*	*B*	*H*	*D*	r_1	f_1	r_2	f_2
Concrete	6	4580	2880	2440	2400	1740	480	2550	300
	8	4880	3180	2540	2700	1960	540	2860	340
	10	5180	3480	2640	3000	2180	600	3180	380
	12	5380	3780	2700	3200	2320	640	3400	400
Lime-Concrete	6	4580	2960	2480	2400	1740	480	2550	300
	8	4880	3260	2580	2700	1960	540	2860	340
	10	5180	3560	2680	3000	2180	600	3180	380
	12	5380	3760	2740	3200	2320	640	3400	400
Lime-Clay	6	4580	3060	2530	2400	1740	480	2550	300
	8	4880	3360	2630	2700	1960	540	2860	340
	10	5180	3660	2730	3000	2180	600	3180	380
	12	5280	3860	3790	3200	2320	640	3400	400

4.3 Fixed Dome Type Plants

In fixed dome type biogas plant which was first developed in China (Fig. 4.8), gas is stored in upper part of the digester. In this design there is no separate gasholder and upper portion of the digester pit itself acts as gasholder. Displaced level of slurry provides requisite pressure for release of gas for its subsequent use [1]. The first pilot plant as per this design was developed by the Gobar Gas Research Station, Ajitmal, Etawah in 1977. Most of the fixed-dome type biogas plants in China have their digester capacity between 6 to 8 m^3 and a small proportion have their capacity upto 12 m^3. Community plants have their capacity between 50 to 100 m^3 depending upon waste availability and gas requirement [11]. The plant works on both animal wastes and plant waste materials and is convenient to work with in both continuous and batch mode. Normally animal wastes are fed daily whereas crop residues are fed in batches. As the plant does not involve any steel parts and it can be built with local materials, its construction costs are low and operation costs virtually nil. Being underground, space needs are also minimal. Materials like lime-clay, lime-concrete, concrete, bricks, stones etc. can be used for plant construction. It, however, requires considerable skills in dome construction. Construction materials required are cheap and easily available (Table 4.9).

In China a number of organisations have done pioneering work in developing and modifying fixed dome plants which include: National Biogas Office, Beijing; Science and Technology Association, Shanghai; Biogas Research Institute, Chengdu; Agricultural University, Hangrhow; South-West Institute for Agriculture, Chengdu, and the Chengdu Institute of Biology. In Chinese

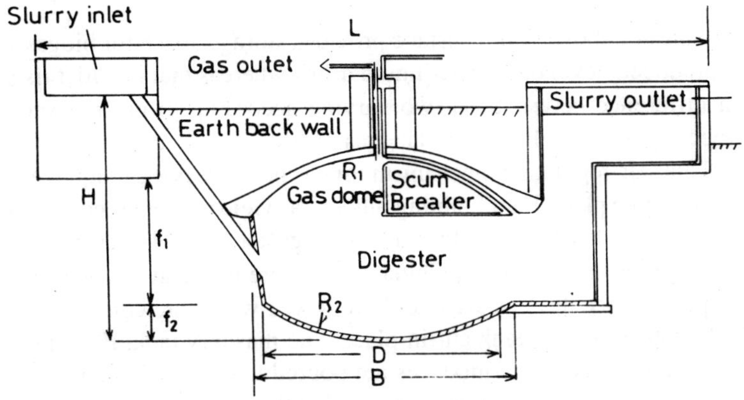

Fig. 4.8 A schematic view of Chinese biogas plant [Source: Bansal, N.K., A Technoeconomic Assessment of Solar – Assisted Biogas Systems, Energy Sources, Vol 10, 1988, p. 217.]

Table 4.9 Bill of Material for Different Capacity Chinese Fixed Dome Type Biogas Plants

List of Materials	*Quantity Required*		
	Digester Volume 6 m³	*Digester Volume 10 m³*	*Digester Volume 50 m³*
Masonry 5.0 m^3	5.0 m^3	6.5 m^3	20 m^3
Bricks (20 × 5 × 10 cm)	5000	6500	20000
Cement			
Lime	280 kg	360 kg	1100 kg
Sand	2700 kg	2100 kg	8400 kg
Internal Plastering	A = 23 m^2	A = 31 m^2	A = 89 m^2
	0.8 m^3	1.0 m^3	2.5 m^3
Cement			
Lime	200 kg	250 kg	600 kg
Sand	1200 kg	1500 kg	4000 kg
Paint	70 kg	95 kg	270 kg
1" gas pipe	1.5 m	1.5 m	1.5 m
Gas tap			

Note : Add an allowance of 10-30% for losses

Source : GATE, Biogas Plants: Building Instructions, German Appropriate Technology Exchange, BORDA, Bremen, Germany (erstwhile FRG), 1980.

designs it is usual to provide a manhole cover at the top. In earlier designs the plant depth varied from 3 to 4 m but present trends limit it to within 2 m. In equivalent capacity moving drum type plant corresponding height varies from 3.5 to 4 m. Inlet and outlets are connected at the mid-point of the digester. Upper part of digester above liquid surface provides storage space for biogas. When gas is produced, level of digester liquid drops whereas that in outlet rises with height difference between the two varying according to gas pressure. This difference in height helps to regulate gas pressure within digester over a wide range. This self-regulating mechanism for gas pressure is the unique feature of fixed dome type plant. For obtaining higher gas yield per unit digester volume, it is desirable to keep free surface area of digester liquid larger in relation to height. In Sichuan Province of China where there is a heavy concentration of biogas plants three designs are mainly followed which are commonly known as 'tank cast-in-place', 'tank cut-in-place' and masonry tank. For tanks 'cast-in-place' suitable concrete mixtures are poured to cast walls, dome and base whereas 'cut-in-place' tanks are dug out directly from suitable hard clay soils with minimum of structure work. For masonry tanks, bricks, stones, precast cement structures etc. are used [11].

Based on Chinese design, Gobar Gas Research Station, Ajitmal (Etawah) developed the first ever fixed dome biogas plant in India in 1977 which has come to be known as the *Janata* Plant. The plant design has, however, undergone several modifications since then [1, 2]. A cross-sectional view of Janata biogas plant is shown in Fig. 4.9. It involves an underground cylindrical digester and a hemispherical dome but without manhole cover. The dome is fitted with a GI pipe through which the biogas is taken out. It is made of bricks, cement and concrete. Following construction, digester walls and dome are cured for several days for imparting requisite strength to the plant [19]. Important dimension and detailed specifications of Janata plants are included in Tables 4.10, 4.11 and 4.12. A comprehensive estimate of material requirement for constructing different capacity Janata plants is given in Table 4.13.

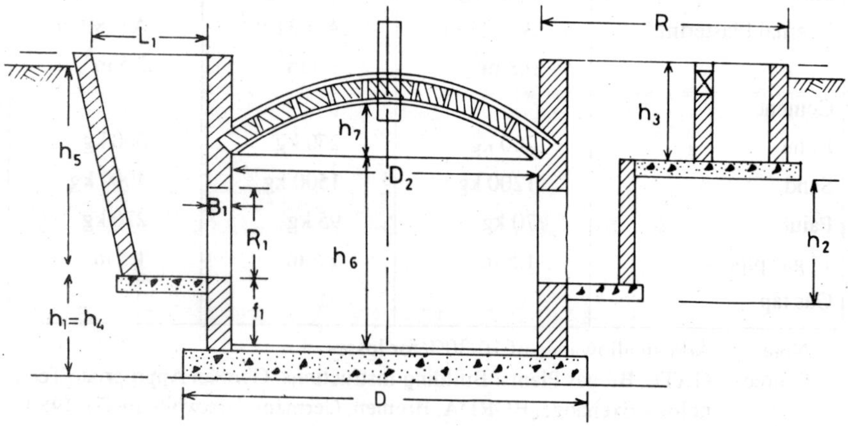

Fig. 4.9 Cross-sectional view of Janata fixed-dome type biogas plant developed in India

Table 4.10 Important Dimensions of Different Capacity Fixed Dome Type (Janata) Biogas Plants

Plant Capacity (m^3)	*Dimensions in mm*														
	D	D_1	D_2	L_1	B_1	f_1	h_1	h_2	h_3	h_4	h_5	h_6	h_7	R	R_1
2	2750	2500	2370	850	610	580	655	765	810	680	1575	1345	520	1610	610 × 610
3	3100	2950	2720	1216	900	746	846	828	880	845	1710	1575	590	1865	610 × 610
4	3610	3400	3000	1372	950	825	1005	990	854	1005	1845	1715	640	2080	610 × 760
6	4110	3080	3420	1520	1350	660	840	1290	1025	840	2315	1950	735	2360	610 × 1000

Table 4.11 Major Dimension of Different Capacity Fixed Dome Type (Janata) Biogas Plants

Dimension	*Dimension for Different Capacity Plants*					
	2 m^3 (70 ft^3)	*3 m^3 (105 ft^3)*	*4 m^3 (150 ft^3)*	*6 m^3 (210 ft^3)*	*9 m^3 (315 ft^3)*	*15m^3 (525 ft^3)*
Diameter (D)	2750 mm (9′ 6″)	3100 mm (10′ 4″)	3610 mm (12′ 0″)	4110 mm (13′ 8″)	4490 mm (15′ 0″)	5350 mm (17′ 3″)
Depth (h_5+h_4)	2255 mm (7′ 6″)	2555 mm (8′ 6″)	2850 mm (9′ 6″)	3155 mm (10′ 6″)	3155 mm (10′ 6″)	3880 mm (13′ 3″)
Inlet length (L_1+115)	965 mm (3′ 3″)	1331 mm (4′ 6″)	1847 mm (4′ 8″)	1635 mm (5′ 6″)	1115 mm (3′ 9″)	1230 mm (4′ 1″)
Inlet width (B_1+230)	840 mm (2′ 9″)	1130 mm (2′ 8″)	1180 mm (3′ 7″)	1580 mm (5′ 3″)	2552 mm (8′ 4″)	3230 mm (11′ 0″)
Outlet length (L_1+115)	965 mm (3′ 3″)	1331 mm (4′ 6″)	1847 mm (4′ 8″)	1635 mm (5′ 6″)	1115 mm (3′ 9″)	1230 mm (4′ 1″)
Outlet width (L_1 + 115)	840 mm (2′ 9″)	1130 mm (2′ 8″)	1180 mm (3′ 7″)	1580 mm (5′ 7″)	2552 mm (8′ 4″)	3230 mm (11′ 0″)

Table 4.12 Detailed Specification of Different Capacity Janata Biogas Plants

Measure	Unit	*Dimensions for Different Capacity Janata Biogas Plants*			
		2 m³ (70 ft³)	*3 m³ (105 ft³)*	*4 m³ (140 ft³)*	*6 m³ (210 ft³)*
1. Volume of digester	m^3 ft^3	5.92 209.03	9.14 322.73	12.12 428.06	17.91 632.40
2. Volume of dome	m^3 ft^3	1.20 42.36	1.80 63.54	2.40 84.72	3.60 127.08
3. Total inside height	m ft	1.86 6.13	2.16 7.12	2.36 7.75	2.69 8.84
4. Internal diameter	m ft	2.37 7.80	2.72 8.95	3.00 9.87	3.42 11.25
5. Wall thickness of digester	m ft	0.12 0.38	0.12 0.38	0.23 0.75	0.23 0.75
6. Foundation dia. of digester	m ft	2.75 9.05	3.10 10.20	3.61 11.89	4.11 13.52
7. Height of dome	m ft	0.52 1.71	0.59 1.94	0.64 2.11	0.74 2.42
8. Thickness of foundation	m ft	0.15 0.49	0.15 0.49	0.23 0.77	0.23 0.77
9. Radius of dome	m ft	1.61 5.30	1.86 6.13	2.08 6.84	2.36 7.75
10. Inlet size of top	(m × m) (ft × ft)	0.85 × 0.61 2.80 × 2.01	1.22 × 0.60 4.00 × 2.63	1.37 × 0.95 4.51 × 3.13	1.58 × 1.35 5 × 6.75
11. Inlet size at bottom	(m × m) (ft × ft)	0.85 × 0.61 2.80 × 2.01	0.61 × 0.61 2.01 × 2.01	0.61 × 0.61 2.01 × 2.01	0.61 × 0.61 2.01 × 2.01
12. Outlet size at top	(m × m) (ft × ft)	0.85 × 0.81 2.80 × 2.67	1.22 × 0.80 4.0 × 2.16	1.37 × 0.95 4.51 × 3.14	1.52 × 1.35 5.00 × 0.61
13. Outlet size of bottom	(m × m) (ft × ft)	0.61 × 0.61 2.01 × 2.01	0.61 × 0.61 2.01 × 2.01	0.61 × 0.61 2.01 × 2.01	0.61 × 0.61 2.01 × 2.01
14. No of bricks	No	1500	2300	3500	4000
15. Cement bags	No	20	25	30	40
16. Sand	m^3 ft^3	2.00 70.00	2.30 80.00	3.66 125.00	4.30 150.00
17. Coarse sand	m^3 ft^3	0.85 30	1.15 40	1.60 50	1.70 60

Contd...

18. Brick ballast	m^3 ft^3	1.00 35.00	1.15 40.00	1.42 50.00	2.00 70.00
19. Pebbles	m^3 ft^3	0.10 3.35	0.10 3.53	0.30 10.59	0.30 14.12
20. Construction cost	Rs.	1393.50	1865.50	2470.75	3282.50
21. Cost of accessories	Rs.	255.00	380.00	435.00	475.00
22. Total cost (1978) (say)	Rs.	1648.50 1650.00	2245.50 2250.00	2905.75 2910.00	3757.50 3760.00
23. Cattle required for generating wastes	No.	3-4	4-5	5-7	7-10
24. Daily requirement of fresh wastes	kg	30-45	45-50	55-60	80-100
25. Number of person served	No.	5-8	8-12	12-16	16-20

Sources: 1. Bahadur, S. and Singh, K.K., Janata Biogas Plants, Planning Research and Action Division, State Planning Institute, Lucknow, 1978 [For S.No. 1 to 22].

2. Khandelwal, K.C., A Janata Model Gobar Gas Plant, Pamphlet No. 104, AFPRO, New Delhi, 1980 [For S.No. 23 to 25].

Scientists at the Tata Energy Research Institute, New Delhi following some modifications have developed three alternative versions of Janata biogas plant which are commonly referred to as the TERI mark I, II and III. All the three TERI designs have a hydraulic retention time of 50 days. The cost of Mark I for capacities of 1-6 m^3 is higher than that of Janata plants by 1 to 27 per cent. For capacities of 1-3 m^3, Mark II models are 9 to 22 per cent costlier than the Janata models. The cost of the Mark III model is lower than that of Mark I and II for all capacities and comparable with the cost of the Janata model of capacity below 3 m^3 [20].

4.4 Miscellaneous other Biogas Plants Developed in India

A number of plant designs have been developed by different organisations and individuals in India which are either modifications of movable drum or fixed type or entirely different new designs [1, 7, 8]. Some broad details and distinctive feature of these plant design are given below. For more precise and exact information, reader is advised to contact concerned organisations and individuals, and consult appropriate building instructions and manuals [10, 24].

Table 4.13 Material Requirement for Different Capacity Janata Biogas Plants

Item	*Unit*	*Material Required for Different Capacity Plants*					
		2 m^3 (70 ft^3)	*3 m^3 (105 ft^3)*	*4 m^3 (150 ft^3)*	*6 m^3 (210 ft^3)*	*9 m^3 (315 ft^3)*	*15 m^3 (525 ft^3)*
Construction Materials							
1. Bricks	No	1500	1500	3500	4000	5500	9500
2. Cement	50kg bags	20	25	30	40	65	95
3. Brick ballast (1 ½ dia) or stone chips	ft^3	35	40	50	70	150	200
4. Fine sand	ft^3	70	80	125	150	200	300
5. Coarse sand/ morrum/fine gravel	ft^3	30	40	50	60	100	200
6. Bit black enamel paint	litre	1 ½	2	2 ½	3	3 ½	4
7. GI pipe with two sockets for gas outlet (1/2″ or 3/4″)	ft	$\frac{3}{4}$	$\frac{3}{4}$	$\frac{3}{4}$	$\frac{3}{4}$	1	$\frac{1}{2}$
8. Inlet and outlet cover (cement concrete slab or wooden cover)		(*X*)	(*X*)	(*X*)	(*X*)	(*X*)	(*X*)
9. Rent for shuttering materials (wooden planks & bamboo-sticks etc.)	Rs.	25	30	35	40	45	50
Labour and Masons							
10. Labour for digging the pit	Labour days	6-7	7-8	9-10	12-15	20-25	30-35
11. Labour for construction work	Labour days	16-20	24-28	32-36	50-54	80-90	100-125
12. Masonry labour	Mason days	8-12	12-16	16-20	25-29	40-50	60-65
Accessories							
13. Alkathaline pipe	m	15	15	15	15	12	20
14. PVC pipe (1/2″ dia)	m	5	5	5	5	5-10	5-10
15. Gate valve	No.	1	1	1	1	1	1
16. Two way connection	No.	1	1	1	1	1	1
17. Burner	No.	1	1	2	2	2	2
18. Gas lamp	No.	–	1	1	1	1	1

Note : (*X*) – Dimension of covers same as that of corresponding inlet and outlet.
Source : Khandelwal, K.C., A Janata Model Gobar Gas Plant, Pamphlet No. 104, AFPRO, New Delhi. 1980.

Deenbandhu Biogas Plant. Action for Food Production (AFPRO), an NGO, in 1984 developed a low cost fixed dome plant called 'Deenbandhu Model' meaning 'friend of the poor.' The history of this model dates back to 1981 when the AFPRO organised a training programme aided by F&AO/UNDP at Varanasi where AFPRO presented 8 alternative designs of 2 m^3 capacity plants which following some modifications in 1984 were later standardised and promoted as Deenbandhu model. It is appropriate for using all types of wastes and minimises biogas losses from inlet chamber and ensures maximum utilisation of digester volume thereby making the plant operate at designed HRT. This model is 30 per cent cheaper than Janata model and 45 per cent less than the KVIC plants of comparable capacities. It is 15-30 per cen cheaper than the Pragati Model developed by BORDA and UNDARP (United Socio-Economic Development of Research Programme). Storage capacity is 33 per cent of daily gas production. It has curved bottom and a hemispherical top which are joined at their bases with no cylindrical portion in between (Fig. 4.10). Displaced slurry following fermentation moves to the outlet displacement chamber as there is no displacement space on the inlet side (Table 4.14). An inlet pipe connects mixing tank with the digester. The plant design is based on gas yield of 0.04 m^3 per kg of fresh dung, and HRT of 40 days in plains and 55 days in hilly regions [29].

Table 4.14 Bill-of Material for Different Capacity Deenbandhu Biogas Plants

Description	*Unit*	*2 m^3*	*3 m^3*	*4 m^3*
1. Bricks (class I)	No	1000	1300	1600
2. Cement bags	No	14	16	22
3. Stone chips	ft^3	40	50	60
4. Sand	ft^3	40	50	60
5. Coarse sand	ft^3	40	50	60
6. G.I 1/2" dia. pipe with sockets	inches	7	7	7
7. AC 6" dia. pipe	ft	6	6	6
8. Iron bans (6 mm dia) for outer tank cover	kg	7	10	12
9. Paint	litre	1	1.5	2

Source : TERI, Biogas: A Source of Rural Employment, Tata Energy Research Institute, New Delhi, 1994, p. 36.

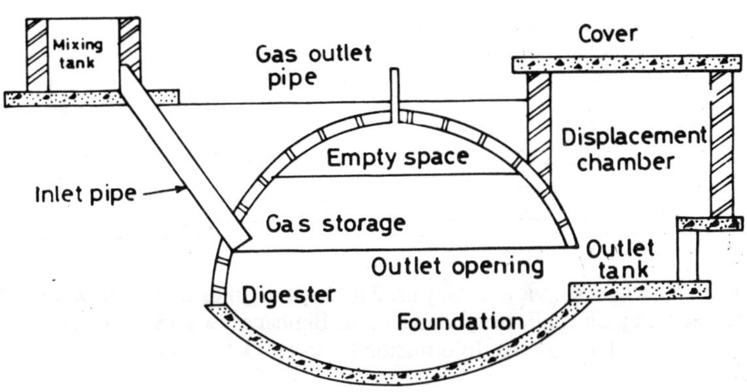

Fig. 4.10 A cross-sectional view of Deenbandhu Biogas Plant

Cattle dung slurry prepared in 1 : 1 ratio with water is fed upto the level of second step in the outlet tank which is also the base of the outlet displacement chamber. As the gas generates and accumulates in the empty portion of the plant, it presses the slurry of the digester and displaces it into the outlet displacement chamber. The slurry level in the digester falls whereas in the outlet chamber it starts rising. This fall and rise continues till the level in the digester reaches the upper end of the outlet opening, and at this stage the slurry level in the outlet tank reaches the height of discharge opening. Any unused gas beyond this stage escapes through the outlet tank [29].

The diameter of the base is almost the same as the diameter of JBPs of identical capacities. For example, in a 2 m^3 plant the bottom with a large curvature has a base diameter of 2.55 m which is the same as in a JBP. Effluent discharge opening in the outlet displacement chamber is at least 150 mm (6 inch) lower than the bottom of the gas outlet which is flushed with ceiling. It helps to prevent entry of gas or its foam into the gas pipeline taken as a trade-off between low discharge opening and high dome. For plant bottom, base diameter to rise of arch ratio is kept as 5 : 1. Design of Deenbandhu model (1, 2, 3, 4 and 6 m^3 capacities) was approved by the DNES in 1986 for countrywide popularisation under the NPBD. Based on AFPRO study whereas gas yield per unit weight of feed is same for both Deenbandhu and JBP the productivity per unit volume of digester is higher by 20 to 40 per cent in Deenbandhu plant for all months [24].

Gayatri Model. *Gayatri* model was developed by the Government Implements Factory, Bhubaneshwar. It eliminates the use of brick dome and minimises the use of cement. In place of brick dome, a pre-fabricated fibre-glass reinforced plastic (FRP) dome of the same dimension is used for gas collection. A pilot-scale Gayatri model plant of 2 m^3 capacity as shown in Fig. 4.11 has been functioning satisfactorily at the Government Implements Factory, Bhubaneshwar [23, 24].

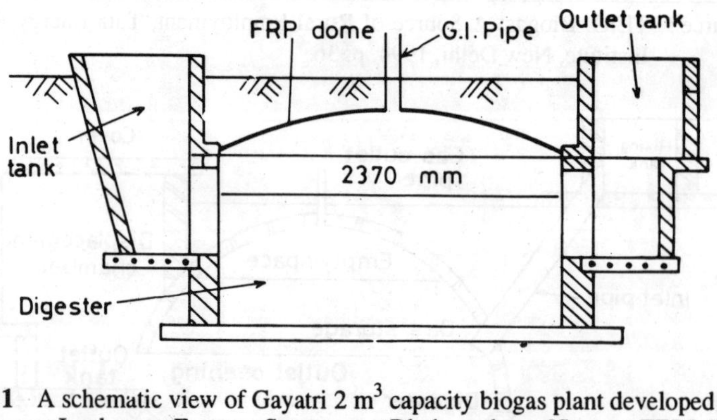

Fig. 4.11 A schematic view of Gayatri 2 m^3 capacity biogas plant developed by the Government Implement Factory, Satyanagar, Bhubaneshwar [Source: TERI, Biogas Technology: An Information Package, Oct., 1985, p. 33.]

TNAU Model. Tamil Nadu Agricultural University at Coimbatore evolved four designs of biogas plants with a view to achieve among others reduction in retention period. One of its model is similar to the fixed dome plant of the Janata type with a flat bottom whereas the other model resembles the Nepalese version of the Chinese model with curved digester bottom and ferro-cement dome. The others two designs are mobile plants while one being cubical the other cylindrical in shape. Special feature of all these plants is low retention period of 15 days which is achieved by periodic addition of micro-organisms which speed up the digestion process.

Krishna Biogas Plant. *Shivsaadan* Renewable Research Institute (Address: 192/3, Industrial Estate, Sangli 416416, Maharashtra) has developed an alternative fixed dome type plant made of prefabricated concrete which it named as Krishna biogas plant (Fig. 4.11a). The plant consists of a digester with its dome and the lower cylindrical portion cast monolithically in reinforced cement concrete (RCC). The inner surface of the structure that comes in contact with biogas under varying pressures is coated with an epoxy paint to prevent possible gas leakage. In addition, a single compensatory slurry displacement tank, also cast in RCC is provided over the dome to account for increase in gas pressure. Besides reducing overall costs, Krishna model overcomes several limitations commonly associated with conventional fixed dome plants such as frequent gas leakage. Use of pre-fabricated concrete structure achieves faster construction and closer adherence to time-bound programme of constructing a specified number of plants through more realistic scheduling of production, transportation and erection of plant sequences. By February, 1991, the Shivasadan Renewable Research Institute had set up 725 Krishna biogas plants. The cost of the Krishna model is claimed to be 15 to 20 per cent lower than the corresponding KVIC model.

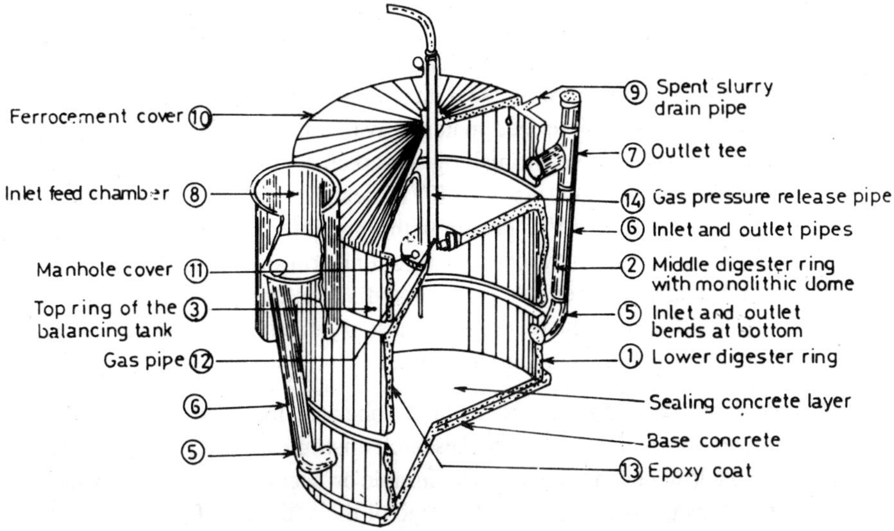

Fig. 4.11a Cut section view of Krishna Biogas Plant

The Krishna plant was originally designed for 30 days HRT as per MNES specifications for the Maharashtra region. It is possible to connect a latrine unit to the biogas plant so that night-soil could be digested along with the cattle dung. Details of digester volume and gas holding volume for the plants of 2, 3, 4 and 6 m^3 gas production capacities are given in Table 4.15. It may be noted while the digesters volume is 1.5 times of this nominal size, the gas holding volume is half the nominal size.

Table 4.15 Major Dimensions of Krishna Fixed-Dome Biogas Plant

Capacity (m^3)	*Digester Tank*			*Gas Balancing Tank*			*Mixing Tank Dia.*
	Diameter	*Effective Depth*	*Volume (m^3)*	*Diameter*	*Effective Depth*	*Volume (m^3)*	
2	1829	1143	3	1829	381	1	457
3	2134	1293	4.6	2134	457	1.6	533
4	2134	1676	6	2134	610	2.2	610
6	2134	2515	9	2134	838	3	762

The gas balancing tank is automatically filled up by the digested substrate twice a day coinciding with the two meals times following increase in gas pressure inside the digester. As the gas is consumed for cooking or heating, the digested substrate returns back to the digester through automatic balancing. Digested slurry is directly fed to the compost pit through spent slurry drain pipe [28].

Manipal Model. This model has features of both Indian and Chinese models. It has a rectangular masonry digester with an inclined RCC roof. There is no separate gasholder unlike in earlier Indian designs and larger inclined space above the slurry level provides storage area for resulting biogas to accumulate. Digester can be built, partly inside or fully above ground depending upon ground conditions. Variation in pressure is achieved by a shallow ferro-cement cover dipped in water seal on sloping roof slabs. Inlet and outlet pipes are selected according to feed material availability and desired mode of plant operation (Fig. 4.12). Resulting biogas is taken out with the help of a pipe. This model is more economical vis-a-vis KVIC design economical as it replaces mild steel gasholder of the KVIC design with a small ferro-cement gasholder. Furthermore, ferro-cement gasholder can be of uniform size regardless of plant capacity which makes it amenable for standardisation and pre-fabrication. In addition sloping RCC roof and rectangular masonry digester achieve sufficient ease in plant construction and maintenance.

Spherical Biogas Plant. A fixed-dome spherical low-cost plant design was developed by the Tata Energy Research Institute (TERI), New Delhi. Besides being a low-cost model it is also durable on account of its shape which can withstand gas pressure (Fig. 4.12a). One of the additional features of this

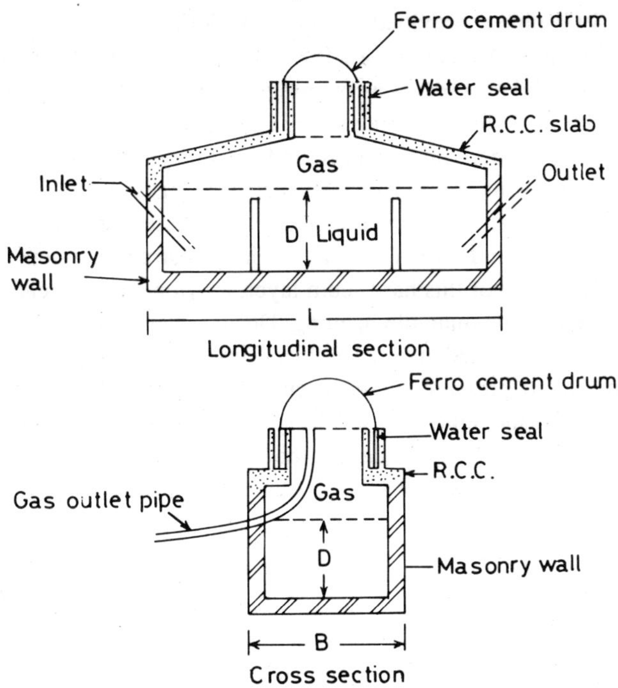

Fig. 4.12 Cross-sectional view of biogas plant developed by the Manipal Institute of Technology, Manipal [Source: TERI, Biogas Technology: An Information Package, Oct., 1985, p. 35]

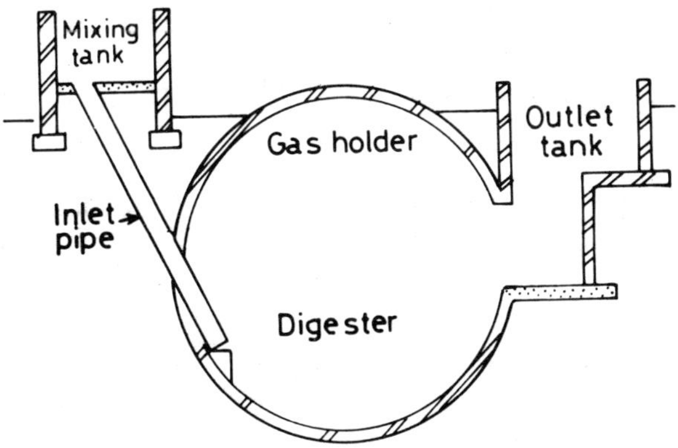

Fig. 4.12a Schematic view of Spherical Biogas Plant

plant is that it uses a nylon net for automatic breaking of the scum formation over the slurry surface within the digester. This plant is designed for a retention period of 50 days with a maximum pressure of 2 ft. 6 inches (75 cm). Digester and gasholder together form a complete sphere in which the top portion serves as the gasholder. Mixing tank and outlet tank are on either side of the digester. The gasholder is protected by an additional layer of tiles which are laid over the bricks. A GI-pipe is provided on top of the dome and a valve is fitted to it to regulate gas flow. A manhole of 2 ft × 2 ft is provided at digester top which facilitates entry and also serves as opening to the outlet tank. Organic wastes are mixed with water for preparing slurry in the mixing tank. Once uniformly mixed slurry is prepared, lid is opened and the slurry flows into the spherical digester through the inlet pipe. The slurry inside the digester may contain lighter particles like grass and other waste materials. These particles float on top surface of the slurry and form a hard layer which obstructs gas generation. For preventing formation of this hard scum layer, a nylon net is spread inside the digester just above the manhole opening. During digestion process, biogas is produced and accumulated in the gasholder. The slurry level within the digester goes down due to pressure of the accumulated gas and rises up with its consumption. This movement causes slurry to pass through the nylon net which breaks up the scum. [For more details the reader is advised to refer to TERI document, Spherical Biogas Plants: A Manual for Extension Workers, New Delhi, 1988].

ASTRA Model. Centre of Applications of the Science and Technology in Rural Areas (ASTRA), a unit of the Indian Institute of Science, Bangalore, developed a plant model which has come to be known as the ASTRA model. It is a floating drum type biogas plant which utilises a solar water heater and a solar still for obtaining higher gas yield. A schematic view of the ASTRA model is shown in Fig. 4.13. The ASTRA model needs reduced retention period, incurs low cost and causes higher gas yield due to the facility for maintaining higher slurry temperature particularly during winter. Digester and gasholder dimensions can be optimally selected that correspond to the minimum plant cost. Mild steel sides of the gasholder are extended by 0.3 m above the black painted roof to form a 0.3 m deep water pond on gasholder top. The pond is covered with a

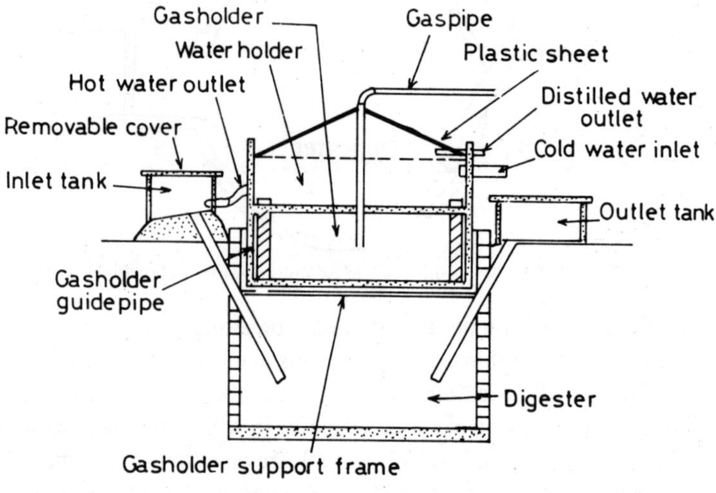

Fig. 4.13 Schematic view of the ASTRA biogas model [Source: TERI, Biogas Technology: An Information Package, Oct., 1985, p. 38.]

polythene sheet. By making the transparent cover assume a tent like shape, the roof-top solar heater serves as solar still to yield distilled water. Two experimental ASTRA plants of 2 and 6 m^3 capacity with their performance analysed over a period of one year were marginally superior to the matching KVIC design with their costs less by 40 per cent.

Belur Math Model. This model was developed by Swami Vimuktananda at Ramakrishna Mission at Belur Math (West Bengal). Its distinctive feature lies in having a separate gasholder independent of digester. A schematic view of this design is shown in Fig. 4.14. Gasholder consists of two cylindrical tanks placed opposite to one another with lower one filled with water. Slurry is normally fed to the digester every day. Resulting biogas following anaerobic fermentation is passed through a pipeline to the gasholder and from there through another pipeline to the place of use.

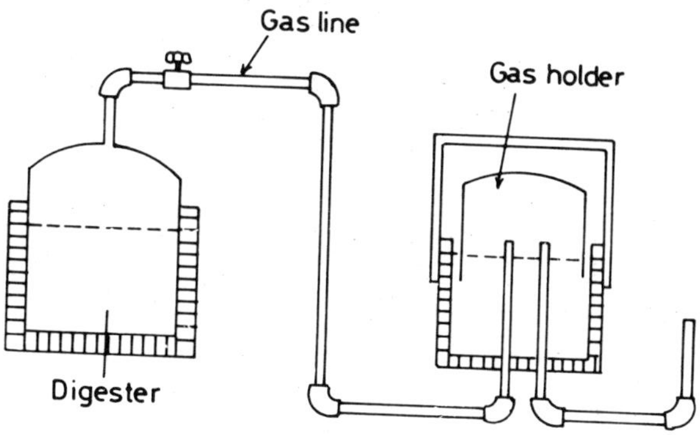

Fig. 4.14 Schematic view of plant model developed by the R.K. Mission, Sardapitha, Belur Math, Howrah [Source: TERI. Biogas Technology: An Information Package, Oct., 1985, p. 40.]

NEERI Model. This model evolved by the National Environmental Engineering Research Institute (NEERI) at Nagpur is a modified version of the movable drum type biogas plant. Its digester made of brickwork partly lies below ground and partly above. It is specially designed to run on night-soil. There is a facility for periodic sludge removal into sludge drying bed where it is dewatered and dried. A schematic view of the plant is shown in Fig. 4.15. Supernatant of the digester and underflow of drying beds are further treated in a stabilisation pond.

Bajwa-KVIC Model. The Bajwa-KVIC model also called the Ganesh model is an improved version of the KVIC model. Whereas the KVIC model uses the mild steel and cement for its construction, the modified Bajwa-KVIC model uses only a few mild steel round bars and no cement. Its digester is made of bamboo strips, mild steel bars and low density polyethylene sheets. Gasholder is made of high density polyethylene sheets. Gasholder is a floating drum guided by a central guidepipe as in the KVIC model. Gasholder is very light and does not require periodical maintenance. It is possible to carry out

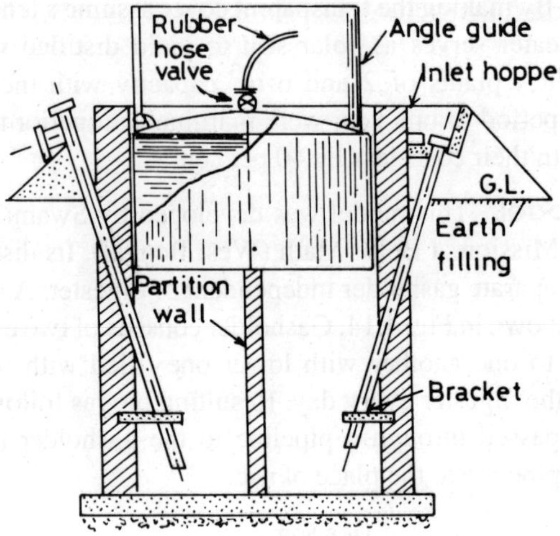

Fig. 4.15 Schematic view of plant model working on night-soil developed by the NEERI, Nagpur [Source: Chawla, O.P., Advances in Biogas Technology, ICAR, New Delhi, 1986, p. 65]

repair locally with the help of kerosene-fired torch. Plant construction is easier since locally available materials like bamboo-split can be used and costly items like cement and mild steel are used bare minimum. LDPE and HDPE sheets are found very effective in checking water seepage than cement or brick apart from ease to carry. Life of the plant however following exposure of LDPE gasholder to ultraviolet rays is comparatively less. A 2-m^3 capacity plant (Fig. 4.16) has been installed at the Government Implement Factory, Bhubaneshwar and is reportedly working satisfactorily.

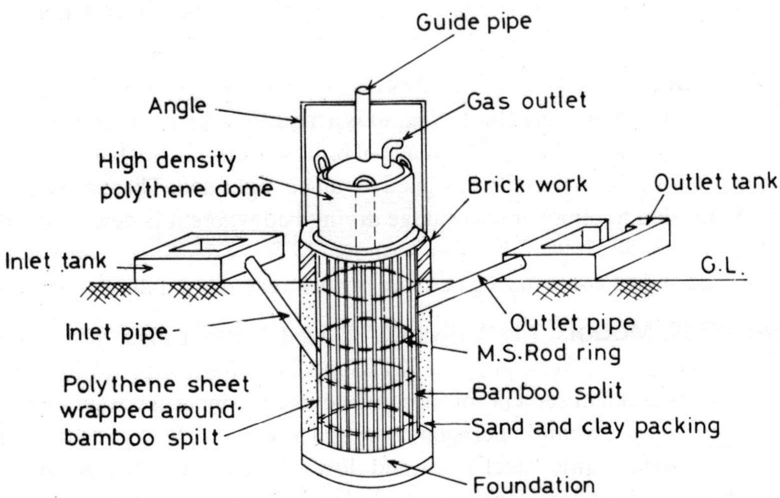

Fig. 4.16 A schematic view of Bajwa-KVIC model initially commissioned at the Government Implement Factory, Satyanagar, Bhubaneshwar [Source: TERI, Biogas Technology: An Information Package, Oct., 1985, p. 34.]

Jyoti Top Loaded Digester. Design of this model (Fig. 4.17) was developed by the Jyoti Solar Energy Institute at Vallabh Vidyanagar (Gujarat) which is particularly appropriate for utilising agricultural waste materials and crop residues. In this model which is a modified version of the KVIC model, wastes are admitted from the gasholder top leading to the slurry surface with the help of a plunger arrangement. Care is taken to make inlet pipe sufficiently long so that it always touches the slurry surface for preventing gas leakage through inlet. For achieving adequate dispersion of wastes as well as avoiding rotation of wastes as a whole, stirrer with paddles near the surface and passing through central guidepipes are provided. Gasholder can move up and down inside the peripheral channel close to upper part of the digester depending upon the gas pressure. Jyoti model is helpful in alleviating the problem of scum formation which is quite severe in digestion of forest and crop residues. The plant design obviates the need for fine shredding and drying of the residues before feeding them.

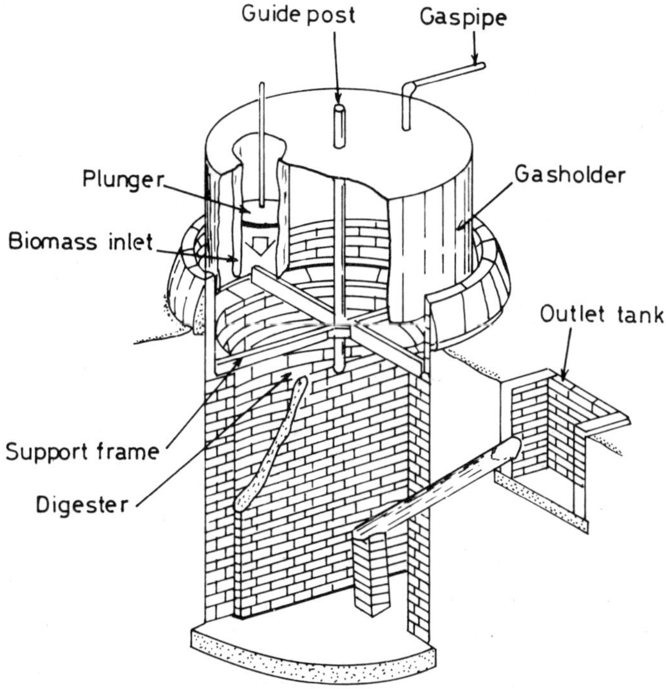

Fig. 4.17 Cross-sectional view of biogas plant developed by the Jyoti Solar Energy Institute, Vallabh Vidyanagar [Source: TERI, Biogas Technology: An Information Package, Oct., 1985, p. 37.]

Jwala Biogas Plant. This design was developed by the AMM Murugappa Chettiar Research Centre, Madras. It has cylindrical digester which is largely kept below ground level. Gasholder is made up of low density polyethylene (LPDE) sheet which is fixed on a geodesic balloon. Composite dome is held inside a water seal to avoid gas leakage. Scum-breaking is achieved with the help of a stirrer which passes through a central guide-tube. It has a partition wall which is tangentially touched by the inlet pipe (Fig. 4.18). This model is relatively inexpensive as no costly materials are involved in its construction.

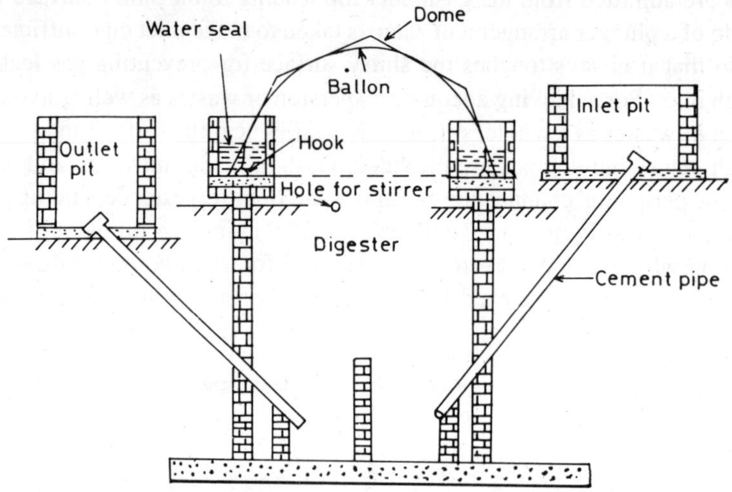

Fig. 4.18 Cross-sectional view of Jwala model developed by the AMM Murugappa Chettiar Research Centre, Madras [Source: TERI, Biogas Technology: An Information Package, Oct., 1985, p. 39]

Mud Jar Biogas Plant. Centre of Science for Villages, Wardha, developed a low cost biogas plant which resembles a '*Surahi*' local name for a mud pot which can be built by a village potter and hardly needs any maintenance [5]. Ten mud jars of 300 litre capacity each connected in series were used as digesters and resulting gas was sufficient for meeting cooking needs of a family of five for four hours use at a stretch. Based on 1982 price level, the plant cost was Rs. 650 which is less than one-fourth of the traditional masonry built biogas plant of the Indian design. Mud jars were connected with three inch thick clay pipes. The first mud jar was attached to a small pot to serve as inlet for feed. Similarly the last jar was connected to a pot to act as the plant outlet. Whereas inlet and outlet pots were placed above the ground, three-fourth portion of the intermediate mud jars (digesters) were embedded within the ground (Fig. 4.19). Partial underground placement of digesters minimises the problems of leaks and plant breakage. Upper one-fourth portion of the jars act as gasholder where gas is stored. For preventing gas seepage through the porous walls of mud jars, their upper exposed portion can be painted with coal tar or locally developed paint made out of sap of trees and lime. Generated gas can either be taken out through flexible tubes or collected in leak-proof balloons. Mud jar plants can prove valuable in areas that prossess good soil properties.

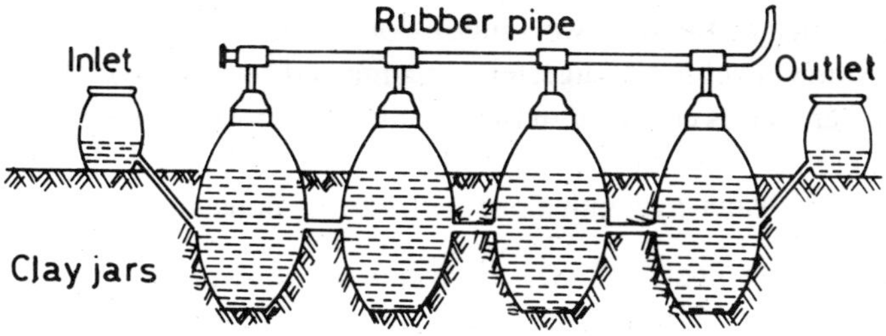

Fig. 4.19 A view of '*Surahi*' Biogas Plant [Source: TERI, Biogas News. Inaugural Issue, Feb., 1982, p. 1.]

'Kachra' Biogas Plant. Dr. K.P. Goswami, an eminent agricultural scientist, developed a plant what has come to be known as '*Kachra*' biogas plant arising out of years of work at the Punjab Agricultural University, Ludhiana and later Gujarat Agricultural University at Anand [12, 13]. The need to develop a special plant arose following difficulty in fermenting straw like oversized agricultural wastes which float and cause clogging in the digester. The '*Kachra*' biogas plant can work on any plant material or animal wastes in any proportion but it works more efficiently with leafy materials than woody materials. Initially a bench scale plant (2-3 m^3 biogas per day capacity) was designed and developed which was later on modified. The earlier design was made of iron sheets and included three adjacent rectangular masonry chambers, 213 cm deep and 183 cm wide with a portion 150 cm deep kept underground. It had two 243 cm long digester tanks separated by a partition wall and having a 40-cm wide gate opening. The two tanks were provided with separate stirrers for vertical stirring of digester slurry. The installation cost of the original double chambered 'Kachra' biogas plant was one and half times more than the conventional plant of the KVIC design of identical capacity. The economics of plant operation was, however, matching and comparable with that of a conventional plant.

Development activities were continued as a result of which double chamber in earlier design was replaced by a single chamber. The modified design consists of a $2.5 \times 2 \times 1.65$ m^3 rectangular brick masonry digester covered with a rectangular gasholder (Fig. 4.20). The outlet is built contiguous to the digester

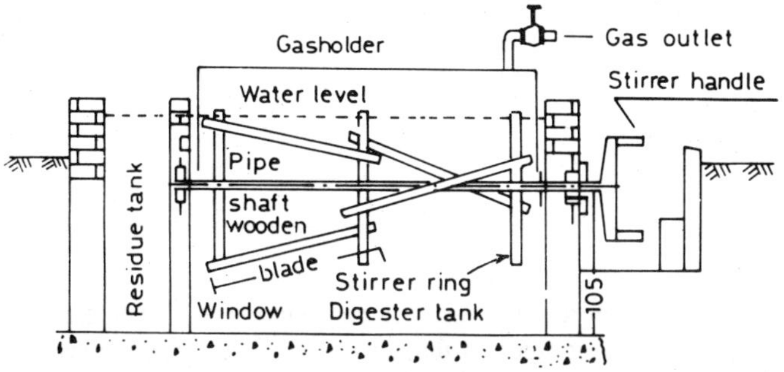

Fig. 4.20 A schematic view of *Kachra* biogas plant

which is connected through two windows of size 45×45 cm^2 near the digester bottom. Slurry level in the digester is maintained by overflow channel in the exit chamber. Gas is collected in free-floating type rectangular gasholder. The plant is preferred to be operated on a single type of feedstock, be it, paddy straw, wheat straw or water-hyacinth alone with stirring done horizontally.

Flexible Biogas Plant (FLXI). Conventional underground digesters are not appropriate for thousand of miles along the coastlines in India, Pakistan and Bangladesh and several other countries with vast coastlines. Recurrent floods in the vicinity of Ganges, Indus and Brahmaputra river banks adversely affect the success of biogas programme in these densely populated regions. In these areas balloon type flexible biogas plants made of neoprene-rubber material could provide an effective solution (Fig. 4.24). Flexi plants can be easily installed above ground, conveniently transported to isolated areas, quickly installed at any time and can be built up at low cost. In India the *Swastik* Company, Pune, supplies flexible Biogas plants briefly called FLXI. Gas production per filling quantity is comparable to the similar range from conventional biogas plants. Playing children or roaming buffaloes may cause leakages but can be easily repaired. Growing or splitting families may carry or replace their balloon plant whereas with conventional plants it becomes a lost investment. Economics of the plant may improve if more plants are made by the Swastik Company. A minimum of five operation years is needed to amortize the investment. However as the average life span of the Swastik plant is barely eight to ten years, long term return on investment for conventional underground digesters is accordingly higher. Hence the Swastik plant is not an alternative to conventional plant where these are found appropriate but it should merely be viewed as a substitute option where conventional plants cannot be built underground.

Ramakrishna Mission Technical Institute, Mylapore (Madras) has recently designed and constructed a small portable biogas plant (Fig. 4.20a). It consists

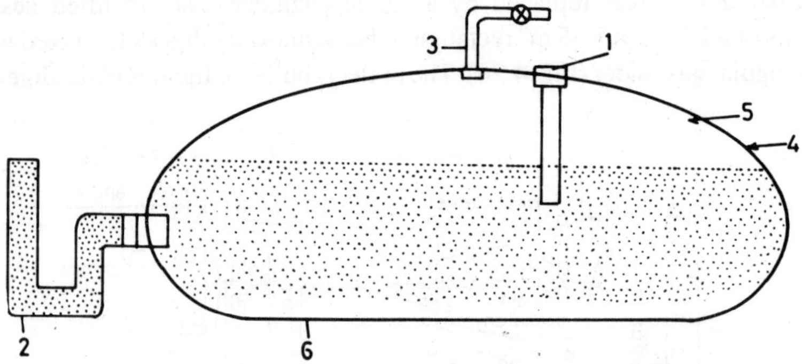

1. Inlet for feeding fresh cowdung slurry
2. Outlet for digested slurry
3. Outlet with valve for biogas
4. PVC Bag
5. Fibre layer
6. Slurry

Fig. 4.20a Portable biogas plant

of a flexible bag made of a layer of PVC black sheet on inside and a layer of woven material often used in a fertiliser bag outside. The volumetric capacity of the bag is 2.2 cubic metres. Input slurry consisting of equal parts by volume of cowdung and water is poured into the bag through an opening. Inlet pipe dips sufficiently into the slurry to prevent gas escape through it. Digested slurry is discharged through bent pipe whose bends help to maintain the desired slurry level. When fresh slurry is poured in, an equal amount of digested slurry is displaced through the discharge or outlet tube. Biogas is collected in the bag which gradually expands it. Gas needs to be continuously utilised to prevent building up of excessive gas pressure as otherwise it may cause gas escape through the inlet or outlet tubes [27].

Plants approved by the MNES under NPBD. Under the NPBD a total six types of biogas plants have been approved by the MNES. This includes KVIC movable drum type, Janata fixed dome type, Deenbandhu fixed dome type, Pragati movable drum type, ferrocement and Swastik biogas plants [26]. Initially only two models, namely, KVIC movable drum type and Janata fixed type were approved for promotion. As the programme made headway, four more models were added out of which Deenbandhu, a low cost fixed dome type has surpassed all others. Whereas other plants have already been described and illustrated, cross-sectional view of Pragati Model is shown in Fig. 4.20b. Material requirement for constructing different capacity Pragati model is given in Table 4.16. The Pragati model though it is of movable drum type but it has a spherical digester with a conical bottom.

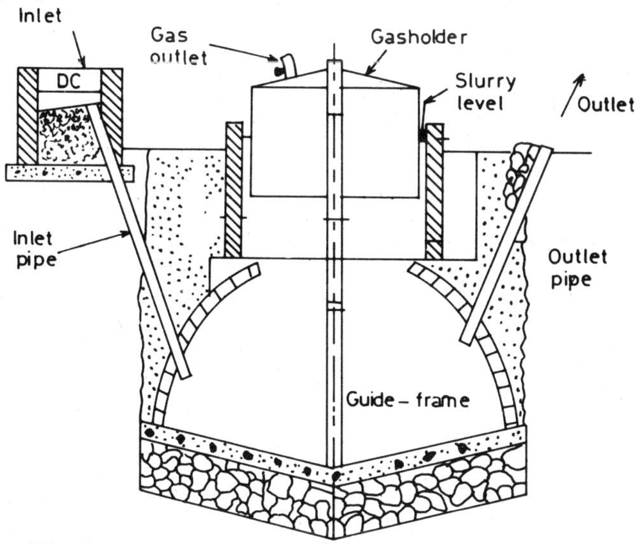

Fig. 4.20b Cross-sectional view of the Pragati Model

Source: TERI, Managing the Biogas programme, A Monograph sponsored by the MNES, Tata Energy Research Institute, New Delhi, 1994, p. 46

Table 4.16 Bill-of-Material for Different Capacity Pragati Biogas Plants (All dimension in mm)

Description	*Unit*	*2 m^3*	*3 m^3*	*4 m3*
1. Bricks	No	700-750	1200-1250	1775-1825
2. Cleared and screen-ed sand	m^3	1.65	1.8	4
3. Stone metal (25 mm to 40mm)	kg	0.95	0.98	1.7
4. Cement gauge bags	No	9.5	18	2.1

Note : Gasholders height is 100 cm.

Source : TERI, Biogas: A Source of Rural Employment, Tata Energy Research Institute, New Delhi, 1994, p. 36.

4.5 An Overview of Select Plant Designs in World

Several plant designs are popular in western world. A schematic view of a typical on-farm biogas plant used in western countries is shown in Fig. 4.21 [6]. In Philippines one of the early plants used is schematically shown in Fig. 4.22. Being a two stage process it ensures more efficient fermentation of fibrous agricultural waste materials like crop-residues and vegetable wastes. Schematic view of an oil drum biogas plant commonly used in Indonesia is given in Fig. 4.23. In several countries including Belgium flexible bag type

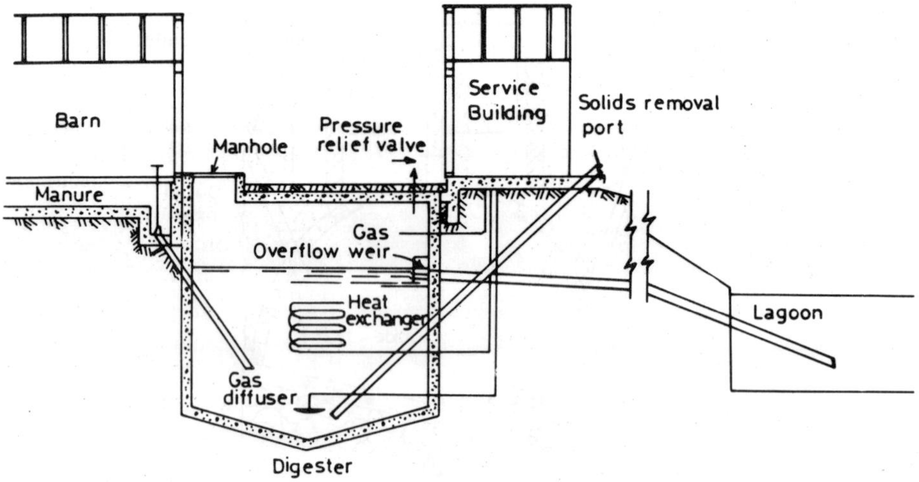

Fig. 4.21 Schematic view of a typical on-farm biogas plant used in western countries. [Source: Biomass Energy Institute, Biogas Production from Animal Manure, Winnipeg, Manitoba, 1978, p. 13]

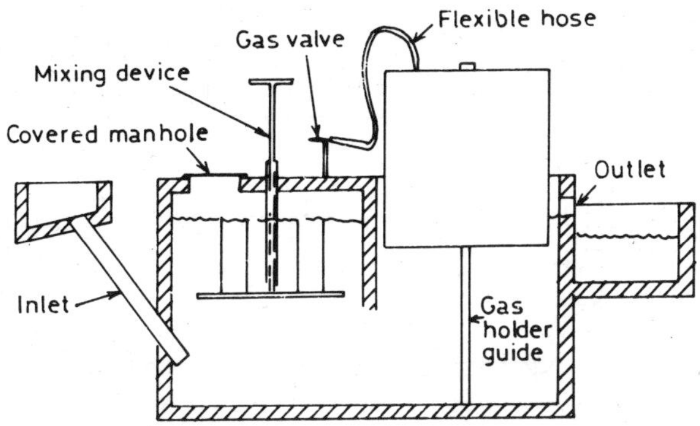

Fig. 4.22 Schematic view of a two-chamber rectangular biogas plant commonly used in Philippines [Source: UN, Guidebook on Biogas Development, New York, 1980, p. 16.]

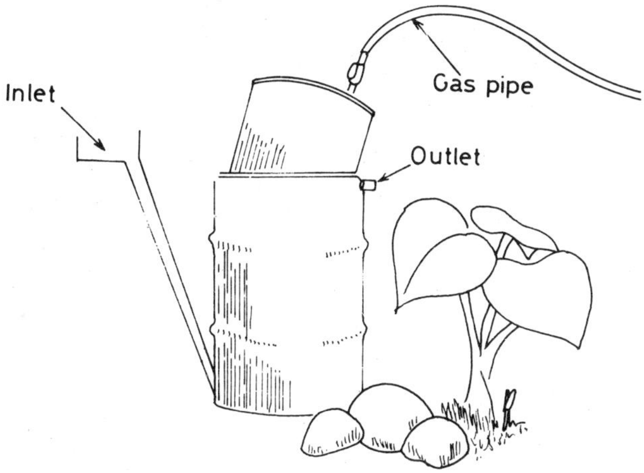

Fig. 4.23 Schematic view of an oil drum biogas plant commonly used in Indonesia [Source: UN, Guidebook on Biogas Development, New York, 1980, p. 16.]

biogas plants (Fig. 4.24) are popular. These are made of strong plastics resistant to ultraviolet rays [25]. Hypalon-Neoprene coated fabric is commonly used in their construction. These plants are easy to build at low costs, are portable and enable constant gas pressure to be built-up. H.A. Hamad *et al.* [15] give an account of a viable model proposed for application of biogas technology in new desert community areas in Egypt. The model involves a novel architectural design and house planning to promote biomethanation for the entire community for meeting its energy, fertiliser and sanitation needs.

A novel continuous-fed biogas plant made of ceramic precast has been designed by the Energy Research Section of the Development Research and Pilot Centre (DRPC), Ho-Chi-Minh City, Vietnam. The plant consists of two crossed and horizontal iron bars in digester which serve as scum breakers. Scum

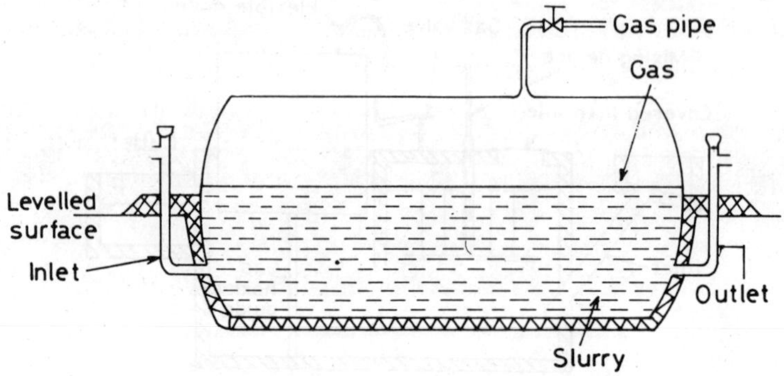

Fig. 4.24 Cross-sectional view of Neoprene bag model developed and commonly used in Taiwan [Source: TERI, Biogas Technology: An Information package, Oct., 1985, p. 25.]

breaking is affected by movement of slurry resulting from pressure variations. Besides outlet valve digested sludge can be removed through equalisation tank. By 1983, DRPC had designed two sizes of this plant. The small-scale plant had a 300 litre digester and 150 litre equalisation tank. Loading capacity of this plant was 70 kg piggery waste. Second model was meant for household use and had a 135 litre digester and 65 litre equalisation tank. Loading capacity of family-size plant is 40 to 50 kg of piggery waste. Retention period in both the plants is 40 days [22].

Recently, animal industry in Japan has expanded rapidly with the increasing demand for animal products following rapid growth in Japanese living standards. This expansion is characterised by a rapid increase in number of animals and significant decrease in number of farmers. Anaerobic fermentation of wastes with a view to obtain both fuel and fertiliser has emerged as an attractive option [14]. Initial studies on biogas production from animal wastes were carried out by the National Institute of Animal Industry using a cylindrical digester made of fibre-glass reinforced plastic which is shown in Fig. 4.25. The size of the digester (200 litres) was adequate for the waste from one pig and it

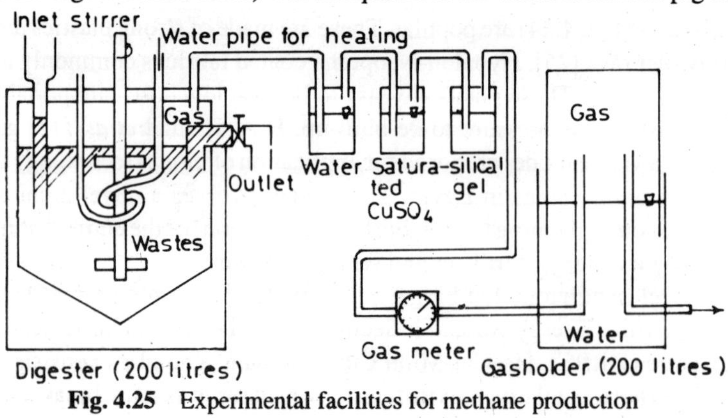

Fig. 4.25 Experimental facilities for methane production

contained about 150 litres of waste slurry. The temperature was kept at 35°C by circulating hot water through a coil in the slurry. The digester was insulated by a 5 cm layer of glass fibre. Methane fermentation followed from a mixture consisting of 44 kg of fresh pig faeces and 88 litres of tap water with 30 litres of anaerobically-fermented cow dung as seed for the fermentation. The daily loading was 7.5 litres of slurry consisting of 2.5 kg of pig faeces and 5 kg of tap water. In steady state it produced 170-200 litres of biogas per day. Kagawa Prefecture Livestock Experiment Station wayback in 1978 used a much bigger digester of 3 m^3 capacity. The digester shown in Fig. 4.26 is equipped with a submersible pump (0.4 kw) and is operated under mesophilic condition (35°C). Biogas generated from pig wastes is stored in a gasholder (1.8 m^3) and used in the experiment involving running a modified petrol engine (5 hp with 2000 rpm) which consumes 24.6 litres of biogas in 1 hour. The capacity of the digester at Nagano Prefecture Livestock Experiment Station (Fig. 4.27) is 200 m^3 which is constructed under a roof for 30 head of dairy cattle. Wastes from the dairy cattle are introduced by gravity into the digester with two submersible pumps for stirring and heating. The pumps are surrounded by screens which prevent clogging caused by coarse organic material. Biogas produced is used to generate steam and provide hot water. Systematic experiments with a larger installation are being carried out at the Hiroshima Prefecture Livestock Experiment Station (Fig. 4.28). Wastes from 200 head of pigs are mixed with 2-3 volumes of tap water and charged into two digesters daily. The digesters with polyurethane insulation are kept under mesophilic conditions by submersible pumps, and produce enough gas to run air-conditioning facilities for about 8 hours each day.

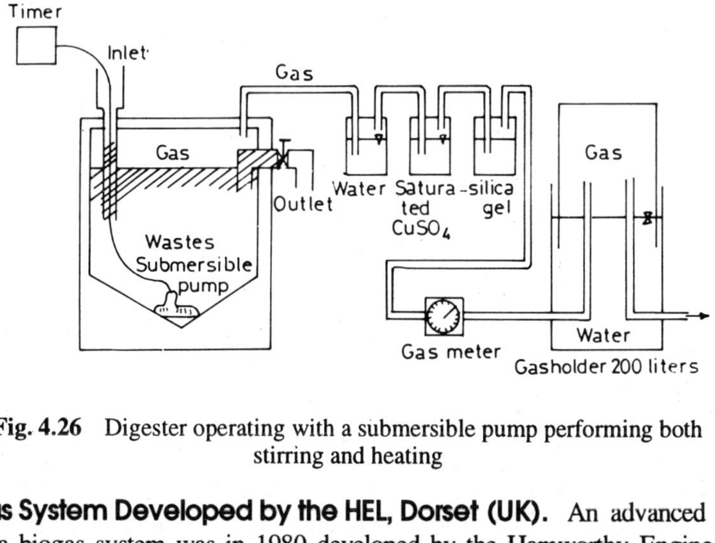

Fig. 4.26 Digester operating with a submersible pump performing both stirring and heating

Biogas System Developed by the HEL, Dorset (UK). An advanced design of a biogas system was in 1980 developed by the Hamworthy Engineering

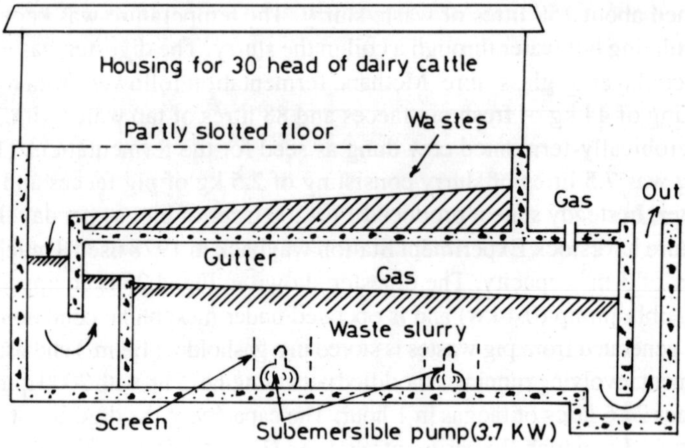

Fig. 4.27 Facilities for methane production at Nagano Prefecture Livestock Experiment Station (Vertical section view)

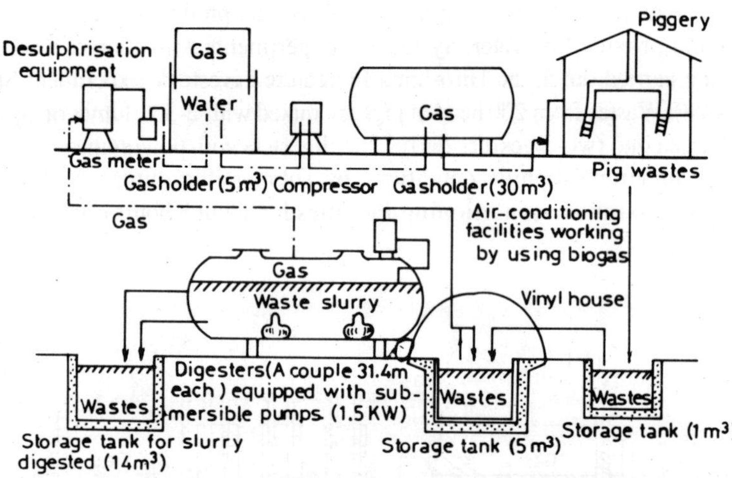

Fig. 4.28 Facilities for methane production and for its utilisation at Hiroshima Prefecture Livestock Experimental Station

Limited, Dorset (Fig. 4.29). A demonstration unit of this system is working at the Walnut Tree Farm, Chediston, Suffolk (UK) to produce biogas and fertiliser containing high proportions of phosphates and nitrogen. This system developed was the result of over six years of collaborative endeavours put up by the Hamworthy Engineering Limited and the Department of Microbiology at University College, Galway (Cardiff) which were responsible for the scientific aspect of the project; and the Polytechnic of Wales where pilot plant evaluation studies were carried out. The biogas plant at the Walnut Tree Farm was designed to treat 18 tons of pig wastes per day with 5 to 6 per cent total solids

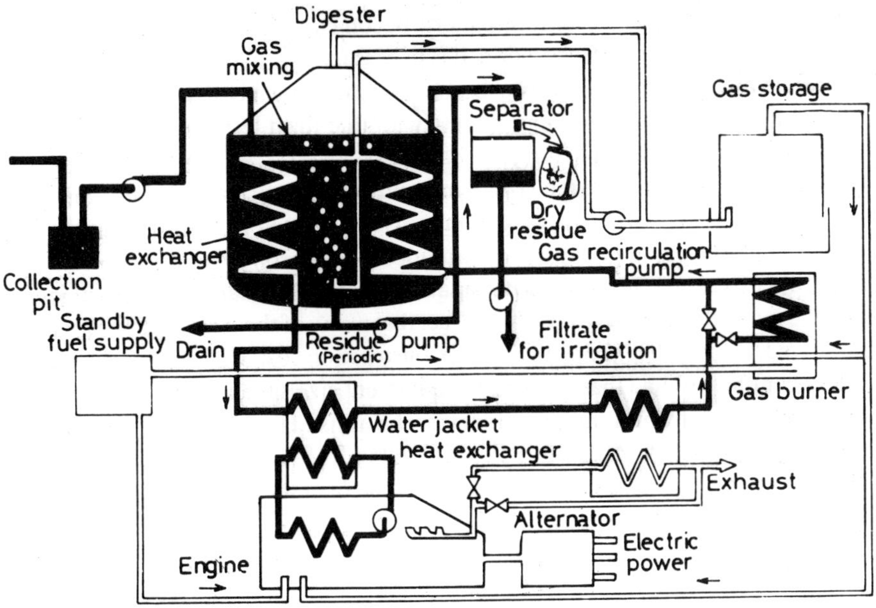

Fig. 4.29 Flow diagram of an anaerobic digester at Suffolk Farm, UK

and 6 tons of cattle waste per day with 8 to 9 per cent total solids and 15 day retention time. The digester has a volume of 350 m^3 and a diameter of 9.15 m with a conical top.

Biogas is collected from the digester top and held in a small 10 m^3 gasholder. The effluent from the plant is passed through a vibratory screen separator to remove any undigested material and the filterate containing about 1 per cent total solids is pumped to a storage lagoon used for irrigation purposes. The process operates on a continuous cycle with minimum retention time. Wastes are fed into the digester by a pump. There is a provision to warm up and agitate the slurry with a mixing device.

4.6 Concluding Remarks

In prevailing circumstances in a developing country like India, it is not practical to feed a digester on continuous basis as it requires expensive feeding and withdrawal arrangement. Instead it is operationally convenient to run the plant on 'fill and draw' basis with filling and drawal being done at regular intervals varying anywhere from one day to a week. A biogas plant can also be operated on 'all in and all out' basis in which case it is charged and emptied when gas production is over. This type of feeding is particularly appropriate in situations where feed materials like crop residues are generated at one point of time such as at the end of harvest period and subsequent additions occur only in limited quantities. In situations where digesters are fed on daily or weekly basis or any other periodic interval, charging pattern becomes semi-continuous or batch operation. In case of batch operation, it is necessary to provide storage facilities for achieving uniform gas supply. Batch operation is generally carried out by emptying two third of digester contents with one third left inside to seed the incoming batch of influent slurry.

In a situation where large size community plant is proposed, there can be two options, namely, to go for several smaller plants or to opt for a single equivalent large plant. Latter option provides higher economies of scale which are ordinarily not achievable with smaller plants. For example, large plants operate more efficiently in view of the fact that many of them are equipped with improved heating and mixing facilities. Construction cost of a community plant is less per unit volume than combined equivalent size several units. Following higher yield, gas from community plant can be utilised for large scale industrial applications which is not feasible from small family units. The main advantage of a family unit is that it offers flexibility of operation and gas utilisation according to individual needs.

REFERENCES

1. Bahadur, S. and Singh, K.K., *Janata Biogas Plants*, Planning Research and Action Division, State Planning Institute, Lucknow, 1978.
2. BABA, Digest No. 10, Dec., 1983.
3. Bansal, N.K., A Techno-Economic Assessment of Solar Assisted Biogas Systems, *Energy Sources*, Vol. **10**, 1988.
4. Anonymous, Noval Biogas Plant, *Biogas News*, Inaugural Issue, Feb., 1982.
5. BEI (Biomass Energy Institute), *Biogas Production from Animal Manure*, Winnipeg, Manitoba, 1978.
6. BIS, *Specification for Gobar Gas Plant and Equipment* IS : 9478-1980, Bureau of Indian Standards, New Delhi, 1980.
7. BORDA, *Biogas Plants: Building Instructions*, GATE, P.O. Box 5180, 6236 Eschborn 1, Germany.
8. BORDA, *Biogas: Manual for the Realisation of Biogas Programmes*, GATE, P.O. Box 5180. 6236 Eschborn 1, Germany, 1980.
9. Buren, A.V., *A Chinese Biogas Manual*, Intermediate Technological Pub. Ltd, London, 1979.
10. Chawla, O.P., *Advances in Biogas Technology*, ICAR, New Delhi, 1986.
11. FAO, *China: Azolla Propagation and Small-scale Biogas Technology*, Rome, 1978.
12. Goswami, K.P., *Development of a 'Kachra' Biogas Plant for Biogas and Manure*, Proc. National Symp. on Biogas Technology and Usage, Nov. 29-30, 1977.
13. Goswami, K.P., Appropriate Technology for Rural India to Produce Biogas from Vegetative Waste, *Energy Sources*, **11**, 59-67 1989.
14. Haga, K., *et al.*, Methane Production from Animal Wastes and its Prospects in Japan, *Agricultural Wastes*, **1**, 45-55 1979.

15. Hamad, H.A., *et al.*, Planning Egyptian Villages in New Reclaimed Areas with Regard to Accommodation of Biogas Technology, *Jour. of Institution of Engineers (India)*, **68**, IDP3, June, 48-52 1988.
16. IARI, *Cow Dung Gas Plant*, Indian Agricultural Research Institute, New Delhi, 1979.
17. Khandelwal, K.C., *Some Informations About Biogas Technology and Development in India,* A Report, Ministry of Agriculture and Irrigation, New Delhi, 1978.
18. Khandelwal, K.C. and Mahdi, S.S., *Biogas Technology: A Practical Handbook,* Tata McGraw-Hill Pub. Co. Ltd, 12/4 Asaf Ali Road, New Delhi 110002.
19. Khandelwal, K.C., *A Janata Model Gobar Gas Plant,* Pamphlet No. 104, AFPRO, New Delhi, 1980.
20. Kishore, V.V.N., Raman, P. and Ramarao, V.V., Fixed Dome Biogas Plants Developed by the TERI, *Rural Technology Journal*, **4(2)**, 26-40 1987.
21. NAS, *Methane Generation from Human, Animal and Agricultural Wastes,* Report of an Ad. Hoc. Panel of the Advisory Committed and Technology Innovation, National Academy of Sciences, Washington DC, 1977.
22. PTI, *Science Service*, March 16-31, 1984, p. 17.
23. Qidwai, M.S., *Invention Intelligence*, **19(1)**, Jan, 9 1983.
24. TERI, *Biogas Technology: An Information Package*, New Delhi, Oct., 1985.
25. UN, *Guidebook on Biogas Development*, United Nations, New York, 1980.
26. TERI, *Biogas Technology: A Promise for the Future*, A Monograph sponsored by the MNES, Tata Energy Research Institute, New Delhi, 1994.
27. Anonymous, Portable Biogas Plant, *The Hindu*, Madras, Sep 21, 1994, p. 27.
28. SRERI, Krishna Biogas Plant: New Innovative Model, *Biogas Forum*, **IV** No. **51**, 9-15 1992.
29. Singh, J.B., Myles, R.M. and Dhussa, A., *Manual for Deenbandhu Biogas Plants*, Tata Mc Graw Hill Pub. Co. Ltd., New Delhi, 1988.

Chapter 5

DESIGN, CONSTRUCTION, OPERATION AND MAINTENANCE OF BIOGAS PLANTS

5.1 Introduction

For obtaining maximum yield from biogas plants, it is necessary that plants have correct design parameters, are duly constructed as per laid-down specifications, and operated and maintained according to the prescribed instructions. It is also important that the plant be set up at an appropriate site. Site selection is particularly important for community plants as they need be located at sites convenient to several users. Family units are generally set up close to the point of use. Plant capacity need to be assessed before setting up a plant. Wastes availability and total gas requirement of a prospective user need be worked out while determining the plant size. Selection of plant type whether it is to be fixed dome or movable drum type or some other is another vital issue. Fixed dome plant involves no steel parts, runs on a batch-continuous process, provides richer fertiliser and ensures greater thermal insulation of the digester. Movable drum type on the other hand is conducive to build uniform gas pressure, offers ease of scum-removal following facility of lifting gasholder and enables easy clearance of blockages due to inlet and outlet being straight. For trouble-free operation of movable drum type plant, it is important that it be fed with homogenous, non-fibrous and sufficiently diluted feed. On the contrary in a fixed dome type plant as digested slurry is cleared manually, even long fibrous materials can be used to feed.

Some kind of checklisting is necessary for making a choice between fixed dome and movable drum type plant. Being different in design, selection of design parameters for a fixed dome and movable drum plants involves altogether different methodology and considerations. TERI has developed methodology for selection of design parameters for Janata biogas plants. IISc, Bangalore in past developed a methodology for designing a moving drum biogas plant. Selection of pipelines and aspects relating to distribution of biogas were studied in past by the Biogas Research Centre of the KVIC (Mumbai). Estimation of heating needs for maintaining the digester at desired temperature is particularly important for plants located in colder regions. A number of heat transfer models have been developed for this purpose. Exploitation of solar energy for warming of digester contents is a practical and economic approach. Procedures for constructing biogas plants have been fairly standardised and a number of comprehensive manuals exist. Several new materials such as plastics and ferrocement have been successfully used for plant construction. In this chapter it has been endeavoured to provide an overview of all aspects pertaining to the design, construction, operation and maintenance of biogas plants.

5.2 Selection of Biogas Plant Sites

Biogas plants can be built either close to the source of waste generation or away from it. In case plants are needed to be constructed afar, this may not be cost-effective in many situations. Normally for family-size plants it is economical to construct plants on site close to households where wastes are generated. For community plants, however, these can be set up at off-site provided economics of plant operation supports such a course. In family units wastes are usually collected from available sources, mixed to form slurry and finally fed to plants. As per the practice followed in China, pigstys, household latrines, poultry sheds etc. are also connected to plants either directly or through mixing chamber. For the Chinese practice to succeed in India, social barriers associated with human faeces and pig-wastes need to be removed. If animal sheds and household toilets are to be connected to plants, it is necessary to provide slanting floors for easy wastes-flow. If animal urine is to be added to feed, animals need to be properly stabled particularly during nights to help store dung and urine. For obtaining higher gas yield, the plant should be so located that it receives direct sunshine for most part of the day. For preventing water contamination, drinking water wells should be at least 15 meters away from biogas plants. Demonstration plants should be located at convenient sites to attract maximum promotional benefit.

In several situations organic wastes need to be kept in storage sheds before use. These sheds can be made of local materials such as bamboo, tree limbs, palm thatch, bricks, tiles, mortar etc. From sheds wastes can be carried to sites manually in shoulder-buckets or animal-drawn carts in close-containers depending upon distances involved. If water-weeds have to be used as feed then these, on small scale, can be collected by two or more persons pulling a net stretched between two bamboo poles across the water surface. Algae and water-hyacinth need be dried before use. Dried matter can also be used as bedding in animal sheds to soak urine and then used [29].

5.3 Checklist for Determining the Size of Biogas Plant

For selecting plant parameters, it is necessary to first assess gas demand. For determining gas requirement biogas consumption norms for cooking, lighting, driving engines etc. need to be known. Common norms for biogas consumption for different applications are as follows:

Gas Consumption Rates for Different Applications:
Cooking : 0.25-0.42 m^3 (8-15 ft^3) per person per day
Lighting : 0.11-0.15 m^3 (4-5.5 ft^3) per hour per lamp
Driving engines : 0.45 m^3 (15 ft^3) per HP per hour
Generating electricity : 0.6 m^3 per kwh

- 1 kg of fresh animal wastes generates 0.05 m^3 of gas.
- 1 m^3 of biogas is equivalent to [7]:
  - 2 kg of fuelwood
  - 0.6 litres of kerosene
  - 0.5 litres of petrol
  - 0.4 litres of diesel
  - Cooking of 3 meals for 3 persons
  - Running of 1 HP IC engine for 2 hours
  - Running of 300 litre refrigerator for 3 hours
  - 1.25 kw of electricity
  - Lighting a gas lamp for six hours or six lamps for one hour
  - Illuminating twenty five 40 w electric bulbs for one hour, equivalent to a 60 w electric light burning for 6 to 7 hours
  - Driving a 3-tonne lorry for 2.8 hours
- The weight of methane is roughly half that of air.

$$\frac{1 \text{ m}^3 \text{ of methane}}{1 \text{ m}^3 \text{ of air}} = \frac{0.716 \text{ kg}}{1.293 \text{ kg}} = 0.554$$

- Approximately 1 m^3 of digester space is occupied by 1000 kg of animal manure and water.
- 1 m^3 of waste materials yield 0.15 to 0.30 m^3 of gas per day depending upon the climatic conditions and the type of material used.
- Digester volume should be at least 30 times the volume of daily feed rate. Considering that feed remains inside the digester for more than six weeks or so, it is generally preferred to keep the digester volume roughly 42 times the volume or daily feed input.
- Volume of the pit can be determined by estimating how much gas will be needed and how it will be utilised. In rural areas a family of five requires 1 m^3 per day for cooking and lighting. In summer each m^3 of digester space produces 0.15-0.2 m^3 gas per day and in winter 0.1-0.15 m^3 per day. While building digester one can go by the norm that 1.5-2 m^3 of volume should be allowed per head. Digester volume can also be calculated by following the norm that for two or less people there should be no more than 3 m^3 digester volume per head; for three to five people, no more than 2 m^3 per head, and for five or more people there should be no more than 1.5 m^3 digester volume per head [39].

Following simple relationship can be used for determining the total gas requirement.

Biogas requirement per unit × Number of units = Total gas requirement.

Gas requirement for cooking =
0.25 m^3 per person per day × No. of person = __________ m^3
Gas requirement for lighting =
0.15 m^3 per lamp per hour × No. of lamps = __________ m^3

- Gas consumption of engines =
0.45 m^3 per HP per hour × No. of operating hours = __________ m^3

Gas requirement for other applications:

- Refrigeration : _______ × _______ = _______ m^3
- Crop drying : _______ × _______ = _______ m^3
- In biogas plant itself for heating digester contents : _______ × _______ = _______ m^3
- Additional gas needs : _______ × _______ = _______ m^3

Total gas needs : = _______ m^3

After assessing gas needs, it is necessary to estimate the amount of digested sludge and biogas obtainable from available wastes. While estimating this, it is necessary to know norms of wastes and gas yield from different animals (Table 5.1).

Table 5.1 Biogas Yield Obtainable from Different Wastes

Animal	*Waste Yield per day (kg)*	*Biogas Obtainable per kg of Different Wastes*		*Biogas Obtainable per Animal per Day*	
		Litre	*ft^3*	*Litre*	*ft^3*
One cattle head	10	36	1.3	3600	13.0
One water buffalo	15	36	1.3	540.0	19.5
One pig (app. 50 kg)	2.25	78	2.8	180.0	6.3
One chicken (app. 2 kg)	0.18	62	2.2	11.2	0.4
One human adult	0.4	70	2.5	28.0	1.0

Source : GATE, Biogas Plants: Building Instructions, German Appropriate Technology Exchange, German Agency for Technical Cooperation, P.O. Box 5180, 6236 Eschborn 1, Germany.

Following simple relationships can be used for estimating biogas yield.

Generation of animal manure

Number of animals or other sources of wastes × rate of waste yield per animal per unit time

Assessment of biogas yield

Number of animals or other sources of wastes × rate of gas generation per animal per unit time = Biogas yield per unit time

Estimation of wastes availability:

Number of buffaloes	______	× 15 kg/day	= ______	kg/day
Number of cows	______	× 10 kg/day	= ______	kg/day
Number of calves	______	× 5 kg/day	= ______	kg/day
Number of pigs (50 kg)	______	× 2 kg/day	= ______	kg/day
Number of horses	______	× 10 kg/day	= ______	kg/day
Number of sheeps	______	× 2 kg/day	= ______	kg/day
Number of chicken	______	× 0.18 kg/day	= ______	kg/day
Number of human adults	______	× 0.4 kg/day	= ______	kg/day
Biomass			= ______	kg/day
Additional wastes availability			= ______	kg/day
Total wastes availability			= ______	kg/day

Estimation of biogas yield

Number of buffalos	______	× 0.540 m^3/day	= ______	m^3/day
Number of cows	______	× 0.360 m^3/day	= ______	m^3/day
Number of calves	______	× 0.200 m^3/day	= ______	m^3/day
Number of pigs (50 kg)	______	× 0.180 m^3/day	= ______	m^3/day
Number of horses	______	× 0.350 m^3/day	= ______	m^3/day
Number of sheep	______	× 0.100 m^3/day	= ______	m^3/day
Number of chickens	______	× 0.011 m^3/day	= ______	m^3/day
Number of human adults	______	× 0.030 m^3/day	= ______	m^3/day
Biomass			= ______	m^3/day
Additional biogas production			= ______	m^3/day
Total biogas yield			= ______	m^3/day

Following assessment of biogas demand and supply at a prospective site, rationality of having a biogas plant at that site can be evaluated by systematically proceeding as in Fig. 5.1.

5.4 Selection of Parameters Affecting Plant Size

Digester temperature and detention period. There are several factors that affect size of a biogas plant. Temperature and detention period which are interrelated are two important parameters. Detention period is defined as the time taken by the feed to flow from inlet to outlet. A set of recommended norms for detention times in different world regions are given in Table 5.2.

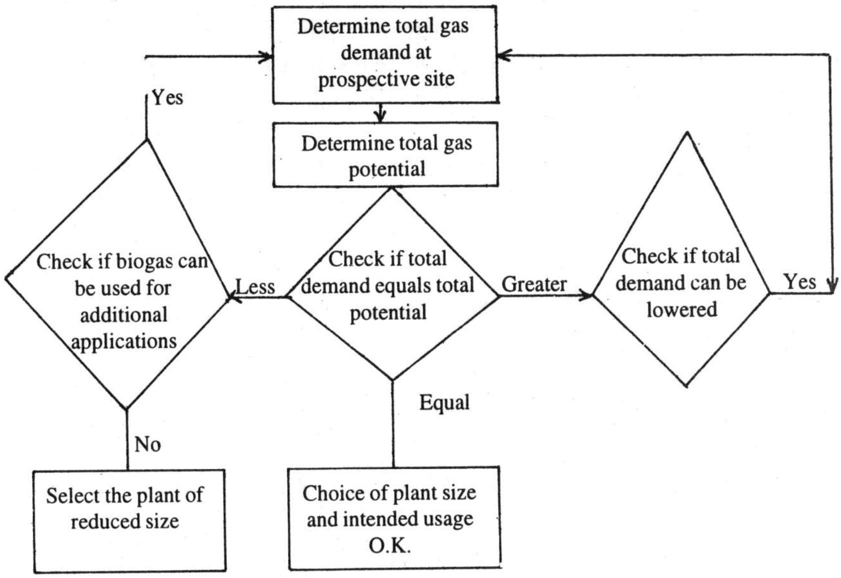

Fig. 5.1 A flow diagram for determining size of a biogas plant

Table 5.2 Recommended Norms for Detention Time in Different World Regions

Regions of the World	*Recommended Detention Time (days)*
Hot, tropical plains climate like Sudan, Cameroon, Sri Lanka, Indonesia, Venezuela, Central America etc.	30-40
Hot regions which do not appreciably cool down during winter such as India, Thailand, Philippines, Kenya, Ethiopia etc.	40-60
Cold climatic regions experiencing very low temperature during winter such as China, Korea, Russia etc.	60-90

Source : GATE, Biogas Plants : Building Instructions, German Appropriate Technology Exchange, German Agency for Technical Cooperation, P.O. Box 5180, 6236 Eschborn 1, Germany (erstwhile), 1979.

Biogas is formed following anaerobic fermentation of organic wastes due to bacterial action. Bacteria are most active when digester temperature is kept uniform and maintained between 25 and 35°C. If it is lower or fluctuates, biogas yield is suboptimal. At temperatures below 15°C, biogas yield almost ceases. Due to this reason, biogas plants are less effective in cold climatic regions. For estimating biogas yield more realistically in such regions, temperature is taken as 5°C less than the actual temperature. Fig. 5.2 shows the relationship between detention period, digester temperature and gas yield. Higher temperature tends to intensify microbial activity leading to increased gas output. The higher is the temperature, the shorter can be the detention period. With shorter detention period, throughput becomes faster, with the result that digester can be made of smaller size. Effect of detention period on gas yield for different feeds is shown in Fig. 5.3.

From the knowledge of climatic conditions at the prospective site, and with the help of information such as given in Figs. 5.2 and 5.3, detention period to suit particular climatic conditions can be selected.

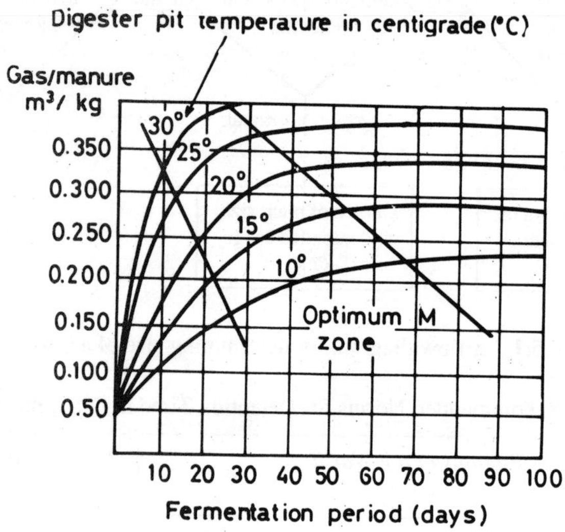

Fig. 5.2 Effect of fermentation period on gas yield at different temperatures

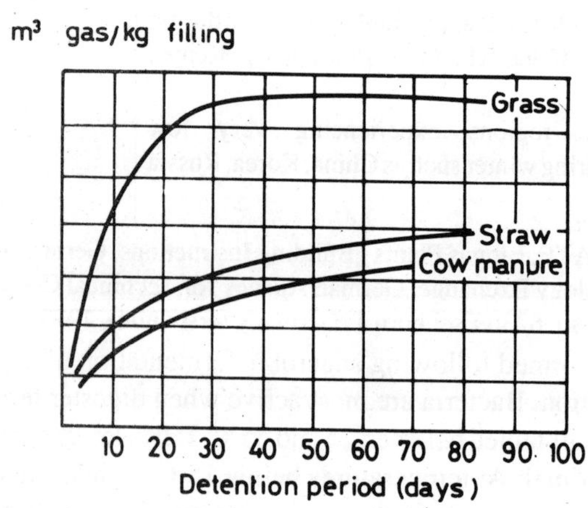

Fig. 5.3 Effect of detention period on gas yield for different feeds.

Digester temperature = __________ °C

Detention period = __________ days

Quantity of feed to be used. Plant slurry to be fed to a biogas plant should have organic wastes and water in proper mix. A set of typical mixing ratios are given in Table 5.3.

Table 5.3 Typical Mixing Ratios for Preparing Plant Slurry

Mixture	*Proportions*
Fresh cowdung : water	1 : 1
Superficially dry cowdung : water	1 : 2
Horse and sheep's dung : water	1 : 1 to 1 : 3
Human faces : water	1 : 1
Green refuse : water	1 : 0.5 to 1 : 2

Source: GATE, Biogas Plants: Building Instructions, German Appropriate Technology Exchange, German Agency for Technical Cooperation, P.O. Box, 5180, 6236 Eschborn 1, Germany (ertwhile FRG) 1979.

Estimation of quantity of waste material and water added per day:

1. Cow dung added per day (litres) ______	+	Water added per day (litres)	=	Total volume (litres)
2. Pig manure added per day litres ______	+	–do–	=	–do–
3. Other animal faeces added per day (litres) ______	+	–do–	=	–do–
4. Human faeces added per day (litres) ______	+	–do–	=	–do–
5. Green matter added per day (litres) ______	+	–do–	=	–do–

Whence,

Total volume added to the plant per day : ______ litres

: ______ kg

(1000 litres = 1 m^3)

Estimation of size of digester and gasholder. Volume of digester can be determined by multiplying the volume of slurry added per day with the detention period. Digester volume is essentially the product of feed rate and detention period.

$$\text{Volume of slurry added per day (litres)} \times \text{Detention period (days)} = \frac{\text{Digester volume}}{1000}\ (m^3)$$

By referring to Table 5.1 or other similar source of information, it is possible to estimate at biogas yield per day in appropriate units such as m^3/day.

Volume of gas production per day from plant = ______ m^3/day

As a rough rule, volume of gasholder is usually taken as half of daily gas production rate.

Volume of gasholder = __________ m^3

5.5 Checklist for Deciding between Fixed Dome and Movable Drum Biogas Plant

Both fixed dome and movable drum type biogas plants have their own merits and demerits. The principal benefit of the former is that it is relatively inexpensive and maintenance problems are fewer with consequent longer working life. On account of being underground and less susceptible to temperature variations, gas yield from the former is more uniform. In view of the fact that in fixed dome type plants, digested slurry is cleared manually, even long fibrous materials can be easily fed to it. However, to facilitate easy feeding and removal of fibrous materials following digestion it is desirable to have inlet and outlet somewhat larger. Fixed dome plants are particularly suitable in colder regions as on account of being underground, effect of low temperatures is largely compensated. However, these plants are unsuitable to fully exploit naturally available solar energy for warming up digester contents. Fixed dome type plan requires no steel sheet, runs on a continuous batch process, provides richei fertiliser and ensures greater thermal insulation of the digester. This type however normally does not have any digester heating and mixing device attached to it. Movable drum type plant on the other hand offers the advantage of building uniform gas pressure, ease in scum removal following facility of lifting gasholder, easy clearance of blockages due to inlet and outlet being straight, and other benefits associated with a continuous process. For trouble free-operation it is necessary that plant be fed with homogenous, non-fibrous and sufficiently diluted feed. In a movable drum unit there occurs higher degree of heat losses due to more exposed surface. This plant mostly works on animal manure. Movable drum type plants can be constructed both above and underground. Underground digester pit construction is usally preferred. Main advantage of underground construction is that incoming slurry can be fed undergravity from ground. A plant above ground on the other hand is easy to maintain and the plant exploits the solar heat for warming up the digester contents to a far greater extent than when underground. The main drawbacks of movable drum plants are that it is costly to construct and suffers from corrosion problems particularly in steel gasholder. A detailed comparison of fixed dome type and movable drum plant is summarised in Table 1.2 in Chapter 1. A checklist for deciding whether a prospetive owner should go in for a fixed dome or movable drum type plant is given in Table 5.4 which is a modified form of the one initially developed by the BORDA [5, 6]. Checklist approach based on relative merit and demerit of each type provides a reasonably accurate guideline for making a choice between fixed dome and movable drum type

Table 5.4 Check-list for Making a Choice between Fixed Dome Type and Movable Drum Type Biogas Plant

S. No.	*Check-list*	*Fixed Dome Type*	*Movable Drum Type*
1.	Which type of plant already exists in the vicinity of intended site?	☐	☐
2.	Is anticipated digester temperature above 15°C?	No ☐	Yes ☐
3.	Are winters normally cold?	Yes ☐	No ☐
4.	Is steel to be used in gasholder construction is readily and cheaply available?	No ☐	Yes ☐
5.	Are binding materials such as cement and lime are cheaply and readily available?	Yes ☐	No ☐
6.	Do facilities for fabricating gasholders exist in the area?	No ☐	Yes ☐
7.	Is necessary expertise locally available to do brick-work in curves?	Yes ☐	No ☐
8.	Does facility for sealing/coating required for dome exist?	Yes ☐	No ☐
9.	Is it possible to paint gasholder every year?	No ☐	Yes ☐
10.	Is plant required to operate only on agricultural crop residues such as grass or bagasse?	No ☐	Yes ☐
11.	Will it be possible to prepare homogeneous slurry comprising animal and human faeces?	No ☐	Yes ☐
12.	Which type of plant you can construct confidently?	☐	☐
13.	Which plant according to your estimates would cost less?	☐	☐
	Total number of crosses	–	–

biogas plants. A plant which scores maximum crosses is most-suited for a particular site under consideration.

Selection of a particular type of plant depends on several factor such as type and particle size of available waste materials; skill level of local artisans and financial and technical capability of individual plant owners. For example, if expertise to construct a fixed dome type plant not locally available and it is quite expensive to arrange a distant place, one could decide in favour of a movable-drum type plant. The choice can also be based on availability of waste materials. For example, if water-hyacinth and crop residues are available locally in abundance, a Janata or Deenbandhu model should be preferred. These are however general remarks and all situation-specific factors should also be taken into account while making a choice.

5.6 Selection of Design Parameters for JBPs

A biogas digester is essentially a chemical reactor which is linked to both inflow and outflow channels. Rate of gas production is dependent not only on the microbial growth rate but also on the effectiveness of dispersion of the solid, liquid and gas phases and on the residence time distribution within the reactor. Dispersion or mixing is achieved by stirring which leads to higher gas yield. The fact that all the feed material may not reside in the digester for the same time underscoves the need to study residence time distribution. If feed is divided into N smaller portions, some parts might stay longer in the digester than others. Residence time dispersion depends upon such factors like geometry of digester or degree of dispersion. All these factors lead to following different design approaches one of which is as follows [21].

The main parameter which governs the cost of digester is its volume (V) which is made up of two parts, namely, active digester volume V_S and gas storage space. V_S represents the volume occupied by the slurry in the digester which can be computed by the relationship,

$$V_S\,(\text{m}^3) = \text{Slurry feed rate (m}^3\text{/day)} \times \text{HRT (days)}$$

Volume of input slurry for a given quantity of organic wastes depends upon the dilution. Slurry is diluted to optimum level as with higher dilution bigger capacity digester is required. A dilution of cow dung with water in the ratio of 1 : 1 is generally preferred. Composition of cattle dung may vary from place to place and thus it is advisable to analyse the TS fraction so that the amount of water to be added to bring the final TS content within 8 to 10 per cent can be determined. The fraction of TS can be measured by weighing a small quantity (5-10 g) of fresh dung, keeping it in an oven at about 100°C till all the water has evaporated, and then weighing the residue.

For operational ease, biogas plants are fed on semi-continuous basis which is generally done daily. Volume of gas storage space is so provided that it is sufficient to accommodate gas generated in 24 hours under normal operating conditions. In movable drum type plants as gas accumulates inside top of the gasholder it rises providing a variable volume for gas storage with pressure remaining more or less constant. On the contrary in fixed dome type plant gas produced pushes the slurry down displacing an equal amount of slurry into the inlet and outlet tanks. This process compresses the gas and raises its pressure.

Fixed dome plants can be made in different shapes: rectangular, cylindrical, spherical or ellipsoidal. However cylindrical shapes are generally more common particularly in India which again can be divided into two types, namely, flat and curved bottom digesters. Important dimensions of the two types are shown in Figs. 5.4 and 5.5 respectively. Digester consists of a cylindrical brickwalled pit with inlet and outlet tanks. When gas pressure is equal to atmospheric pressure, no gas can be withdrawn for use under this case as slurry level in the digester is equal to that in the inlet and outlet tanks. Resulting gas from the surface of the slurry accumulates in the space above and gets compressed slightly. As the compressed gas pushes down the slurry level in the digester, the slurry level in the inlet and outlet tanks rises to a point such that the pressure of the gas is balanced by the pressure of the water column in the inlet and outlet tanks. When the storage space is completely filled with gas, slurry level inside digester is pushed to the lowest position and the slurry level in the inlet and outlet tanks rises to the highest position. Volume of slurry in excess of active digester volume is pushed out through an opening in the outlet

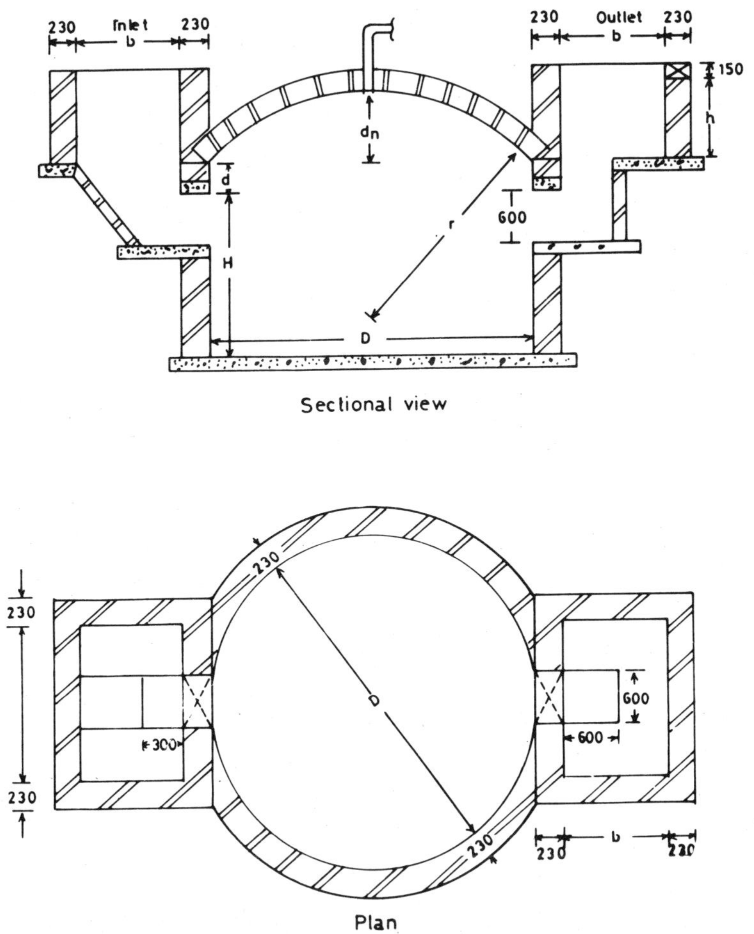

Fig. 5.4 Cross-sectional view of fixed dome Janata biogas plant with flat bottom [Source: TERI, Fixed Dome Biogas Plants, New Delhi, June, 1987, p. 6]

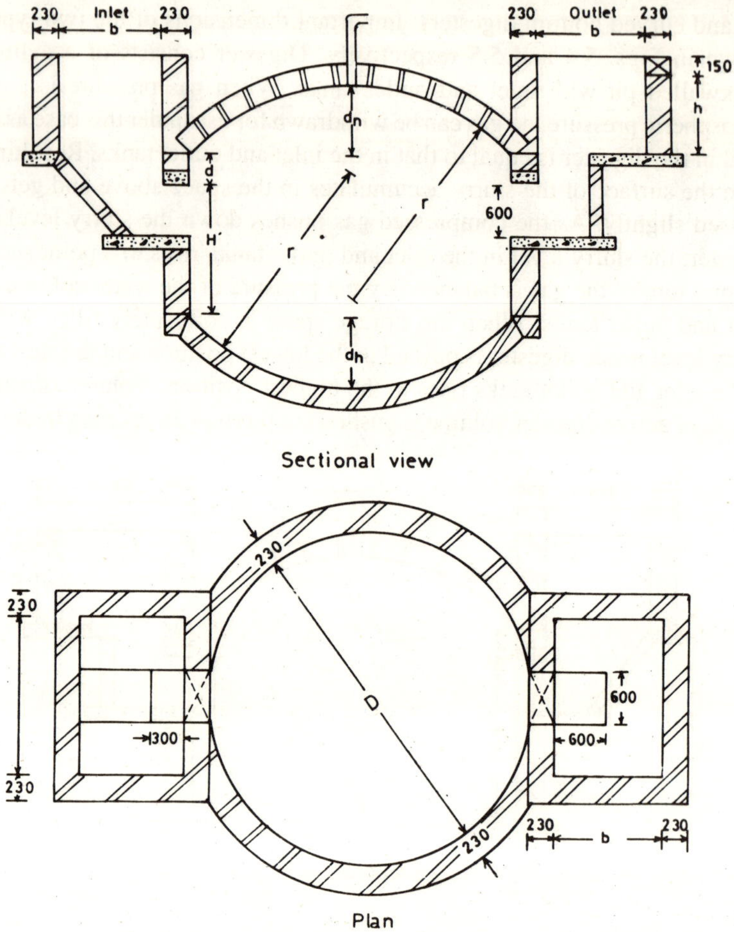

Fig. 5.5 Cross-sectional view of fixed dome Janata biogas plant with curved bottom [Source: TERI, Fixed Dome Biogas Plants New Delhi, June, 1987, p. 7]

tank. Thus under ideal operating conditions, volume of slurry flowing out of the digester in a day equals that flowing in [21].

Design of a biogas plant starts from the information concerning availability of wastes per day. Following is the description of various parameters used by TERI for this purpose [21].

W = Weight of animal wastes fed per day (kg/day)

G = Gas production rate (m^3/day)

V_s = Active slurry volume in the digester (m^3)

V_{sd} = Slurry displacement volume (m^3)

V_d = Dome volume (m^3)

H = Height of the cylindrical portion of the digester upto the top edge of the inlet/outlet opening (lintel), for flat bottom digesters (m)

H' = Same as H, but for curved bottom digesters (m)

D = Diameter of the digester (m)

d = Slurry displacement inside the digester (m)

d_h = Height of the dome (m)

r = Radius of the dome (m)

h = Slurry displacement in inlet and outlet tanks (m)

lb = Length and breadth of the inlet and outlet tanks (m)

Various design parameters can be estimated with the help of the following relationships.

1. Gas production rate *G*. One kg of undiluted cattle dung roughly yields about 0.04 m^3 of the gas. Hence,

$$G = 0.04\ W$$

2. Active slurry volume *Vs*. V_s which is related to the HRT can be determined by the following relationship,

$$V_s = \text{HRT} \cdot \frac{2\ W}{1000}$$

If HRT is assumed as 50 days, V_s can be expressed in terms of G by the following relationship.

$$V_s = 2.5\ G$$

3. Estimation of *H* and *D*. There are no specific rules but generally D is kept twice of H. Knowing the active slurry volume from the previous equation, H can be determined in the following way.

$$(\pi/4)\ D^2 \cdot H = V_s \text{ and } D = 2\ H$$

Whence,

$$H = (V_s/\pi)^{1/3}$$

4. Estimation of slurry displacement parameter. Selection of a val-ue of D depends upon gas utilisation pattern. Assuming 3-hour cooking for evening and morning, variable gas storage V_{sd} can be expressed as,

$$\frac{3}{24} G + V_{sd} = 0.5\ G$$

Whence,

$$V_{sd} = 0.375\, G \approx 0.4\, G$$

$$(\pi/4)\, D^2 \,.\, d = V_{sd} \text{ but } (\pi/4)\,.\, D^2 = \frac{V_s}{H} = \frac{2.5\, G}{H}$$

On substitution,

$$\frac{2.5\, G}{H} \cdot d = 0.4\, G$$

Thus,

$$d = 0.16\, H$$

Computation of slurry displacements in inlet and outlet tanks. Maximum pressure exerted on gas is equal to the pressure of the water (slurry) column above the lowest slurry level in the inlet and outlet tanks. This pressure is normally selected to be 0.85 m water gauge as a safe limit for brick/concrete domes. Thus,

$$h + d = 0.85$$

i.e., with the knowledge of d, h can be determined from the above relationship.

Length (*l*) and breadth (*b*) of the inlet and outlet tanks. There is no specific criterion with regard to the choice of the shape of cross-section of inlet and outlet tanks, but generally a rectangular shape with $l = 1.5\, b$ is preferred. For a case when inlet and outlet tanks have identical shapes, volume of slurry displaced downwards inside the digester is equal to the total volume of slurry displaced upwards in the inlet and outlet tanks. Thus,

$$2\, lbh = V_{sd} = 0.4\, G$$

Therefore on substituting $l = 1.5\, b$, we get

$$2\,(1.5\, b)\, bh = 0.4\, G$$

Whence,

$$b = [0.26/1.5\, h]^{1/2}$$

By using the values of h and G obtained earlier, b can be estimated from the above relationship and l can be selected as 1.5 times of b.

In some cases, an inlet pipe of 6 to 8 inch diameter is provided instead of inlet tank. In this situation, area of the outlet should be doubled to accommodate the slurry displacement. For this

$$b_{outlet} = [0.4\, G/1.5\, h]^{1/2}$$

$$l_{outlet} = 1.5\, b_{outlet}$$

Computation of the dome height. Volume of dome which is a part of sphere is given by the following relationship.

$$V_d = (\pi/6) \,.\, d_h \,.\, [3\,(D/2)^2 + d_h^2]$$

Recalling that the total volume of gas space is taken as equal to G, V_{sd} as 0.4 G, the remaining gas space volume which is the volume of the dome becomes (G – 0.4 G) or 0.6 G. Substituting this in the above equation, we get

$$0.6\,G = (\pi/6) \,.\, d_h \,.\, [3\,(D/2)^2 + d_h^2]$$

Value of d_h can be obtained by solving the above equation which is cubic. This, among others, can be solved by iteration. An algebraic solution also exists which is obtained by computing series of parameters as follows:

$$p = 0.75\,D^2$$

$$q = -\,0.6\,(6/\pi)\,G$$

$$R = (p/3)^2 + (q/2)^2$$

$$A = [(-\,q/2) + \sqrt{R}]^{1/3}$$

$$B = [(-\,q/2) - \sqrt{R}]^{1/3}$$

Then d_h is given by

$$d_h = A + B$$

Estimation of the radius of dome. Dome radius is determined with the help of the following relationship.

$$r = [(D/2)^2 + d_h^2]/2\,d_n$$

Determination of *H′* for curved bottom digesters. For digesters with curved bottom, the lower base is equal to the upper dome. H' is thus computed by the following relationship.

$$(\pi/4) \,.\, D^2\,H = (\pi/4) \,.\, D^2 \,.\, H' + 0.6\,G$$

By simplification, one gets

$$H' = \frac{1.9\,H}{2.5}$$

Determination of other parameters. The size of the openings in inlet and outlet tanks are normally kept of size 0.5 m × 0.6 m regardless of digester capacity. This particular size is selected so that a man can go inside the digester during the construction period. Digester walls 9 inch thick (full brick) and the walls of the inlet and outlet boxes (below the lintel) are 4.5 inch thick (half brick). For curved bottoms, two brick layers are provided, the lower layer being 4.5 inch and the upper layer 3 inch thick. Concreting whether plain or reinforced should be 4 inch thick.

Selection of design parameters for 2 m³ capacity plant. For G = 2 m³, W = 50 kg/day, different parameters can be computed in the following manner

$$G = 0.04\,W = 0.04 \times 50 = 2\,\text{m}^3/\text{day},$$

$$V_s = 2.5\,G = 2.5 \times 2 = 5\,\text{m}^3$$

$$H = (V_s/\pi)^{1/3} = (5/\pi)^{1/3} = 1.17\,\text{m}$$

$$D = 2H = 2 \times 1.17 = 2.34\,\text{m}$$

$$d = 0.4\,(H/2.5) = 0.4\,(1.17/2.5) = 0.19\,\text{m}$$

$$h = 0.85 - d = 0.85 - 0.19 = 0.66\,\text{m}$$

$$b = (0.2\,G/1.5h)^{1/2} = (0.2 \times 2/1.5 \times 0.66)^{1/2} = 0.635\,\text{m}$$

$$l = 1.5 \times b = 1.5 \times 0.635 = 0.953\,\text{m}$$

$$p = 0.75\,D^2 = 0.75\,(2.34)^2 = 4.107$$

$$q = -\,0.6\,(6/\pi)\,G = -\,0.6\,(6/\pi)\,2.0 = -\,2.292$$

$$R = (p/3)^2 + (q/2)^2 = 2.566 + 1.313 = 3.879$$

$$\sqrt{R} = \sqrt{3.879} = 1.9696$$

$$A = [(-q/2) + R]^{1/3} = (1.146 + 1.9696)^{1/3} = 1.4605$$

$$B = [(-q/2) - R]^{1/3} = (1.146 - 1.9696)^{1/3} = -\,0.9374$$

$$d_h = A + B = 1.4605 - 0.9374 = 0.523\,\text{m}$$

$$r = [(D/2)^2 + d_h^2]/2d_h = (1.17^2 + 0.523^2)/2 \times 0.523 = 1.57\,\text{m}$$

Finally summarising these parameters, we get

G = 2.000 m³/day	h = 0.660 m
V_s = 5.000 m³	b = 0.635 m
H = 1.170 m	l = 0.953 m
D = 2.340 m	d_h = 0.523 m
d = 0.190 m	r = 1.570 m

By performing the calculations in the above manner, parameters and dimensions of large capacity plants can similarly be worked out complete summary of which is summarised in Table 5.5. In India it is customary to

Table 5.5 Important Dimensions of Fixed Dome Biogas Plants
(All dimensions in metres)

G (m^3/day)	H	D	d	h	b	l	d	r	H'
1.000	0.927	1.854	0.148	0.702	0.436	0.654	0.416	1.240	0.704
2.000	1.168	2.335	0.187	0.663	0.634	0.951	0.524	1.562	0.887
3.000	1.336	2.673	0.214	0.636	0.793	1.189	0.600	1.787	1.016
4.000	1.471	2.941	0.235	0.615	0.931	1.397	0.661	1.967	1.118
5.000	1.584	3.168	0.253	0.597	1.057	1.586	0.712	2.119	1.204
6.000	1.683	3.367	0.269	0.581	1.174	1.761	0.757	2.251	1.279
7.000	1.772	3.544	0.284	0.566	1.284	1.925	0.796	2.369	1.347
8.000	1.853	3.705	0.296	0.554	1.388	2.082	0.833	2.477	1.408
9.000	1.927	3.853	0.308	0.542	1.488	2.232	0.866	2.576	1.464
10.000	1.995	3.991	0.319	0.531	1.585	2.378	0.897	2.668	1.517
11.000	2.060	4.120	0.330	0.520	1.679	2.518	0.926	2.754	1.565
12.000	2.120	4.241	0.339	0.511	1.770	2.655	0.953	2.835	1.611
13.000	2.178	4.355	0.348	0.502	1.859	2.788	0.979	2.911	1.655
14.000	2.232	4.464	0.357	0.493	1.946	2.919	1.004	2.984	1.696
15.000	2.284	4.568	0.365	0.485	2.032	3.047	1.027	3.053	1.736
16.000	2.334	4.667	0.373	0.477	2.116	3.173	1.049	3.119	1.773
17.000	2.381	4.762	0.381	0.469	2.198	3.298	1.071	3.183	1.810
18.000	2.427	4.854	0.388	0.462	2.280	3.420	1.091	3.244	1.844
19.000	2.471	4.942	0.395	0.455	2.361	3.541	1.111	3.303	1.878
20.000	2.514	5.027	0.402	0.448	2.440	3.660	1.130	3.360	1.910

Source : Tata Energy Research Institute, Fixed Dome Biogas Plants: A Design, Construction and Operation Manual, New Delhi, 1987, p. 12.

express the size or capacity of the plant in terms of designed gas production rate per day whereas in China it is expressed in terms of the volume of the digester. Thus if the capacity is mentioned as 2 m^3 or 70 ft^3, it implies that the plant is designed for producing 2 m^3 (70 ft^3) of gas per day. It is to be emphasised that design parameters are by no means 'optimum' ones. For instance, if it is possible to maintain temperature of digester contents at 35°C throughout the year, HRT can be significantly reduced and other parameters will accordingly change. After fixing the design parameters it is also necessary to work out the estimates of items like cement, sand, brick and steel which are used in plant construction. For further details reader is advised to refer to the TERI document (1987) entitled 'Fixed Dome Biogas Plants: A Design, Construction and Operation Manual' which provides excellent coverage of the subject [21].

5.7 Selecting Design Parameters Using a Cost Minimisation Model

D.K. Subramanian, P. Rajabapaiah and A.K.N. Reddy [34] evolved a cost optimisation model for optimising design parameters of movable drum type biogas plants. Volume V' of a digester is product of volume v_{sl} of slurry which is charged daily and retention time t_d of slurry. Thus,

$$V' = V_{sl} \cdot t_d$$

Necessary experiments for developing a theoretical approach for optimising design parameters were carried out at ASTRA, IISc., Bangalore on a KVIC type conventional plant having a rated capacity of 200 ft^3 of biogas/day (5.66 m^3/day). This plant has the following dimensions; diameter and height of gasholder 6 ft (1.85 m) and 4 ft (1.22 m) respectively; and diameter and depth of digester 6.5 ft (1.98 m) and 16 ft (4.88 m) respectively. In KVIC design, mild steel floating gasholder accounts for nearly 40 per cent of the total cost of biogas plant. If r, R_s and R_r are capital costs of gasholder, its sides and roof respectively, then

$$R = R_s + R_r$$

$$= \pi\, Dht\rho u + (\pi\, D^2/4) \,.\, t \,.\, \rho \,.\, u$$

where D is diameter of gasholder, h its height, t its thickness, ρ its density and u its unit cost in Rs/kg (comprising cost of steel, transport, fabrication, welding and painting). Height h of gasholder can be estimated by

$$h = 4\, v/\pi\, D^2 = 4\, \gamma\, C/\pi\, D^2$$

where v is the volume of gasholder and γ is the maximum fraction of daily gas production, i.e. actual plant capacity C which is intended to be stored in gasholder.

After substitution and denoting $\alpha = t\rho u$, we get

$$R = (4\,\alpha\,\gamma\,C/D) + (\pi\,\alpha\,D^2/4)$$

The objective function R can be minimised by differentiating it w.r.t. variable D and equating the derivative to zero, where from the diameter D_k, corresponding to this minimum cost gasholder is

$$D_k = (8\,\gamma/\pi)^{1/3} \cdot C^{1/3}$$

It is to be seen that D_k increase linearly with $C^{1/3}$, the slope being $(8\,\gamma/\pi)^{1/3}$ The corresponding height h_k of the gasholder is

$$h_k = 4V_k/\pi\,D_k^2 = 4\,\gamma\,C/\pi\,D_k^2$$

After substituting the value of D_k in the cast equation, the minimum cost becomes

$$R = (3\,\alpha\,\pi^{1/3} \,.\, \gamma^{2/3}) \,.\, C^{2/3}$$

From the equation for total plant costs, it is to be seen that plant costs do not increase linearly with plant capacity. In other words there are definite economies of scale in constructing several plants at a place. As a first approximation excluding excavation costs, the total costs of N plants each of capacity (C/N) is $N^{1/3}$ times the cost of one plant of capacity C. Based upon the cost optimisation approach for selecting design parameters, plants were constructed as per the modified dimensions and their performances were compared with the corresponding size of the KVIC model. D.K. Subramanian, *et al.* [34] carried out cost optimisation studies for two cases, (i) when cost of gasholder alone is to be minimised, and (ii) when cost of gasholder plus digester is to be minimised. Minimisation of the cost of gasholder alone leads to narrow and deep digesters of conventional plants. It was found that if total capital cost of gasholder plus digester is to be minimised, the optimisation leads to wide and shallow digesters which are less expensive. For evaluating this alternative two prototype plants were designed, constructed and operated which were both less expensive and efficient in performance [34]. It was found that if the cost of the digester and gasholder is considered, the least expensive plant is one in which the diameter of the digester equals its depth [34]. Field tests of modified experimental models indicated that gas output in warm weather exceeded that from movable drum type KVIC plants of seventies although these are still not as suitable for colder areas due to larger surface heat losses.

G.J. Mohanrao [27] carried out experimental studies to analyse the effect of digester diameter to depth ratio on biogas yield. He measured the biogas yield in litres per m^3 of digester volume and in litres per kg of volatile solids added. He reported that gas yield is maximum when digester diameter to depth ratio is 1 and remains reasonably high till the ratio is 0.66. Hence, if his experimental results are to form the basis, digester diameter to depth ratio should be kept within the desirable range while designing the digester and it should not be allowed to be less than half in any case (Table 5.6).

Table 5.6 Effect of Digester Diameter to Depth Ratio on Biogas Production at a Loading of 20 kg of Dung/m^3/day and Keeping a Detention Period of 24 days

S. No.	*Diameter to Depth Ratio*	*Biogas Yield Litres per m^3 of Digester Volume*	*Biogas Yield Litres per kg of Volatile Solids Added*	*Slurry pH*
1	1.50	340	106	7.00
2	1.00	510	158	7.05
3	0.66	480	150	7.00
4	0.50	410	128	7.00

5.8 Supply and Distribution of Biogas

Distribution of biogas from plant to its place of use forms an important part of biogas system. For satisfactory operation of biogas appliances, it is necessary to have a uniform gas pressure of around 7.6 cm W.G. (3 inch W.G.) at its point of use. For carrying gas to households for cooking and lighting, generally GI pipes of 25.4 mm (1 inch) diameter are used and within homes GI pipes of 12.7 mm (1/2 inch) diameter are regarded appropriate. For efficient gas flow it is necessary to select the pipeline of right length and diameter. While following inside pipe there occurs drop in pressure (h) which is directly proportional to the length of the pipeline, specific gravity of gas, square of gas flow rate (q^2) and coefficient of friction; and inversely proportional to the 5th power of the pipe diameter. Hence

$$h \propto \frac{lsq^2 f}{d^5}$$

whence,

$$q = \text{constant} \sqrt{\frac{h\,d^5}{slf}}$$

Assuming f as 0.0065, the relationship for rate of gas flow (Pole's equation) assumes the following form

$$q = 1350 \sqrt{\frac{h d^5}{sl}}$$

By using the Pole's formula G.L. Patankar, *et al.* [12] computed gas flow for varying pipe length and diameter combinations complete details of which are summarised in Table 5.7. A similar data in which rate of gas flow is expressed in ft^3/hr instead m^3/hr is given in Table 5.8 contents of which are self explanatory. By referring to these tables or by making independent calculations for a particular situation, one can determine needed pipeline diameter [31].

Central Salt and Marine Chemicals Research Institute (CSMCRI), Bhavnagar, developed a flowmeter for measuring volumetric flow upto 2.5 litres at normal temperature and pressure [1]. Besides measuring flow, it can also help to estimate methane content of biogas within 5 minutes. It consists of a graduated gas collecting cylinder, manometric reservoir and a wash bottle. Capacity of gas collecting cylinder can be adjusted as per the requirement. Biogas is made to collect in a cylinder. Gas flow is measured at atmospheric pressure after equalising levels in cylinder and manometric reservoir. Gas flow from gasholder to the cylinder is then stopped using the stopcock. Biogas is then made to pass through baryte solution kept in a bottle. It absorbs carbon dioxide and hydrogen sulfide from biogas. The purified methane then enters the aspirator containing water which displaces equal volume of water into the measuring cylinder which indicates the volume of methane generated.

5.9 Estimation of Heating Needs for Maintaining the Digester at Desired Temperature

Temperature affects microbial activity of the two types of bacterias that cause anaerobic fermentation. Mesophiles perform best in the temperature range of 29 to 38°C and the thermophiles are most active between 49 and 57°C. Gas yield substantially slows down at 15°C and ceases altogether below this temperature. The rate of gas generation nearly doubles for every 10°C rise in temperature variation between 15 and 35°C. A biogas plant supposed to operate at a temperature of 15°C needs to have digester volume four times greater than one at 35°C. Comparative biogas yield from cattle dung in different states of India based on mean ambient temperatures is summarised in Table 5.9. As expected gas yield in climatically cold states like Jammu and Kashmir is substantially low as compared to the other states. In view of particular climatic conditions mesophilic digestion is more convenient and economical for most parts of India. For mesophilic fermentation a single temperature range of 30-35°C is considered most convenient and economical to work with. As gas yield depends on digester temperature, choice of detention period is governed by ambient temperatures. Length of detention period is inversely related to temperature. If temperature is low, a longer detention period is necessary which is met by designing larger digester and hence more expensive biogas systems.

Table 5.7 Recommended Pipe Diameters for Varying Flow Rates and Distances between Biogas Plant and Points of Use

Gas Flow Rate (ft^3/hr)	*Distances*							
	25 m	*50 m*	*100 m*	*150 m*	*200 m*	*300 m*	*400 m*	*500 m*
16	½"	3/4" (for 25 m) ½" (for 25 m)	3/4"	3/4"	1" (for 150 m) 3/4" (for 50 m)	1" (for 200 m) 3/4" (for 100 m)	1" (for 350 m) 3/4" (for 50 m)	1"
24	½"	3/4"	3/4"	1" (for 100 m) 3/4" (for 50 m)	1" (for 150 m) 3/4" (for 50 m)	1"	1 ½" (for 200 m) 1" (for 150 m) 3/4" (for 50 m)	1 ½" (for 200 m) 1" (for 300 m)
32	3/4"	1"	3/4"	1" (for 100 m) 3/4" (for 50 m)	1"	1 ½" (for 200 m) 1" (for 50 m) 3/4" (for 50 m)	1 ½" (for 200 m) 1" (for 200 m)	1 ½" (for 350 m) 1" (for 150 m)
48	3/4"	3/4"	1" (for 75 m) 3/4" (for 25 m)	1"	1"	1 ½" (for 150 m) 1" (for 150 m)	1 ½" (for 300 m) 1" (for 150 m)	1 ½" (for 400 m) 1" (for 100 m)
64	3/4	1"	1 ½" (for 50 m) 1" (for 50 m)	1 ½" (for 50 m) 1" (for 50 m)	1 ½" (for 150 m) 1" (for 50 m)	1 ½"	1 ½"	2" (for 150 m) 1 ½" (for 350 m)

Source: Dhussa, A., Designing Biogas Distribution System, Bioenergy Reviews, 2 (1 and 2), 71 (1983).

Table 5.8 Rate of Gas Flow for Varying Combinations of Length and Diameter of Pipelines

Length of Pipe (m)	*Rate of Gas Flow for Different Pipe Diameters (m^3/hour)*		
	12.7 mm (1/2 inch)	*19.2 mm (3/4 inch)*	*25.4 mm (1 inch)*
1	4.99	12.65	25.98
3	2.65	7.31	14.99
6	1.87	5.17	10.60
9	1.53	4.22	8.66
12	1.32	3.65	7.50
15	1.18	3.27	6.71
18	1.08	2.98	6.12
21	1.06	2.76	5.67
24	0.94	2.58	5.30
27	0.88	2.43	5.00
30	0.84	2.31	4.74
33	0.80	2.20	4.52
36		2.11	4.33
39		2.03	4.16
42		1.95	4.00
45		1.88	3.87
48		1.83	3.75
51		1.77	3.64
54		1.72	3.53
57		1.68	3.44
60		1.63	3.35
63		1.69	3.27
66		1.56	3.20
69		1.52	3.13
72		1.49	3.06
75		1.46	3.00
80		1.41	2.90
90		1.46	2.73
100		1.41	2.60
125		1.33	2.32
150		1.26	2.12
175		1.13	1.96
200		1.03	1.84
250		0.96	1.64
300		0.89	1.50
350		0.80	1.38
400			1.29
450			1.22
500			1.16

Source : Patankar, G.L., Recent Developments in Biogas Technology, *A* Report, Gobar Gas Development Centre, Borivli, Mumbai, 1977.

As for the fixed dome Janata biogas plants, temperatures in underground plants are mainly influenced by earth temperatures. The mean gas yield from these digesters is reported to be 0.15 m^3/day cubic meter of the digester volume though the rate of gas production varies substantially from north to south and from summer to winter [2, 3].

Table 5.9 Comparative Biogas Production from Cattle Dung in Different States of India

State	*Monthly Mean Temperature (°C)*		*Rate of Biogas Production (m^3/tonne dung)*
	Minimum	*Maximum*	
Andhra Pradesh	22.8	33.7	75
Assam	16.8	20.4	64
Bihar	17.0	31.7	68
Gujarat	20.7	32.6	71
Haryana	13.5	34.2	66
Himachal Pradesh	10.7	28.7	52
Jammu and Kashmir	3.2	21.0	23
Karnataka	21.9	29.4	69
Kerala	26.6	28.8	74
Madhya Pradesh	17.7	33.3	67
Nagaland	16.8	29.4	64
Orissa	20.6	31.3	72
Punjab	13.5	33.2	66
Rajasthan	15.8	33.7	69
Tamil Nadu	24.7	31.6	77
Uttar Pradesh	15.1	32.9	61
Delhi	13.5	34.2	66

The total heat requirement (Q_T) for maintaining digester slurry at a desired temperature can be estimated as a sum of the heat quantities in the following manner [15, 16].

$$Q_T = Q_L + Q_W + Q_g + Q_I - Q_r$$

where,

Q_L = heat loss through digester walls, floor and its top

Q_W = heat loss following evaporation

Q_g = heat loss due to the gas leaving the digester

Q_I = heat required raise temperature of influent slurry the desired level

and,

Q_r = heat of reaction resulting from anaerobic fermentation

The heat loss through the digester walls (Q_L) is the sum of heat loss through top, side walls, and its bottom which can be calculated from the overall heat transfer coefficients of top, side walls and bottom of the digester, respectively. It is reasonable to assume that all losses from digester take place through conduction which can be estimated with the help of the following fundamental equation.

Heat loss through digester walls below liquid level ($Q_{W.}$) is given by

$$Q_{W_1} = \frac{A_{W_1} \cdot (T_1 - T_2)}{\left[\dfrac{x}{k_1} + \dfrac{x}{k_3} + \dfrac{1}{h_0}\right]}$$

where,

Q_{W_1} = rate of heat loss through digester walls below the liquid level (energy/time)

A_{W_1} = wall area below liquid level (distance^2)

Heat loss through digester walls above liquid level (Q_{W_2}) is given by

$$Q_{W_2} = \frac{A_{W_2} \cdot (T_1 - T_2)}{\left[\dfrac{1}{h_j} + \dfrac{1}{K_a} + \dfrac{x}{k_1} + \dfrac{x}{k_3} + \dfrac{1}{h_0}\right]}$$

where,

Q_{W_2} = rate of heat loss through digester walls above the liquid level (energy/time)

A_{W_2} = wall area above liquid level (distance^2)

Heat loss through flow (Q_f) is given by

$$Q_f = \frac{A_f \cdot (T_1 - T_2)}{\left[\dfrac{x}{k_1} + \dfrac{x}{k_2} + \dfrac{x}{k_3} + \dfrac{1}{h_0}\right]}$$

where,

Q_f = rate of heat loss through the flow of digester contents (energy /time)

A_f = area of flow (distance^2)

Total digester heat loss (Q_L) is, therefore,

$$Q_L = Q_R + Q_{W_1} + Q_{W_2} + Q_f$$

The heat loss due to evaporated water (Q_W) is the sum of the sensible heat loss of the steam and the heat of evaporation of water. The sensible heat loss with the dry biogas leaving the digester is the sum of the sensible heat in methane and carbon dioxide.

The heat required to raise the influent slurry to the digester operating temperature (T_1) can be calculated from:

$$Q_1 = W \cdot C_P (T_1 - T_I) \cdot E_F$$

where T_1 is the digester operating temperature, T_I is the influent slurry temperature, W is the total weight of slurry to be added to the digester per day, C_P is the specific heat of the influent slurry and E_F is the efficiency factor.

$$Q = U.A. (T_2 - T_1)$$

where,

Q = rate of heat loss (energy/time)

U = overall coefficient of thermal conductivity (energy/time-area-temperature)

A = area normal to the direction of heat flow

T_2 = air temperature outside the digester

T_1 = digester operating temperature

Overall coefficient of thermal conductivity U is a function of the unit surface conduction inside and outside plus the thermal resistance of each of the materials as governed by the following relationship.

$$\frac{1}{U} = \left[\frac{1}{h_j} + \frac{x}{k_1} + \frac{x}{k_2} + \ldots + \frac{1}{h_0} + \frac{1}{K_a}\right]$$

Thus,

$$U = \frac{1}{\left[\frac{1}{h_j} + \frac{x}{k_1} + \frac{x}{k_2} + \ldots + \frac{1}{h_0} + \frac{1}{K_a}\right]}$$

where,

x = material thickness (length)

k = coefficient of thermal conductivity of the material (energy/time length-temperature)

h_j = inside unit surface conductance (energy/time-area-temperature)

h_0 = outside unit surface conductance (energy/time-area-temperature)

K_a = coefficient of conductivity of air and gas (energy/time-area-temperature)

With the help of above relationship, different heat loss quantities can be estimated in the following manner.

Heat loss from digester roof (Q_R) is given by

$$Q_R = \frac{A_R (T_1 - T_2)}{\left[\frac{1}{h_j} + \frac{1}{K_a} + \frac{x}{k_1} + \frac{x}{k_3} + \frac{1}{h_0}\right]}$$

where,

Q_R = rate of heat loss through digester roof (energy/time):

A_R = roof area (distance^2)

For a digester working on continuous feeding, the heat taken away by the removed sludge is given by [23].

$$Q = \frac{V}{RT} \cdot C_{pl} \cdot (T_1 - T_{in})$$

where V is the volume of slurry present in digester, RT is the retention time, C_{pl} its specific heat, T_{in} its inlet temperature and T_1 is the temperature of the digesting slurry

Total heat removed is given by

$$Q_{tot} + Q = \alpha_1 T_1 - \alpha_2 T_2 - C$$

where T_2 is the ambient temperature and C is trivially connected to all thermal constants.

In energy units, biogas production P may be written as:

$$P = P_c - V M_g f(T_1) = f(T_1)$$

where P_c is the calorific value of biogas, M_g is the specific biogas production at a reference temperature (e.g. 35°C) and $f(T_1)$ is an experimental digestion temperature modulating function.

Q_{tot} is required to be supplied to the fermenting mass to keep it at the desired temperature, consuming Q_{tot}/η energy (η is an efficiency of the heating process, e.g. burner plus pipes). The net gas production thus shall be

$$N_P = P - Q_{tot}/\eta = f(T_1) - (\alpha_1 T_1 - \alpha_2 T_2 - C)/\eta$$

The above equation is written with a view to facilitate conceptual understanding of the heating process.

C.R. Prasad and S.R.C. Sathyanarayan [34] developed thermal model to study heat-transfer from slurry and gasholder to the surrounding earth and air respectively for the case of a movable-drum type biogas plant. Computations were made for two conditions: slurry at an ambient temperature of 20°C and slurry at 35°C, the latter being ambient temperature of mesophilic fermentation. Gasholder is a major source of heat-loss under both conditions. It was endeavoured to estimate quantity of heat which must be supplied externally to compensate for net heat loss which occurs if the slurry is maintained at 35°C. For a case when part of biogas produced is used for heating the digester contents, it was found that there occurs a net increase in gas output following heating and therefore a net benefit by operating the plant at 35°C. The results of the thermal analysis can be used to develop a strategy for operating the plant at optimum temperature or at temperature higher than the ambient.

J. Kishor, *et al.* [22] described approaches for thermal balance of fixed dome biogas plants and enhancement of gas yield during winters. They presented a thermal model of a fixed dome biogas plant with or without hot charging and having the ground surface above the plant blackened and glazed. The rate of heat transfer between the ground and plant was determined by electrical simulation by using laboratory scale model of the plant made of copper and suspended in copper sulphate solution. Studies were carried out keeping in view monthly variation in temperatures at Madras, New Delhi and Srinagar, and suggest strategies to alleviate problems of low temperatures corresponding to 1, 4, 8, 14 and 30 m^3 capacity biogas plants. It was reported to be possible to maintain the slurry temperature between 28 and 35°C in a place like Delhi by blackening and single glazing of the ground above the plant. In places of colder climates even double glazing may not be enough. Hot charging from a shallow solar pond alone was not found sufficient to maintain desired temperature of digester contents during winters in Delhi and Srinagar. In places of hot climates where temperature is sufficiently high, hot charging from a shallow point or blackening/glazing of the ground was found to be sufficient to meet the objective. In colder places like Srinagar using part of the gas produced to heat digester contents and blackening and double glazing of the ground above dome was suggested to counter the effect of low temperatures.

G.P. Devkota [17] studied the effect of composting on biogas yield which is based on the fact that yield changes with temperature of the digester contents. In regions where the temperature drops significantly at night and for certain seasons of the year, composting is a useful method for generating heat for smaller dome-type biogas digesters. Tests were carried out in Butwal (Nepal) throughout the cold season with the result that temperature of digester contents insulated by compost was found to be 2°C higher than for uninsulated digesters. Sixty kg of buffalo dung was daily fed into each digester, and it was found that digesters insulated with compost produced 10 to 37 per cent more gas than without.

5.10 Example Showing Steps for Selecting Plant Parameters including Heating Needs

This example illustrates the steps for computing design parameters of a 100 ft^3 capacity movable drum biogas plant. Based on a set of assumptions it is endeavoured to calculate biogas requirement, manure requirement, digester, capacity, gasholder size, heat requirement, heat losses, and digested slurry withdrawn every day from plant. This exercise is based on a similar study reported in NAS publication [29] but with an altered design parameters.

Following is the summary of various assumptions and conversion factors used.

- Biogas is used to run engine-generator set to produce electricity.
- Thermal efficiency of the engine is 40 per cent.
- Electrical mechanical efficiency of generator is 75 per cent.
- One cubic foot of biogas containing 55 per cent methane provides 550 BTU of heat.
- Forty per cent of fuel energy is transferable to cooling water.
- 1 kwh = 3411 BTU.
- Ten pounds of dung plus water is needed to prepare slurry which occupies a volume of 2/7 ft^3. This is equivalent to 1.8 litre of slurry per kg of dung.
- Digester is expected to operate at 95°F (35°C).
- For meeting village needs 125 kwh of electricity is needed to be generated daily.

Estimation of biogas requirement. Heat needed to generate 125 kwh of power per day

$$= 3411\left(\frac{\text{BTU}}{\text{kwh}}\right) \times 125\left(\frac{\text{kwh}}{\text{day}}\right) = 4{,}26{,}375 \text{ BTU/day}$$

Actual heat needed per day with assumed engine efficiency of 40 per cent

$$= \frac{4{,}26{,}375}{0.40} = 1{,}065{,}937 \text{ BTU}$$

Daily heat requirement with generator efficiency assumed as 75 per cent

$$= \frac{1{,}065{,}937}{0.75} = 1{,}421{,}250 \text{ BTU}$$

Daily biogas requirement for the village with the assumed rate that 1 ft^3 of gas provides 550 BTU of heat

$$= 1{,}421{,}250\left(\frac{\text{BTU}}{\text{day}}\right) \times \frac{1}{550}\left(\frac{\text{ft}^3}{\text{BTU}}\right)$$

$$= 2585 \text{ ft}^3\text{/day } (73.2 \text{ m}^3\text{/day})$$

Estimation of manure requirement. With an assumed norm that 1 kg (2.20462 lb) of cattle manure yields 1.3 ft^3 of biogas (Table 5.1) or 1 lb of cattle manure produces 0.60 ft^3 of gas

Dung requirement per day

$$= \frac{2585}{0.60} = 4308 \text{ or, say, } 4310 \text{ lb } (1955 \text{ kg})$$

Number of cattle needed to generate 4310 lb (1955 kg) of dung per day

$$= \frac{4310}{33} = \frac{1955}{15} = 130$$

Selection of digester capacity. It is assumed that 4310 lb (1955 kg) or dung is fed to the digester daily which should remain inside for 50 days (Table 5.2) for undergoing anaerobic fermentation.

With the assumed norm that 10 lb of dung occupies 2/7 ft^3 of space and 4310 lb of dung is fed to the digester, daily volume needed becomes

$$= \frac{2}{7} (ft^3) \times \frac{1}{10 \text{ (lb)}} \times 4310 = 123.14 \text{ ft}^3 \text{ } (3.48 \text{ m}^3)$$

For a 50 days detention period, total digester volume needed is

$$= 123.14 \left(\frac{ft^3}{day} \right) \times 50 \text{ days} = 6157 \text{ ft}^3 \text{ } (17.34 \text{ m}^3)$$

If a cylindrical shaped digester is chosen with diameter (*D*) to depth or height (*h*) ratio as 1 : 1, requisite height and diameter of digester to match estimated digester volume can be calculated as follows:

$$\frac{\pi D^2}{4} \times (h = D) = 6157 \text{ ft}^3$$

or,

$$D^3 = \frac{6157 \times 4}{3.14} = 7843.3 \text{ ft}^3$$

Whence,

Height (*h*) = 19.8 ft (6.06 m)

Diameter (*D*) = 19.8 ft (6.06 m)

Estimation of gasholder size. Based on presumption that gas is to be used regularly and withdrawn more or less at constant rate, gasholder needs to have only half the volume of estimated daily gas production. Thus for daily gas production of 2585 ft^3/day (73.2 m^3/day), gasholder needs to have a capacity of only 1292.5 ft^3 (36.6 m^3). For a proper alignment in a movable drum type plant, it is an appropriate design practice to keep gasholder diameter smaller by 6 inches (15.24 cm) than that of digester. Diameter of gasholder should therefore be 19.3 ft (5.9 m). The height of gasholder is thus given by:

$$\text{Height of gasholder} = \frac{\text{Volume}}{\text{Area}} = \frac{1292.5 \times 4}{3.14\,(19.3)^2} = 4.42 \text{ ft}$$

$$\text{Gasholder diameter} = 19.3 \text{ ft } (5.9 \text{ m})$$

$$\text{Gasholder height} = 4.42 \text{ ft } (1.35 \text{ m})$$

It is to be noted that on account of its weight, gasholder tends to compress biogas accumulated with the result that actual volume of gas collected above the slurry surface is less than the gasholder capacity.

Estimation of heat requirement. Estimates of heat requirement for warming up the plant slurry should be based keeping coldest month of the year in view. For tropical and subtropical regions the minimum average temperature for the country as a whole is assumed as 20°C (68°F) and for extremely cold regions like Laddakh the minimum average ambient temperature is assumed as 1°C (35°F). As the available digester temperature is less than the optimum desired temperature, for obtaining higher gas yield it is desirable to raise the digester temperature from 20°C (68°F) to 35°C (95°F). As already estimated, 4310 lb (1955 kg) along with equal quantity of water is to be added daily which means that a total of 8620 lb (3911 kg) of slurry is to be fed every day. If water and dung are assumed to have same specific heat, requisite heat to raise the temperature of input slurry from 68°F to 95°F is

$$= 8620 \text{ lb} \times \frac{1 \text{ BTU}}{\text{lb} \times {}^\circ\text{F}} (95 - 68)^\circ\text{F} = 2{,}32{,}740 \text{ BTU/day}$$

$$= 2{,}45{,}540 \text{ kJ/day}$$

Estimation of heat losses. Amount of heat losses from digester to the environment depends upon the ambient temperature. For estimating heat losses following temperatures are assumed.

Mean ambient temperature (coldest) = 1°C (35°F)

Mean ambient temperature close to the digester wall = 20°C (68°F)

For estimating heat losses, following area calculations are made.

$$\text{Roof area of gasholder} = \frac{\pi\,(19.3)^2}{4}$$

$$= 292.4 \text{ ft}^2 \ (27.2 \text{ m}^2)$$

$$\text{Area of gasholders walls} = \pi \times 19.3 \times 4.4$$

$$= 266.64 \text{ ft}^2 \ (24.8 \text{ m}^2)$$

Area of digester bottom
+ area of digester walls

+ area of annular space between digester and gasholder

$$= \frac{\pi\,(19.8)^2}{4} + \pi \times 19.8 \times 19.8 + \frac{\pi}{4}(19.8^2 - 19.3^2)$$

$$= 1569.44 \text{ ft}^2 \ (145.95 \text{ m}^2)$$

Values of different thermal conductivities which we have denoted by q are as follows:

Thermal conductivity (q_r) from steel gasholder roof top and walls is given by

$$q_r = 0.16 \text{ BTU/hr/ft}^2/^\circ\text{F} \ (3.27 \text{ kJ/hr/m}^2/^\circ\text{C})$$

Thermal conductivity (q_w) from digester floor, walls and annular clearance between digester and gasholder is given by

$$q_w = 0.12 \text{ BTU/hr/ft}^2/^\circ\text{F} \ (2.45 \text{ kJ/hr/m}^2/^\circ\text{C})$$

Heat loss from gasholder top

$$= 0.16 \frac{\text{BTU}}{[\text{hr. ft.}^2\ ^\circ\text{F}]} \times 292.4 \text{ ft}^2 \ (95 - 36)^\circ\text{F}$$

$$= 2760.25 \text{ BTU/hr} \ (2912 \text{ kJ/hr})$$

Heat loss from gasholder walls

$$= 0.16 \frac{\text{BTU}}{[\text{hr. ft.}^2\ ^\circ\text{F}]} \times 266.64 \text{ ft}^2 \ (95 - 68)^\circ\text{F}$$

$$= 1151.88 \text{ BTU/hr} \ (1215.23 \text{ kJ/hr})$$

Heat loss from digester bottom and digester surface area including clearance between digester and gasholder

$$= 0.12 \frac{\text{BTU}}{[\text{hr. ft.}^2 \cdot ^\circ\text{F}]} \times 1569.44 \ (95 - 68)^\circ\text{F}$$

$$= 5084.98 \text{ BTU/hr} \ (5364.65 \text{ kJ/hr})$$

Thus, total heat loss per day becomes

$$= 24\,(2760.25 + 1151.88 + 5084.98)$$

$$= 215930 \text{ BTU/day} \ (240335 \text{ kJ/day})$$

The total heat to be supplied is the sum of heat losses at operating temperatures and heat needed to raise slurry to the desired operating temperature

$$= 232740 + 215930 = 4{,}48{,}670 \text{ BTU/day} \ (4{,}73{,}347 \text{ kJ/day})$$

The actual amount of heat produced by the biogas run diesel engine coupled generator which is not converted to electricity based on an overall efficiency of 30 per cent (75 per cent × 40 per cent) is 1,421,250 – 4,26,375 = 9,94,875 BTU (10,49,593 kJ) per day. This is the same as 70 per cent of 1,41,250 BTU per day. With a total heat requirement of 4,48,670 BTU/hr (4,73,347 kJ/day), the amount of heat available for transfer to cooling water and that which will somehow go to the environment will be 9,94,875 – 4,48,670 = 5,46,205 BTU/day (5,76,246 kJ/day).

Fifty per cent of this heat quantity is assumed to be available for transfer to the cooling water. This implies that 2,73,102.5 BTU (2,88,123 kJ) will be available daily to heat the cooling water. As this quantity of heat is less than the total amount of heat needed to maintain the slurry at operating temperature 4,48,670 BTU/day, (4,73,347 kJ/day), heat available from the engine generator set is not entirely sufficient to warm the digesters content and some kind of supplementary heating from solar energy or other sources shall be necessary.

Estimation of Quantity of Slurry Withdrawn from Digester Daily

Quantity of slurry withdrawn per day from digester is usually less than added each day following certain amount of volatilisation that keeps occurring during the interval. As already estimated total amount of slurry added per day to the digester is 8620 lbs (3911 kg) comprising 4310 lbs of manure and 4310 lbs of water. It is commonly observed that 10 per cent of slurry weight is generally volatilised. Thus the weight of slurry withdrawn per day becomes 8620 × 0.90 = 7758 lbs (3519 kg)

5.11 An Overview of Solar-assisted Biogas Systems

Solar energy is plentiful in most regions of the country. It can be gainfully utilised to keep the temperature of digester contents close to 35°C. Plant slurry can be warmed up in a number of ways. Active and passive methods can be employed to heat up the digester contents using solar energy as external heat source. Active systems involve heating the feedstock or directly heating the digester contents. Pre-heating of the slurry is only marginally beneficial considering that daily input of slurry is only about 2 to 4 per cent of the total volume. Direct heating of the digester contents runs the risk of killing the bacteria and hence altogether eroding the process efficiency. In passive techniques which involves no moving parts, a greenhouse (solar canopy) is usually built over a biogas digester to arrest the solar radiation. Solar flat plate collectors integrated with the biogas plant can also be used to warm the digester contents. Breadbox and solar pond are the main types of solar collectors used for slurry heating. A breadbox solar collector consists of a radiation-absorbing tank within a box fitted with a glass cover. A solar pond consists of a large container kept horizontally on the ground or on insulation, covered with a glazing and filled with liquid. A solar water heating system can also be alternatively used to produce hot water at 50°C which can be circulated into the bed of the digester with the help of heat exchanger to maintain higher temperatures [37, 41].

M.S. Sodha, *et al.* [38] carried out studies concerning evaluation of a new concept of greenhouse-coupled biogas plant for increasing biogas yield especially during winter months. Installation of polyvinyl chloride greenhouse type structure over a biogas plant allows solar heating of the content from 18 to about 37° which is optimal temperature requirement for anaerobic fermentation. Experiments with the two biogas plants of 8 m^3 and 85 m^3 capacity set up at Masoodpur village near New Delhi indicated that plastic enclosures helped to raise gas yield by 15 to 20 per cent. A mathematical model was also developed which helped to predict plant performance in presence and absence of greenhouse under varying climatic conditions [3]. Based on calculations made in respect of a biogas plant installed in the vicinity of Delhi, G.N. Tiwari, *et al.* [42] showed that it was possible to obtain substantial increase in gas yield on a typical winter day by covering the gasholder with a transparent polyethylene sheet during sunshine hours and using a movable insulating material during the off-sunshine hours.

Shallow Solar Pond (SSP) is an inexpensive solar collector which generates hot water. The hot water thus produced can be gainfully utilised in slurry preparation which helps to improve biogas yield particularly during winters. A diagram of the SSP is shown in Fig. 5.6. Its important components are, (i) shallow brick walled container, (ii) a black rexin liner to act as an absorber of solar radiation, (iii) water filled to a depth of 5-10 cm in the container, (iv) a thin transparent PVC sheet floating on water to prevent evaporation of water, (v) a second thin transparent PVC cover above the first one providing an air gap between the two PVC covers, and (vi) inlet and outlet pipes for admitting and discharging cold and hot water respectively. Water is generally filled in morning around 8 AM which gets warmed up to the maximum by 3-4 PM. On a sunny day in winter, water temperatures of 40-45°C can be obtained for a water depth of about 5 cm which corresponds to hot water production rate of

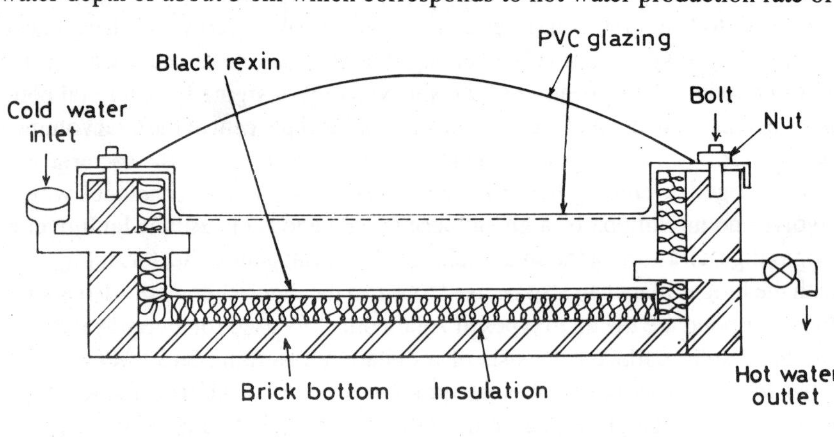

Fig. 5.6 Cross-sectional view of a Shallow Solar Pond [Source: TERI, Fixed Dome Biogas Plants, New Delhi, June 1987, p. 41]

50 litres/m^2 of pond area. It is thus possible to estimate the area of solar pond for a given capacity of digester. As per rates prevailing in 1987, a solar pond costs about Rs 300/m^2 and it becomes less expensive if it is constructed along with the digester. Efficiency of the pond can be further improved if insulation is provided below the layer of black rexin [21].

C.K. Desai, *et al.* [11] studied aspects relating to the design of solar ponds for storing and supplying heat to a biogas plant digester for obtaining added gas yield. It was found that if the temperature of digester content can be maintained at 40°C, retention period for the slurry can be reduced by over 40 per cent. Solar ponds can be helpful in preventing heat losses during night and in maintaining digester temperature at desired level. A small solar pond of 0.65 m depth and 670 litre capacity was used to supply heat to a plant of 0.25 m^3 capacity. Experimental results confirmed that it was possible to reduce retention period by over 45 per cent and obtain much higher and steady gas yield.

R.A. Gupta, *et al.* [13] provided a detailed account of solar-assisted biogas plant which incorporates water heater concept for achieving higher gas yield particularly during winter months. Further improvement in the performance of the system was investigated by using insulation on the walls and base of the digester. They also developed a mathematical relationship which expressed slurry temperature as a function of time. G.N. Tiwari and A. Chandra [41] studied the effectiveness of solar-assisted biogas system particularly during winter months for obtaining higher gas yield. For this purpose biogas unit is exposed to solar radiation to increase the slurry temperature and covered with movable insulation to avoid heat losses from the top and sides of the dome. A. Beba [4] carried out the performance analysis of a simulated hybrid system incorporating biogas fermentor, waste heat recovery system and solar collectors. It was reported that total annual contribution of heat recovery unit and solar systems were approximately 29 and 37 per cent respectively of maximum attainable limit.

Schematic views of insulated and solar-heated digesters involving solar flat plate collector are shown in Figs. 5.7 and 5.8. Their functioning is self explanatory [44].

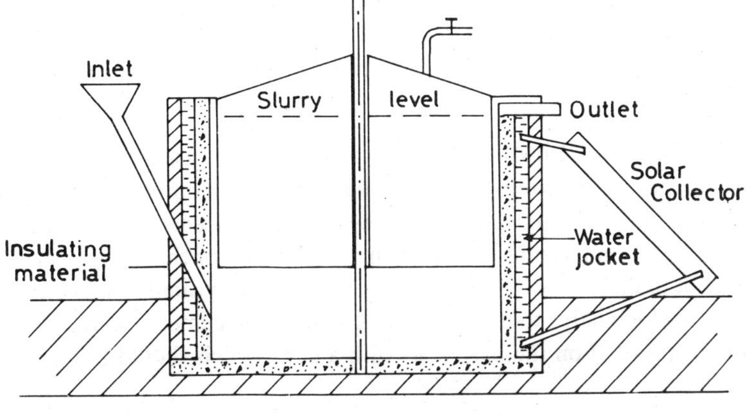

Fig. 5.7 Schematic view of Solar-heated digester [Source: UN, Guidebook on Biogas Development, New York, 1980, p. 80]

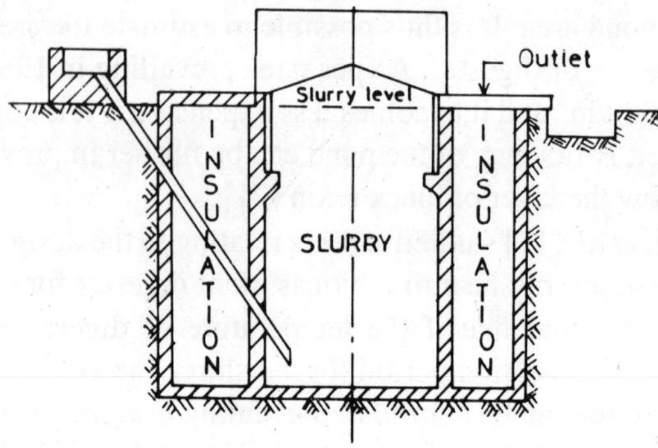

Fig. 5.8 Schematic view of Insulated Digester [Source: UN Guidebook on Biogas Development, New York, 1980, p. 79]

5.12 Procedure for Constructing Movable Drum Type Plant

Biogas plant should be constructed in consonance with the technical specifications as any imperfection will result in low gas yield, reduce plant life, and may even require far more work to remedy defects. The choice of a suitable base is a key factor for the final quality of the pit. Before construction a study should be made with regard to soil conditions and the underground water level. For constructing a movable drum type of biogas plant, generally the following steps need be followed sequentially.

Collection of material. Collect all the materials required, including bricks, stones, cement, steel, mortar etc. Care should be taken to store cement in a dry place. Ensure availability of labour input required including the self-help to be provided by the family members of the plant owner.

Marking of digester pit. Mark the outline of the pit on the ground with the help of a chord and marker.

Excavation of digester pit. Dig the digester pit within the marked ring. Sometimes it may happen that sides may collapse if the soil is not firm. To avoid collapsing, it is customary to dig a 30-50 cm deep ring outside the edge of the pit to be filled with concrete comprising stones, sand and cement in the ratio 10 : 3 : 1.

Pouring the foundation slab. The pit should be straight and the floor horizontal. Before concreting, loose sand and loam is required to be removed. If possible, line with a plastic sheet. The concrete is needed to be poured in a moist condition comprising a mix of 30 mm size gravel or chippings, clean sand and cement in the ratio of 6 : 3 : 1. Carry out necessary stamping to eliminate trapped air which helps to increase strength and save cement. Mark the inside face of the pit-wall on the base slab with a cord which should be tied at the centre.

Pit-wall brickwork. The upper edge of the first layer should be laid horizontally. Surface irregularities should be filled up with extra mortar. Mortar should be sufficiently smooth and sticky. Mixing ratio although dependent on sand quality is generally taken as 1 part of cement to 6 part of sand, or lime, cement and sand mixed in proportions of 1 : 1 : 12. Bricks or stones to be used in construction are required to be dipped in water before use.

Back-filling of pit-wall. Every morning earth or sand should be filled into the space behind the section of the wall built the previous day followed by proper ramming. The partition wall should be built when the outer pit wall has reached the height of the projecting ledge. Install the central guide frame for the gasholder with metal cross in the ledge ring after ensuring its perpendicularity. The ring ledge consists of stones or bricks laid at right angle to the rest of the wall. Complete the rest of the pit masonry work. Cut the outlet pipe to the desired length and insert it into the wall of the composting pit. The inlet pipe should be approximately 50 cm higher than the outlet pipe. Build the mixing tank. It is to be ensured that connections to toilets on one hand should not be lower than the mixing chamber and on the other hand should be higher than the outlet pipe.

Plastering of surfaces. Round off all edges and corners. Apply plastering material in two layers comprising 1 portion of cement with 3 portions of sand and two parts of water. The floor of the mixing chamber should be made with correct slope. Apply rust-proofing paint to the guide frame and its metal cross. Wherever possible, bitumen coating should be applied to all the plastered surfaces.

Construction of the gasholder. Independent of pit-construction, construction of gasholder can simultaneously start in a workshop. Sheet metal of which gasholder is made should not be less than 2 mm thick and weld seams must be gastight. The holder is cylindrical. The outer wall is 3.14 times the diameter in length. The height of the outer wall is the same as the overall height of the holder. The cover of the gasholder is slightly conical. It is formed by marking out and cutting the sheet metal in steps namely, by marking out a width of radius 1.5 cm greater than the radius of the gasholder; measuring and marking a length of 9.5 cm on the circumference of the circle; cutting out a wedge, cutting out a ring for the guide tube plate (10 cm radius); welding together the cut edges of the wedge; and welding the cover to the outer panel. This is to be followed by application of rust-resistant paint to the inside and outside surfaces of the gasholder. Before applying paint, existing patches of rust must be removed with a wire brush. This is to be followed by application of bitumen coating. Following operations also need to be carried out.

Grease the guide-frame sufficiently. Mount the gasholder after closing the main gas tap of the gasholder, if leaking, it should be repaired. Open the main gas tap to exhaust air from the gasholder. Gas tap should remain open till the gas has started coming out. Normally gas pipes vary from 2.5 to 5 cm in diameter. Gas pipes can be either buried in ground or made to lie above grouna. However, it is preferred to keep them underground for keeping them cool. At the lowest point of gas pipe there should be provision for trapping and removing moisture condensate.

5.13 Procedure for Constructing Fixed Dome Type Biogas Plant

For constructing a fixed dome type biogas plant, generally the following steps need be followed sequentially [5, 6].

Collection of material. Collect all the materials required including bricks, stones, cement etc. Care should be observed to store cement in a dry place. Ensure availability of labour input required including the self-help to be provided by the family members of the plant owner.

Marking of digester pit. Mark the outline of the digester pit the inlet and outlet tanks on the ground.

Excavation of digester pit. Dig the digester pit and inlet and outlet tanks according to the marked profile. Sometimes it may happen that sides may collapse if the soil is not firm. To prevent collapsing it is sometimes advisable to taper off the digester walls. It is also possible to excavate the inlet and outlet after the digester pit has been lined. In rocky ground or when the ground-water level is high, the base of the biogas plant can be raised accordingly. For achieving desired insulation, soil around the digester pit should be appropriately packed.

Pouring the foundation slab. The floor can be either of concrete or brick-work and is generally 20 to 40 cm thick. Before concreting loose sand and loam is required to be removed. The floor should be provided with necessary slope as per the design.

Digester pit brickwork. Brickwork should be performed accurately with adequate filling in the joints. For this purpose, bricks, concrete block, or stones can be used. Thickness of digester walls generally varies from 15 to 30 cm. The floor of the digester pit should be strong enough to bear the entire water pressure.

Construction of inlet and outlet tanks. Generally pipes of diameter 25 to 40 cm are appropriate for connecting the inlet and outlet tanks to the digester pit. While making inlet and outlet of brickwork, connecting part to the pit should be most carefully made and if need be, it can be reinforced with concrete. Inlet and outlet pipes are normally inclined at an angle of 50 to 60 degrees.

Back-filling of digester pit wall. Every morning earth or sand should be filled into the space behind the section of the wall built the previous day followed by proper ramming.

Building of dome. Dome-building is a difficult phase in construction of fixed dome type biogas plant as it requires considerable skill and experience. One easy construction method is to use a full wooden shuttering. With necessary skill, a mason can construct a spherical roof without shuttering. Care should be taken to ensure that transition between the dome and vertical wall is adequately reinforced and supported against the surrounding soil. Back-filling is generally recommended to be done every day with the completion of 3 to 4 brick-courses.

Plastering of surfaces. Round off all edges and corners. Apply plastering material in two layers comprising 1 portion of cement with 3 portions of sand and two parts of water. After plastering seating coats need to be applied. Plastering and application of special seating paints are required to make the structure gas and water tight. The plastering need to be carried out in 2-3 successive layers to a total of 2 to 4 cm thickness. Plastered surfaces should be smooth and even and finally given a coating of pure cement slurry.

If cement is not available, mortar can be used in its place which can be categoried as masonry mortar and plastering mortar. Both masonry and plastering mortar can have varying compositions. Following are the standard mortar mixes normally used in China.

Masonry mortar. 3-component mortar 2/2/8 (lime/washed sand/alumina), 4-component mortar 1/2/2/4 (lime/fine ash/alumina/washed sand), 4-component mortar 1/8/8/42 (cement/lime/fine ash/washed sand) or 4-component mortar 0.1/1/2/6 (cement/lime/alumina/washed sand).

Plastering mortar. For the first layer '*A*' 1/2/3 (lime/fine ash/medium sand), for the second layer '*B*' 1/1/2 (lime/fine ash/medium sand) with a thickness of less than 1 cm, for the third layer '*C*' 1/1/1/2 (cement/lime/fine ash/fine sand) with a thickness of 0.5 cm. If in the third layer cement is not to be used, 1/1/1 (lime/fine ash/ fine sand) with a thickness of 0.5 cm can be used instead.

A week after the plastering is done, three layers of bitumen or three coats of a synthetic sealant to provide insulating cover are applied.

Assembly of hatch cover. The hatch cover must be cast in concrete with a step all round. Water is digester slurry keeps the seal moist which makes it unusable after some time. As a result seal is required to be replaced after some time. For breaking the fibrous material a stirrer passes through gas dome. Lower end is pivoted onto the floor of digester. It is sealed by an outer tube which extends down below the minimum slurry level so that gas cannot escape. Upper part of this tube is anchored and sealed in the fixed dome brickwork.

5.14 Common Materials Used in China for Constructing Fixed Dome Biogas Plants

In China for constructing these plants following materials are commonly used [7].

Lime clay. This is commonly used building material in China which after mixing with water in right proportions (21-24 per cent) and requisite curing imparts needed compactness and hardness. Clay easily reacts with lime to form hydrate calcium silicate and hydrate calcium aluminate. Water-hardening capability of lime-clay essentially depends on quality and mixing proportions of lime and clay used, compactness, curing conditions and water content. Moisture content in lime-clay lubricates its grains and makes it plastic. In lime-clay proportions of lime and clay vary from 1 : 9 to 1 : 19 by weight which after curing in water for 5 to 7 days imparts in needed hydro-stability (i.e. it does not give in or collapse in water). Addition of 20 to 30 per cent of coal slag helps to counter shrinking characteristics of lime clay. Moist climate is considered more conducive for efficient curing than arid-climate.

Lime concrete. It is another building material commonly used in China. It consists of mixture of lime, sand and gravel in proportion of 1 : 3 : 6 by volume in water.

Low-strength concrete. This is obtained by mixing concrete with a compressive strength of 30 kg/cm^2 in lime and clay. Concrete mixture is prepared with cement-to-concrete ratio within a desired range of 1.25 to 4 and with some slaked lime powder added to it which enables low quality cement to be used without any loss of strength. Other additives include coal ash, lime kiln slags and clay. Efficiency of various additives used to lower cement grade is represented by a factor called β. Value of β for four different additives commonly used is summarised in Table 5.10. Compressive strength of concrete in kg/cm^2 and achievable after a curing period of 28 days is given by

$$R = 0.5\,(0.82 - a) \cdot R_c \left[\frac{c}{w} - 4\right]$$

where,

a = Additive proportion in cement in per cent used for lowering its grade.

$\frac{c}{w}$ = Cement water ratio by weight.

R_c = Grade number of cement.

Table 5.10 Efficiency Index of Different Additives (Factor β) used for Lowering Cement Grades

Proportion of Additives in Cement (per cent)	*Value of Factor β for different Additives*			
	Coal Ash	*Lime Kiln Slags*	*Clay*	*Slaked Lime Powder*
20	0.66	0.75	0.75	0.64
30	0.50	0.60	0.60	0.53
40	0.38	0.46	0.46	0.40
50	0.22	0.35	0.33	0.31
60	0.13	0.23	0.22	0.20

Source : SPOBD, Biogas Technology and Utilisation, Chengdu Seminar, Sichuan Provincial Office of Biogas Development, Sichuan, China, 1979.

Strength of concrete essentially depends on properties of individual constituents used, water to lime ratio, degree of compactness and size of sand. For retaining properties in low-strength concrete, additive proportions should not exceed 40 per cent limit.

In China concrete is also prepared in clinker-free-cement which is made of lime, gypsum and Pozzolanic (Cinder) material. Concrete prepared from this cement is called clinker-free-concrete. Coal slags, like-kiln slags, crushed bricks and tiles from part of Pozzolanic (Cinder) materials. In this concrete water, clinker-free cement, fine-sand and gravel are mixed in ratio of 0.67 : 1 : 1.85 : 4.07 by weight. Clinker-free-concrete acquires strength with the passage of time and is known to be stronger in moist than arid climate.

Besides use of lime-clay, lime-concrete, low-strength concrete, clinker-free concrete for plant construction, digesters in China are also made of bricks, stones, rocks etc. as is done in India, with their specific choice being governed by local availability of a particular item. Based on Chinese experience for constructing 6 to 8 m^3 capacity plants, 500 kg of cement, 1200 kg of gravel, 1200 kg of sand, 50 bricks weighing 25 kg are consumed. If bricks and cement are used for dome construction; 5 per cent lime 95 per cent clay is used for digester construction, then roughly 200 kg of cement, 300 kg of lime and 500 bricks are needed.

5.15 Various Tests Needed for Fixed Dome Type Biogas Plants

Hydraulic testing for water and gas tightness. Before the plant is put to operation, it is normally advisable to carry out hydraulic testing under water pressure for maintaining the desired level of water and gas-tightness.

Testing for water leakage. For carrying out this test, digester is required to be half-filled with eater and water level noted each day for a period of 6 to 7 days. Significant drop in water level is an indication of water leakage.

Testing for gas leakage. For carrying out this test, all digester outlets are required to be closed to render it gas-tight. This is followed by complete filling in of digester with water to its full capacity and noting down water level each day for a period of 6 to 7 days. If there occurs more than 5 to 10 per cent drop in water level, it is an indication that gas dome, seating is not perfect and needs to be corrected.

Safety pressure gauge helps to detect gas leakage from digester, estimate the gas availability within the digester, and releases extra unutilised gas automatically formed within the digester. The gauge is made of two 1-m long glass tubes (internal diameter: 0.5 cm) and one of rubber, in U-shape, and the glass tubes are filled with coloured water (Fig. 5.9). Tube *A* is connected with the gas pipe attached to the digester through rubber tube and tube *B* is connected to a 100-ml (or larger) bottle. A 2-holed rubber stopper is inserted in the mouth of the bottle, and two glass tubes, one short and one long are placed through

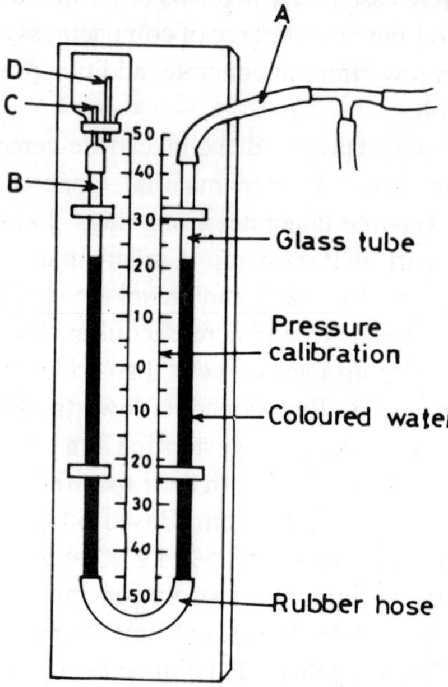

Fig. 5.9 Safety pressure gauge used in biogas systems [Source: IDRC, Compost, Fertilizer, and Biogas Production from Human and Farm Wastes in the People's Republic of China, Ottawa, 1978, p. 92]

the holes of the stopper, the short tube is connected to tube *B* by a rubber hose, and the bottle is inverted and attached near the safety gauge. When the pressure on the water column is high from the biogas plant, the coloured water is forced into the bottle through tube *C* and any extra biogas is released through tube *D*. When the pressure inside the tank is reduced, the coloured water in the bottle flows back into the U-shaped tube through tube *C*. This mechanism enables the water in the gauge to adjust automatically to changes in the digester's internal pressure [7, 17].

5.16 Biogas Plant Operation: Some Observations and Recent Developments

Production of gas normally starts within a week, the gas accumulates inside the holder causing it to float (in movable drum type) and rise. The first gas after start-up may not burn due to high content of carbon dioxide. In such a case accumulated gas needs to be released while observing all safety precautions. The process is to be re-started with drum settling down and again rising with fresh gas which may be similarly tested. After the test is positive the clear gas can be used. In summer the slurry on top of the digester is prone to become so thick that the drum does not easily sink when released for use. The flame obtained is O.K. to begin with but later becomes low. For alleviating this difficulty, a few buckets of water need to be put into the digester and stirred with a bamboo poles until the slurry becomes sufficiently thin. The drum can be shaken few times a day and flame tested. The drum (gasholder) should be taken out and given a protective coating of paint at least once a year. Fibrous materials such as straw, grass etc form a floating scum and can cause blockage of the plant. They should therefore be chopped up to a length of 1-3 cm before being fed to the plant. Bacterial activity should not be overloaded by adding too much material to the pit. Biogas plants should be loaded regularly and uniformly, and the fermentation slurry well stirred to ensure high gas production.

Biogas plant performance can be controlled by studying and monitoring the variation in factors like pH of digester contents, volatile solid concentration, temperature, nutrient availability, concentration of toxic material, feed composition, retention period, loading rate, agitation, heating and heat balance. A digester which works with minimum of control and operates under equilibrium conditions is referred as to 'balanced digester'. However, if any one of the controllable parameters is suddenly changed or if some toxic materials are introduced, the equilibrium is disturbed and an 'unbalanced digester' result which functions at less than the normal efficiency [29]. When there occurs a drop in gas production, it can be due to digester imbalance although other factors could have also contributed to it. For example, it can be due to fall in temperature or the rate at which the feed material is being added. Hence, for identifying the true cause, it is necessary that uniform loading of the digester is maintained and digester temperature closely monitored. If there occurs decline in pH and/or associated increase in volatile acid concentration, it could be due to digester imbalance which needs to be corrected. However, this requires laboratory facilities, equipment and trained personnel to monitor most of the control parameters affecting the anaerobic process. This creates difficulties to farm owners who do not have access to technology or testing facilities for performance evaluation of their plants. As a short-cut villagers can adopt simple commonsense rules. If based on their 'feel' there is no drop in ambient temperature and they are uniform and regular in loading, it is safe to assume that decline could be due to drop in pH. This can be corrected by adding lime to the digester slurry and if the gas yield improves, the course is justified. However, by adding lime if the yields does not improve, then surely some toxic substances might have contributed towards decline. In such a situation as a practical remedy it is advised to dilute the normal feed material with other digestible wastes like straw, hay, gass clippings, urine, night-soil during the period of reduced gas production which in the routine course may not be used. Finally, if all else fails and digester remains non-producing, the only course of action left is to empty the digester and start all over again [28].

While starting up a digester it is a common practice to seed it with an adequate population of both acid forming and methane forming bacteria. S.K. Sharma [36] on the basis of experimental studies carried out at the University of Roorkee found that it is beneficial to start the plant by first filling it with effluent slurry taken from another biogas plant operating on the same or different feedstock. It has been found experimentally that if the wet cattle dung is kept for anaerobic digestion after mixing with water in 1 : 1 ratio (v/v) at 37°C, the usual gas production starts on 7th day but if the influent is mixed with effluent slurry obtained from another biogas plant operating on same feedstock then the gas production starts earlier. The gas production rate increases with increase in proportion of effluent slurry mixed with influent slurry. If the plant is to be fed with crop/agro-residues only, then mixing of effluent from other plant becomes necessary. It was found that biogas with more than 50 per cent methane could be obtained within 2 days of starting the digester when the influent was mixed with effluent slurry in 1 : 1 ratio (v/v) obtained from a cattle dung fed biogas plant.

According to L. Sasse [30] long retention period is one of the effective ways of increasing gas yield. This is backed by experimental results that a 'second push' in gas production occurs after a prolonged retention time. For small farmers with few cattle, use a local material in plant construction, prolonged retention time achievable with somewhat large-size digesters are advisable. Ferro-cement gasholders are preferred to steel types considering that in the case of former construction materials can be transported to remote areas by cycle. Plastic gas pipes have proven to be cheap and long-lasting as compared to iron-pipes. S. Pal, *et al.* [30] studied the effect of loading rates on the performance of three types of biogas plants, i.e. KVIC, Janata and polyethylene-type (flexi) under shallow water-table conditions. The performance was evaluated on the basis of daily gas production, volatile solid reduction, temperature profile of slurry in the digester and NPK concentration in effluent slurry. While gas production was found to be least in the Janata plant at highest of the feeding rates, its performance was found to be most efficient in terms of some of the above parameters.

G.H. Dar and S.M. Tandon [9] studied the effect of pre-treatment of crop residues on biogas yield. Addition of pre-treated (1 per cent NaOH for 7 days) wheat straw, lantantana residue, apple and peach leaf litter to cattle dung was found to improve microbial digestibility and biodegradability during anaerobic fermentation at ambient temperature (28-31°C). Pre-treatment was found to nearly double the gas output including improvement in methane content in biogas. Z. Pechan, *et al.* [33] studied the impact of high ammonium nitrogen concentration on biogas yield from anaerobic digestion of poultry manure. Based on laboratory fermentation of poultry manure conducted at mesophilic temperature corresponding to average total solid concentration of 11-14 per cent in influent and retention period of 27-58 days, no adverse impact on biogas yield was seen when mean ammonia nitrogen concentration in the effluent rose to 4.07-5.85 g/litre. Biogas yield was however found to be directly related to volatile solid loading rate.

M.A. Hamad, *et al.* [14] studied the effect of digester pressure on yield and composition of biogas. A fixed dome type plant fed by buffalo dung with retention period as 40 days was used in carrying out the study. Increase in digester pressure was accompanied by decrease in quantity of gas produced and some increase in methane content.

T. Kate and S. Karanjekar [20] evaluated comparative performance of floating and fixed-dome biogas plants in India. Based on evaluation of one year performance, it was concluded that fixed-dome Janata biogas plants are more efficient in gas production, are cheaper to construct and maintain than floating-dome type plants.

P.H. Liao, *et al.* [24] studied the effect of liquid-solid separation on biogas yield during mesophilic anaerobic digestion of dairy-cattle manure at 35°C in a laboratory-scale digester. Screening of coarse solids from the manure before digestion had a significant effect on biogas production. Biogas yield and methane content in biogas from screened manure were consistently higher than from unscreened manure. A liquid-solid separation pre-treatment step reduced digester volume for a dairy farm without decreasing biogas yield. K.V. Lo and P.H. Liao [25] studied anaerobic digestion of screened dairy manure in anaerobic rotating biological contact reactors at 35°C using rotating discs made of cedar wood and acrylic plastic. Digester using cedar wood discs had a shorter start-up period than acrylic plastic disc. Within an operational period of 3 months, the highest methane yield of 1.67 litre per day per litre was obtained at 1 day hydraulic retention time for the reactor with wood media. On the other hand in case of anaerobic rotating biological reactors using plastic media the highest methane yield of 1.89 litre per day per litre was obtained within an operational period of 10 months at 1 day hydraulic retention time.

Based on their experimental studies, T.H. Chen, *et al.* [8] examined batch digestion of fresh and dry dairy manure and showed that methane yields expressed as m^3 methane/kg volatile solids fed from fresh and dry manure were statistically the same. However fresh manure produced greater total volume of methane than the dry manure following loss in organic materials during drying.

For obtaining more information on construction and plant operation reader may refer to the following sources which provide excellent coverage of the subject matter.

(i) BORDA, Biogas: Manual for the Realisation of Biogas Programmes, GATE, P.O. Box 5180, 6236 Eschborn 1, Germany, 1980.

(ii) BORDA, Biogas Plants: Building Instructions, GATE, P.O. Box 5180, 6236 Eschborn 1, Germany.

(iii) Buren, A.V. (ed.), A Chinese Biogas Manual, Intermediate Technology Pub. Ltd., 9 King Street, London WC 2E 8HN, 1979.

(iv) IRDC, Compost, Fertiliser and Biogas Production from Human and Farm Wastes in the People's Republic of China, Ottawa, 1978, p 92.

(v) Khandelwal, K.C. and Mahdi, S.S., Biogas Technology: A Practical Handbook, Tata McGraw Hill Pub. Co. Ltd, 12/4 Asaf Ali Road, New Delhi 110 002, 1986.

(vi) Kishore, V.V.N., Raman, P. and Ranga Rao, V.V., Fixed Dome Biogas Plants: A Design, Construction and Operational Manual, TERI, Habitat Place Lodhi Road, New Delhi 110 003.
(vii) SPOBD, Biogas Technology and Utilisation Proc. Chengdu Seminar, Sichuan Provincial Office of Biogas Development, Sichuan, China, 1979.
(viii) UN, Guidebook on Biogas Development, New York, 1980.
(ix) Constant, M., Naveau, Ferrero; G.L., and Nyns, E.J., Biogas : End Uses in the European Community, Elsevier Applied Sci., London and New York, 1989.
(x) Singh, J.B., Myles, R.M. and Dhussa, A., A Manual on Deenbandhu Biogas Plants, Tata McGraw Hill Book Pub. Co. Ltd., New Delhi, 1988.

5.17 Protection of Plant Metallic Parts from Corrosion

Corrosion can be viewed as an interactive process with environment which tends to destroy desired properties of the material susceptible to it. Apart from environmental impact, corrosion can also be caused due to direct contact with liquids such as electrolyte. Biogas plants of movable drum type large number of which exist in India are sufficiently prone to corrosion as several of its parts such as gasholder are made of steel. Walls of gasholder are generally found to be more corrosion prone as compared to its top portion. Corrosion depends on several factors such as nature and purity of metal used, level of humidity and oxygen in environment, ambient temperature and type of electrolyte used in situations where electrolyte is the main contributing factor.

A mildsteel gasholder acts as an anode which forms Fe^{++} and Fe^{+++} ions and releases electrons ($2e^-$). Iron and water are ionised in the following manner:

$$Fe \longrightarrow Fe^{++} + 2\,e^-$$

$$H_2O \longrightarrow H^+ + OH^-$$

$$2\,H^+ + 2\,e^- \longrightarrow 2\,H \longrightarrow H_2$$

When hydrogen is liberated, OH^- ions tend to move towards anode, Fe^{++} and Fe^{+++} ions move towards cathode after passing through water film. It is during these journeys in opposite directions that iron and OH^- ions react forming iron hydroxides [Fe $(OH)_2$ and Fe $(OH)_3$]. Carbon dioxide of biogas reacts with these hydroxides to form hydroxy carbonate. This hydroxy carbonate finally degenerates into ferric oxide along with a water molecule [Fe_2O_3. H_2O] signalling the onset of corrosion process. Lack of adequate oxygen also contributes to corrosion by forming certain galvanic cells on metallic surface. In a movable drum type plant, amount of oxygen varies with the position of gasholder.

Degree of corrosion can be minimised in a number of ways which include correct choice of plant materials, removal of corrosive agents and application of an electrical field at phase boundary like cathodic or anodic protection, formation of new interfaces by applying metallic inorganic and organic coatings. Surface coating among them is largely practised and is one of the earliest methods of surface protection. A specific coating well-formulated and applied protects surface either by one, two or even three mechanisms concurrently. Barrier coatings provide protection against the penetrating agents like water and oxygen. By providing sufficiently high thickness of the film, the ingress is arrested. Asphalt and bitumen type of coatings belong to this class. Inhibitive coatings normally contain corrosion inhibitive heavy metal pigments like zinc chromates, and phosphates. They control corrosion by changing the electrode potential at the anodes so that iron does not oxidise. However, this method has limitation that when inhibitor is consumed by combining with other ions, protection breaks down. The third category of coatings, namely, sacrificial coatings contain zinc metal powder or duct particles as the principal component which makes the film conductive. Protective feature depends on sacrificial nature of zinc which protects the surface from corrosion.

Corrosion in biogas plants is usually prevented by applying protective coatings comprising corrosion-resistant materials and making improvements in plant design. While selecting protective coating, it should be ensured that chosen paint is non-porous, adherent, quick-drying and anodic. Synthetic enamel or coal tar are one of the most commonly used paints for this purpose. Coating is done by first cleaning the surface, applying two coats of red oxide or red lead primer, applying a phosphatising agent and a zinc-chromate coat, and finally a synthetic enamel or coal tar coat. It is also sometimes suggested to protect the surface by coating it with chlorinated rubber, epoxy and polyster resins which are usually applied after cleaning the surface. Applying phosphate coat is another effective way for containing corrosion. After coating the surface with zinc chromate, a coat of phosphoric acid is applied to form a protective tenacious film of ferric phosphate which protects the surface.

Apart from using protective coating it should be endeavoured to use corrosion-resistant materials in plant-construction wherever feasible. Digesters made of fiberglass, concrete, masonary brick, stone etc. generally have longer service life than those made of steel. Use of dissimilar metals in plant construction should generally be avoided to minimise corrosion due to electrolysis. Unless economically prohibitive, pipes of stainless steel or similar corrosion-resistant materials should be used for gas supply. All corrosion-prone surfaces should be periodically inspected to minimise chances of corrosion. Feedstock should be freed from non-biodegradable components like sand by installing screening devices, which unnecessarily cause abrasive action on metallic surfaces [31].

5.18 Applications of Plastics in Plant Construction

Plastics like in many other products have also been used in biogas plants. Plastics apart from having energy saving attributes are characterised by their low weight, easy mouldability into any desired shape, insulating, non-toxic and non-corroding characteristics. Plastics can be broadly divided into two categories, namely, thermoplastics and thermo-setting plastics. Thermoplastics can be processed into any desired form and shape by application of heat and pressure by a mechanical process. They retain their moulded form in cold conditions, and can be resoftened any number of times by applying heat and pressure. Thermosetting plastics on the other hand are processed under the influence of heat and pressure but retain their moulded form only when hot and cannot be resoftened easily once set even by applying heat. Thermoplastics are tougher but less rigid and brittle than thermo-setting plastics. Polyvinyl chloride (PVC) and high density polyethylene (HDP) are the two commonly used thermoplastics. The chemical formula of PVC is C_2H_3Cl and its molecular weight is 62.5. In rigid form PVC can be extruded into the shape of hosepipe, conduit, or sheet while in flexible form it can be used as a substitute for rubber such as low voltage cables and household wiring. PVC has a high coefficient of linear expansion, fire-resistant, not easily attacked by alkalies and acids, and has good insulation properties except at very high voltage. HDP is another commonly used thermoplastic. It is interesting to note that both PVC and HDP are finding growing applications in biogas plant construction.

PVC, HDP, nylon laminated sheets or synthetic rubber can be used in making pillow type bag digesters. Experiments on a PVC bag type digester were carried out way back in 1977 at Biogas Research Centre at Etawah. A 100-ft^3 capacity gasholder was built by mounting PVC sheet on a 5′-4″ dia and 4 ft height steel frame. Gasholder was found in good working condition even after a year's working. Slurry was in general found to be non-sticking to the gasholder surface and vertical movements of gasholder did not pose any problems. Similar to PVC gasholder, it was attempted to build a 150 ft^3 capacity bag digester with a flexible bag of size 10′ × 5′ × 3′ made of synthetic rubber/nylon laminated flexible sheet, and attached to a 6-ft long and 3 inch diameter PVC tube. Digester was placed in a small depression on the ground to prevent lateral shifting. Following tendency of digester contents to spread on ground within digester under gravity, this depression helped to build necessary pressure on gas to be released for use. In another experiment, a balloon shaped gasholder of 10 ft^3 capacity made of PVC/nylon laminated sheet was mounted on a 25-ft^3 capacity digester built under the ground; and the second gasholder of the same capacity was mounted on a 48-gallons capacity steel drum. Requisite gas pressure in the latter case was built by fixing a wooden disc within it. The disc moves up on iron pipe fitted inside it as a guideway when gas is formed and comes down when it is consumed. Gas was transferred to its place of use in a 10-ft^3 capacity PVC laminated bag of cylindrical shape.

A.K. Kalia and R.B. Singh [19] reported that plastic gasholders hardly posed any maintenance problems and gave somewhat higher gas output during winters as compared to steel gasholders of similar capacity.

A number of experiments have been carried out in past at KVIC Biogas Research and Development Centre in Mumbai for using plastics in plant construction. Plastic products are generally recommended to be used for constructing parts like gasholder and pipelines, as insulator for preventing heat losses; and sometimes for corrosion-protection of parts like gasholder with the help of thin PVC sheet cover. G.L. Patankar [32] in his experiments at the Biogas R&D Centre Mumbai studied the impact of using PVC sheet on corrosion-prone surface as a substitute for enamel paint which is expensive. Sheet is generally affixed after cleaning the surface and successive coatings of red-oxide and coal tar. While affixing the PVC sheet, it should be ensured that no air-pockets are entrapped between the PVC sheet and gasholder surface. Application of plastics in gasholder construction is now fairly established. High density polyethylene, rigid PVC and glass reinforced polyster sheets are commonly used plastic products for gasholder construction. However, their low weight poses some problems of building requisite gas pressure which is countered by placing artificial weight like brick pieces or sand bags. S.M. Dastur [10] provided an overview of the role of reinforced polyester (RP) in India's biogas programme. Use of RP to replace masonary and steel in movable drum biogas plants can make significant impact on the success of biogas programme in the country as it seeks to alleviate problems of corrosion and poor workmanship associated with existing designs. Problem of fire hazard associated with the use of plastics can be overcome by covering gasholder with fire-resistant materials like asbestos mat [32]. Plastics are also being used in pipeline construction which are corrosion resistant and ensure longer service. Plastic lines are easy to lay, low in cost and exhibit improved flow characteristics. Plastics are particularly advantageous in low pressure gas pipelines, low temperature water lines and for transportation of slurry into and out of plant. R.L. Zhang [45] described the technical characteristics of the soft and hard polyvinyl chloride pipes and polyethylene pipes which are being used in the Chinese countryside and the red-mud plastic pipes which are expected to be used in the near future. Application of plastics in valves, gas-meters and water-remover manufacture is also picking up fast.

Polyethylene is an important plastic product which can be classified as low, medium and high density type (Table 5.11). High density polyethlene is more heat resistant and chemically inert than low density type. Low density resins are softer, more flexible and less heat resistant. A high density polyethylene gasholder was successfully designed and manufactured by Helioplastics, Bombay, in collaboration with the Directorate of Gobar Gas Scheme, KVIC, Bombay. Polyethylene gasholder is claimed to be corrosion, rust, and leak proof, and free from maintenance problems. It is light in weight and possesses strength, stiffness and toughness. It does not crack or break while handling and can be transported over long distances without damage [26].

Table 5.11 Classification of Polyethylenes

Type	*Density, g/cc*	*Category*
1	0.910-0.925	Low
2	0.926-0.940	Medium
3	0.941-0.965	High

Source: Milby, R.V., Plastics Technology, McGraw Hill Book Co, NY, 1973.

Red-mud Plastic (RMP) developed by the Union Industrial Research Laboratories (UIRL), Taiwan, after years of experimentation and improvement has been demonstrated to be superior to several other materials used in digester construction and may have considerable potential in India as well in future. RMP is a mixture of red-mud, PVC, plasticizer, stabiliser and other ingredients [43]. It is elastic, provides considerable resistance to ultraviolet light, hydrogen sulfide gas, strong alkali and acid and can withstand tropical weather for 10-20 years unlike PVC. Red-mud is the material left along with sodium hydroxide, after the aluminium oxide is extracted from Bauxite during production of aluminum metal. For every ton of aluminium metal produced, half ton of red-mud is generated. Red-mud contains SiO_2 (14.6%), Al_2O_3 (22.6%), Fe_2O_3 (35.6%), TiO_2 (7.2%) and Na_2O (9.1%). RMP is produced by thorough mixing red mud with PVC, lubricating oil wastes from industries, and other ingredients. These ingredients are mixed in a Banbury mixer and dumped at temperature of 170 to 180°C which helps formation of carbon-carbon double bonds following loss of hydrogen chloride from PVC. Double bonds make oxidation easier with a resultant drop in molecular weight. These bonds help salts and oxides present in red mud to absorb hydrogen chloride which in turn makes their bonding stronger. Blending with lubricating oil helps to make RMP an ultraviolet absorbent filter, and corrosion and flame resistant. RMP can be brought into both flexible and rigid shape depending upon the amount of blending and temperature employed.

A RMP digester is easy to manufacture, install and transport. It can be easily cleaned and nearly no maintenance is required. Scum if any formed during the fermentation process settles to the digester bottom and hence no stirring is needed. Construction cost of RMP digester is low. In Taiwan several RMP digesters have been inflated and deflated over 60,000 times per year with no reported damage. The digester maintains the initial pressure charge over a long period of time. The estimated life of these digesters is more than 10 years. Physical damage to the digesters is easy to repair. A simple patch of the same material may be applied over the cut and affixed by fusion. Based on experiments carried out in Taiwan, comparative biogas yield from varying size of

RMP digesters while keeping other parameters unchanged are given in Table 5.12. These digesters are not only popular in Taiwan, country of their origin, but have also made considerable headway in Philippines, Tahiti, Paraguay, Hong Kong, Singapore, Australia and Brazil.

Table 5.12 Comparative Biogas Yield from varying Sizes of Red-mud Plastic Digesters

Digester Capacity (m^3)	*Digester size diameter × length (m × m)*	*Weight (kg)*	*No of full grown Pigs required to feed the Digester*	*Biogas Yield (m^3/day)*
20	2 × 6.5	80	8-10	4-5
30	2 × 9.5	150	12-15	6-7.5
50	2 × 16	275	20-25	10-12.5
100	2.5 × 20	550	40-50	20-25
500	5 × 40	2500	200-250	100-12.5

Source: UIRL, Red Mud Plastic, Union Industrial Research Laboratories, Hsinchu, Taiwan, 1981.

Speaking in general terms, though the share of plastic products in plant construction is growing over the years, some difficulties associated with their large scale use for commercial success need be alleviated. These include, among others, problems of non-uniform inadequate gas pressure, gas-leakage near outlet, fire hazard and chances of digester bursting under excessive gas pressure.

5.19 Application of Ferrocement in Plant Construction

Ferrocement is a highly versatile form of reinforced mortar in which closely spaced layers of wire mesh reinforcements are impregnated with rich cement mortar mix. This material can be cast into sections as thin as 1 cm and is therefore suited for precast products. Extensive studies have been carried out in past at the Structural Engineering Research Centre (SERC), Roorkee, on the development and use of ferrocement for rural applications including its use in the construction of gasholder of movable drum type biogas plant (Fig. 5.10). Design and fabrication techniques for ferrocement gasholders suitable for family-sized biogas plants have been developed by the SERC under the 'All India Coordinated Project on Biogas'. Later Cement Research Institute of India (CRII), Ballabgarh, Haryana, also developed in early eighties ferrocement gasholder for medium size biogas plants. Ferrocement unlike reinforced concrete possesses high resistance to cracking and permeability which is essential for the satisfactory functioning of gasholder. A semi-mechanised process for casting thin cylindrical units of ferrocement has been developed and patented by the SERC. By using this process, effective control of thickness and high

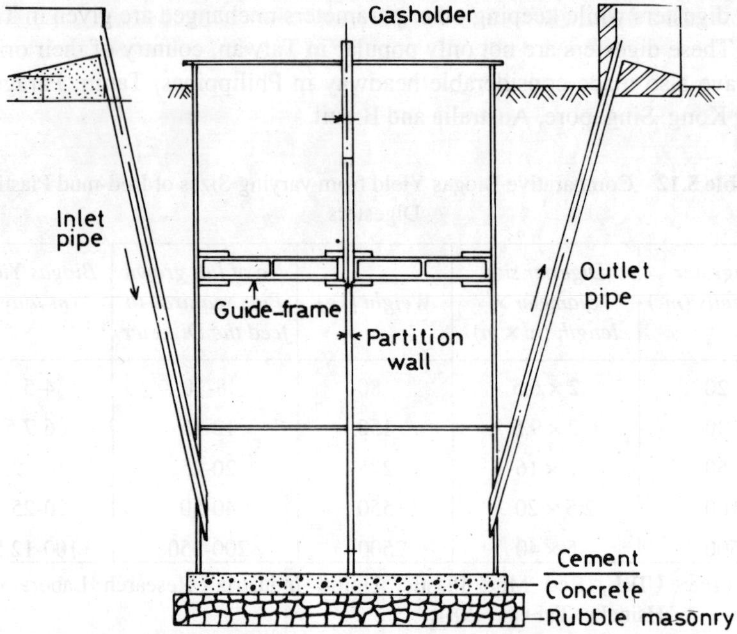

Fig. 5.10 Schematic view of the Ferrocement Digester
[Source: TERI, Biogas Technology: A Promise for the Future, A Monograph sponsored by the MNES, Tata Energy Research Institute, New Delhi, 1994, p. 19.]

degree of compaction with high impermeability to gas is achievable. The process is labour-intensive and does not require power or expensive machinery and therefore ideally-suited for work in the rural areas of the developing countries. Ferrocement has low thermal conductivity and accordingly the rate of gas production is fairly uniform in all seasons. It has high impact strength, cause less maintenance problems, and can withstand the rigours of transport and handling. Damages caused are however repairable easily [401].

The ferrocement gasholder developed at the SERC is suitable for a plant of capacity 2 m^3 of gas per day and plant sizes in medium range. It consists of a wall unit and a roof unit cast separately and assembled. The wall unit is cast on a horizontal wooden mould lined with galvanised iron sheet. The mould is made in several segments to facilitate easy assembling and is mounted on the process stand which enable rotation and stretching of the wire mesh. The reinforcement for the wall unit consists of two layers of galvanised iron woven wire mesh and one layer of mild steel welded with wire fabric in between. The dome shaped roof unit is cast on a masonry mould. After curing the wall and roof unit thoroughly for 14 days, the roof and wall units are joined together to form the gasholder. After joining the wall and roof units, the gasholder is cured thoroughly for 14 days with water after which the unit is ready for testing for leakage of gas. After the surface of the gasholder has fully dried up, suitable surface coating is applied to its inside and outside to improve its impermeability. The scum breaking arrangement is attached to the inserts provided near the top and bottom edges of the wall unit. It consists of three mild steel bars at top and bottom placed radially and connected by vertical bars. The radial bars function in addition as bracings for the gasholder. The scum formed on the surface of the slurry is broken by the vertical bars on rotation of the gasholder once in a few days. Ferrocement gasholder is very economical as compared to a steel gasholder, the savings in cost being as much as 47 per cent [35].

5.20 Concluding Remarks

Design, construction, operation and maintenance are very important aspects of a biogas system. In view of its proven utility, biogas technology has come to stay and rapid advances are taking place in all spheres. Biomethanation on commercial scale in future shall be based not only on agricultural wastes and animal manure but on host of other biomass such as sewage, water-hyacinth, algae, bagasse, waste cotton dust, distillery wastes, textile-mill wastes, press-mud, canteen wastes, cassava leaves, banana peels, palm oil, mill-effluents, industrial-waste waters, willow-dust, fruits and vegetable wastes, household wastes and municipal refuse. In view of their proven potential, it is natural that future digester designs should take into account these sources of biomass availability. Furthermore in order to provide an economically viable energy source from all available sources of biomass, it is essential that future designs are based on the characteristics of biomass with emphasis on low cost materials, improved efficiency of biomethanation and simpler operation.

Sources of biomass for future digester designs can be grouped into four main categories: (1) dilute substrates such as sewage and microscopic algae, (2) organic slurries such as sewage sludge, homogenous animal manures and seaweeds, (3) highly viscous animal waste slurries, and (4) dry crop biomass [18]. There are a number of substrates for which there exists considerable scope to improve the digester design. For instance, design needs to process sewage are entirely different. Sewage is known to have dry organic content as low as 0.02 per cent of total weight as compared to 25 per cent in several animal manures and 80 per cent in field dried crops such as straw and corn stalks. Dilute substrates cause 'washout' of the bacteria faster than they can reproduce; and variations in temperatures and flow-rate inhibit growth of methanogens in traditional designs [18]. The same handling and digester design is not appropriate for all types of diverse organic materials. The flow of materials should be taken into account before designing the system. The availability of raw material, various factors that affect the rate of gas production and local availability of trained manpower needed for plant construction should be taken into account while selecting a particular plant design. In a movable drum type plant corrosion is a serious problem following constant exposure of the metal parts of the system to hydrogen sulfide and organic acids that mav either be present or formed during digestion process. It is thus necessary to paint these parts periodically with corrosion-resistant paint. Use of ferrocement and plastics in plant construction apart from minimising costs also help to reduce corrosion problems. In Janata and Deenbandhu biogas plants which have no moving parts, wear and tear and consequent maintenance problems are few, and this is why their number is on constant rise in the country.

REFERENCES

1. Anjaneyulu, K. and Mehta, D.J., Manometric Glass Device for Measuring Biogas and Methane, *Invention Intelligence*, **19(1)**, Jan., 29-30 1984.
2. Bansal, N.K., *et al.*, A study of Greenhouse Concept on Conventional Biogas Systems for Enhancing Biogas Yield in Winter Months, *Jour. Energy Research*, **9**, 119-128 1985.
3. Bansal, N.K., A Techno-Economic Assessment of Solar-Assisted Biogas Systems, *Energy Research*, **10**, 216 1988.
4. Beba, A., Analysis of a Solar-Heated Biogas Fermenter, *Solar Energy*, **40(3)**, 281-287 1988.
5. BORDA, *Biogas Plants: Building Instructions*, GATE, P.O.Box 5180, 6236 Eschborn 1, Germany, 1980.
6. BORDA, *Biogas: Manual For the Realisation of Biogas Programme.* GATE, P.O. Box 5180, 6236 Eschborn 1, Germany, 1980.
7. Buren, A.V. (ed.), *A Chinese Biogas Manual*, Intermediate Technology Pub. Ltd, 9 King Street, London WC 2E 8HN, 1979.
8. Chen, T.H., *et al.*, Methane Production from Fresh Versus Dry Dairy Manure, *Biological Wastes*, **24(4)**, 297-306 1988.
9. Dar, G.H. and Tandan, S.M., Biogas Production from Pre-treated Wheat Straw, Lantana Residue, Apple and Peach Leaf Litter with Cattle Dung, *Biological Wastes*, **21(2)**, 75-83 1987.
10. Dastur, S.M., Reinforced Polyester Plays Key Role in India's Biogas Programme, *Modern Plastics*, **63**, Oct., 130-133 1986.
11. Desai, C.K., *et al., Use of Solar Pond for Thermal Control of Biogas Plant*, Proc. of the National Seminar on Solar Energy and Rural Development, Kolhapur, May 29-31, 1987, Shivaji Univ., 83-88 1988.
12. Dhussa, A.K., Desigining Biogas Distribution System, *Bio-energy Review*, **2 (182)**, 71 1983.
13. Gupta, R.A., *et al.*, An Improved Solar-Assisted Biogas Plant (Fixed Dome Type): A Transient Analysis, *Energy Conserv. Management*, **28(1)**, 53-57 1988.
14. Hamad, M.A., *et al.*, Rural Biogas Technology: Effect of Digester Pressure on Gas Rate and Composition, *Jour. of Engg. and Applied Science*, **2(1)**, 49-57 1983.
15. Hills, D.H. and Stephens, J.R., Solar Energy Heating of Dairy-Manure Anaerobic Digesters, *Agricultural Wastes*, **2(2)**, 103-118 1980.
16. Hills, D.J. and Stephens, J.R., *Solar Energy, Heating of Methane Digester*, Research Leaflet Series, Dept. of Agri. Eng. UC Appropriate Technology Programme. Univ. of California, Davis, Jan., 1981.
17. IDRC, *Compost, Fertilizer, and Biogas Production from Human and Farm Wastes in the People's Republic of China*, A Monograph Ottawa, 1978, p. 92.

18. Jewell, W.J., *Future Trends in Digester Design*, paper presented at the First Inter. Symp. on Anaerobic Digestion, Univ. Industry Centre, Univ. College, Cardiff, Wales, Sep 21, 1979.
19. Kalia, A.K. and Singh, R.B., Introduction of Plastics in Biogas, *Popular Plastics*, **23(4)**, 52-54 1978.
20. Kate, T. and Karanjekar, S., Comparative Performance of Floating and Fixed Dome Biogas Plants, *Science for Villages*, **70**, 5-7 1983.
21. Kishore, V.V.N., Raman, P., and Ranga Rao, V.V., *Fixed Dome Biogas Plants: A Design, Construction and Operation Manual*, TERI, Habitat Place, Lodhi Road, New Delhi, 1987.
22. Kishore, J., *et al.*, Solar Assisted Biogas Plants and Enhancement of Production in Winters, *Inter. Jour. of Energy Research*, **12(4)**, 711-737 1988.
23. Lavagno, E., *et al.*, On Some Engineering Aspects of Biogas Production, in A. Strub, *et al.*, (eds.), *Energy from Biomass*, Proc. 2nd E.C. Conf., Applied Science Publishers, London, 1983.
24. Lio, P.H., *et al.*, Effect of Liquids-Solids Separation on Biogas Production from Dairy Manure, *Energy in Agriculture*, **3**, 61-69 1984.
25. Lo, K.V. and Lio, P.H., Performance of Anaerobic Rotating Biological Contact Reactors Using Wood and Plastic Support Media, *Energy in Agriculture*, **6(1)**, 77-84 1987.
26. Milby, R.V., *Plastics Technology*, McGraw-Hill Book Co., NY, 1973.
27. Mohanrao, G.J., Scientific Aspects of Cow Dung Digestion. *Khadi Gramodyog*, **29(7)**, 340-347 1974.
28. Nagar, B.R., Biogas Plant Based on Night-Soil and Animal Dung, WHO Inter. Ref. Centre for Wastes Disposal, *IRCWD News*, No. **8**, June, 1975.
29 NAS, *Methane Generation from Human, Animal and Agricultured Wastes*, Report of an Ad. Hoc. Panel of the Advisory Committed on Technology Innovation, National Academy of Sciences, Washington, DC, 1977.
30. Pal, S., *et al.*, Effect of Loading Rates on the Performance of Different Types of Biogas Plants Under Shallow Water-table Conditions in India, *Energy in Agriculture*, **6(3)**, 215-230 1987.
31. Patankar, G.L., *Recent Developments in Biogas Technology*, A Report, Gobar Gas Development Centre, Borivili, Mumbai, 1977.
32. Patankar, G.L., Plastics in Setting Up of a Biogas Plant, *Popular Plastics*, **23(4)**, 55-59 1978.
33. Pechañ, Z., *et al.*, Anaerobic Digestion of Poultry Manure at High Ammonium Nitrogen Concentration, *Biological Wastes*, **20(2)**, 117-313 1987.
34. Prasad, C.R. and Sathyanarayan, S.R.C., Studies on Biogas Technology, Part III, *Proc. Indian Academy of Sciences*, **C2**. 377, 386 1979.
35. Raman, N.V., *et al.*, *Ferrocement Gasholder for Biogas Plants*, Proc. Inter. Conf. on 'Materials for Construction of Developing Countries, Bangkok,' Aug., 749-764 1978.
36. Sharma, S.K., Starting a Biogas Plant, *Urja*, **23(1)**, 31-34 1988.
37. Singh, D., Singh, K.K. and Bansal, D.K., Heat Loss Reduction from the Gas Holder/Fixed Dome of a Community Size Biogas Plant, *Jour. Energy Research*, **9**, 417-430 1985.
38. Sodha, M.S., *et al.*, Effect of PVC Greenhouse in Increasing Biogas Production in Temperate Cold Climatic Conditions, *Energy Conserv. Management*, **27(1)**, 83-90 1987.

39. Subramanian, T.V., Gobar Gas Appliances, *Khadi Gramodyog*, Sep., 532-535 1980.
40. TERI, *Biogas News*, Inaugural Issue, Feb., 1982, p. 2.
41. Tiwari, G.N. and Chandra, A., Solar-Assisted Biogas System: A New Approach, *Energy Conserv. Management*, **26(2)**, 147-150 1986.
42. Tiwari, G.N., A Simple Analysis of Conventional Biogas Plant, *Energy Conserv. Management*, **28(1)**, 1-4 1988.
43. UIRL, *Red Mud Plastic*, Union Industrial Research Laboratories, Hsinchu, Taiwan, 1981.
44. UN, *Guidebook on Biogas Development*, New York, 1980.
45. Zhang, R.L., *Selection of Biogas Pipeline*, Proc. of the Fourth Inter. Symp. on Anaerobic Digestion, Gungzhou, P.R. China, Nov. 11-15, 1985.

Chapter 6

ANALYSIS OF FACTORS AFFECTING BIOGAS YIELD

6.1 Introduction

For obtaining increased biogas yield, anaerobic fermentation should proceed efficiently with minimum of external control. This is possible when affecting parameters in their natural course lie within the desired range and external control becoming necessary only in the event of their falling outside the desired limits. Anaerobic fermentation process is balanced when affecting parameters remain within their desired range and when these fall beyond the limits, the process becomes imbalanced. For example, if volatile fatty acids in digester contents or its pH value lies beyond the desirable limits, a kind of imbalance in the process exists. Imbalances are caused due to retardation in bacterial activity following inhibitions, improper feeding and withdrawal rates and fall in digester temperature. Volatile fatty acid concentration is an important parameter which needs to be controlled for increased biogas yield. For monitoring digester performance, facilities for measuring performance parameters should exist. Such facilities are generally hot available with individual plant owners and exist only in laboratories. To overcome this problem some practical tips can be employed. For instance, in the event of fall in gas yield, impact of factors like digester temperature and loading and withdrawal rates could be analysed. If no apparent temperature variation was felt to occur over the period, improper loading rate could be the possible reason for low yield. With steady temperature and loading rates drop in gas yield could then be either due to decline in pH or due to the presence of toxic materials. As a result of the systematic analysis, if low pH of digester contents is the contributing factor, then in such situations addition of some quantity of lime can counter decline in pH. If this does not lead to increased gas yield, then it is logical to assume that some toxic materials are present in digester which might be responsible for low gas yield. To counter the effect of toxic materials, it is advisable to add easily digestible waste materials like straw, hay, grass clippings, animal-urine, night-soil etc. which are available everywhere. Finally, if all fails and gas output still remains low without showing any sign of improvement, then final option is to empty the digester and charge it afresh without repeating the mistakes of previous charging [14].

6.2 Effect of Agitation on Biogas Yield

Mixing greatly helps to ensure intimate contact between micro-organisms which leads to improved fermentation efficiency. Mixing can be carried out in a number of ways. For instance, if slurry is fed every day instead of feeding periodically at a certain interval, it causes more frequent contact between micro-organisms thus giving desired mixing effect. It can also be achieved by carrying out certain alterations in inlet and outlet pipes of a plant. Mixing can also be carried out by installing certain stirring or mixing devices in the plant. It is also possible to achieve mixing effect by incorporating a nozzle for flushing slurry as provided in the German design of Schmidt-Eggersgluss type biogas plant. Alternatively, mixing can be achieved by installing wooden conical beams which help to break down scum following up and down movement of slurry surface at the time of filling and emptying of digester, as in case of German Weber type biogas plant. Where no mixing device is provided, stratification can be reduced by using a horizontal displacement digester which stimulates a plug flow. In digesters which utilise municipal refuse, part of the gas produced can be recirculated to mix digester contents. It is possible to introduce this feature in large sized biogas plants such as community type. It does not seem to be practical to introduce this feature in family sized plants mainly due to unaffordable costs and skilled supervision needed. A number of experiments have been carried out to analyse the effect of mixing on gas yield. One of the early attempts to study the effect of gas recirculation as an aid to mixing, on biogas yield was made by G.J. Mohanrao [12] important findings of which are reported in Table 6.1. It is to be seen that gas recirculation helps to improve gas yield. Effect of varying degree of mixing of digester contents on per cent total solids formation at different levels in digester as reported by E.R. Coppinger [3] is given in Table 6.2.

6.3 Start-up: Seeding of Biogas Plant

For starting anaerobic fermentation, it is often necessary to introduce enriched seeding bacteria into the digester for starting the process. As is known, anaerobic fermentation proceeds with the help of acid-forming and methane-forming bacteria. Among these two groups of bacteria, the first category of acid formers is more abundant whereas the second category of methane formers is less abundant, more fastidious, and more susceptible to adverse environmental changes. In starting – up a biogas plant, it is a common practice to seed it with adequate population of both acid-forming and methane-forming bacteria. Generally digested sludge from a biogas plant, actively digesting

Table 6.1 Effect of Gas Recirculation on Biogas Yield

Experiment No.	*Biogas Yield*		*Digester Temperature (°C)*	*Ambient Temperature (°C)*
	Fortnightly m^3	*Daily litres*		
	Without Recirculation			
1	1.96	130	30.5	31.8
2	3.89	260	21.0	30.1
3	2.57	170	28.7	28.1
Mean	2.81	187	30.0	–
	With Recirculation			
4	5.35	356	26.7	27.8
5	4.69	313	27.0	26.8
6	5.34	356	27.7	27.7
7	5.81	387	27.8	27.5
Mean	5.29	353	27.3	–

Source : Mohanrao, G.J., Scientific Aspects of Cow Dung Digestion, Khadi Gramodyng, 20(7), 340-347, 1974.

Table 6.2 Effect of Varying of Mixing of Digester Contents on Per Cent Total Solids Formation at Different Levels in a Digester

Level	*Per cent Total Solids*						
	Constant Mixing		*50% Mixing*		*33% Mixing*	*17% Mixing*	
	Jan 21	*Feb 18*	*Mar 4*	*Apr 4*	*Apr 29*	*Jun 1*	*Jun 10*
Perimeter							
top	7.4	8.2	8.2	8.0	8.1	7.3	7.87
	7.4	8.2	8.2	8.0	8.1	7.3	7.87
	7.5	8.4	8.2	8.0	8.2	7.6	7.82
middle	7.2	8.4	7.8	7.9	7.9	7.5	7.87
	7.2	8.3	8.2	8.1	8.1	7.3	8.10
	7.1	8.3	–	7.9	8.2	7.3	8.17
Interior							
top	8.5	9.0	8.4	8.8	8.4	8.7	8.6
bottom	8.0	8.5	8.1	8.6	8.4	7.9	8.2

Source : E.G., The Operation of a 50,000-Gallon Anaerobic Digester at the Monroe State Dairy Farm, Ecotope Group, 2332 East Madison, Seattle, WA, June, 1979.

sludge from a municipal digester, material from a well-rotted manure pit, or cow-dung slurry are used as seed. As a guideline, in continuous operation, the seed material should be at least twice the volume of fresh manure slurry during the start-up phase. The addition of seed is decreased over a 3-week period. Successful operation of biogas plant depends on establishing and maintaining proper balance between acid and methane formers. If during the operation, digester accumulates volatile acids due to overloading, this can be corrected by re-seeding and temporarily suspending feeding of digester or by adding lime in requisite quantities [14, 21].

Addition of inoculum tends to improve both the gas yield and methane content in biogas. Based on studies carried out in China, effect of pig wastes as feed on biogas composition for varying lengths of fermentation period is summarised in Table 6.3.

Table 6.3 Effect of Adding Inoculum to Feedstock on Biogas Composition at Different Time Intervals During Fermentation

Feed Composition	*Biogas Composition at Different Time Intervals during Fermentation*								
	3	*6*	*9*	*12*	*18*	*21*	*24*	*27*	*33*
Anaerobically fermented grass (20%) and a mixture of biogas plant sludge and anaerobically fermented pig wastes (10 %)	5.8	20.5	25.1	28.2	35.2	37.8	43.8	48.1	64.7
Fresh grass (20 %) and fresh pig wastes (10 %)	0	0	0	0	0	0	0	0	0

Effect of adding inoculum to night-soil for analysing its effect on biogas composition and yield is given in Table 6.4. By adding inoculum, it is possible to reduce retention period and increase gas yield. In India, S. Neelakanthan *et al.* [15] carried out experiments to analyse the impact of adding inoculum to cattle wastes as feed both with and without agitation (Table 6.5). The slurry comprised 100 g of dry and 500 g of wet manure with 900 ml of water.

Table 6.4 Effect of Adding Inoculum in Varying Proportions to Night-Soil based Feedstock on Composition and Yield of Biogas

Per cent of Inoculum Added.	*Average Biogas Yield (ml/day)*	*Per Cent Methane Content*
10	17.9	48.2
20	60.0	56.4
30	103.7	66.3

Table 6.5 Effect of Digested Slurry Inoculum and Agitation on Biogas Production

S. No	Period	*Biogas Production in Litres from 500 g Cattle Dung With Varying Inoculum Percentage*									
		Without Agitation					*With Agitation*				
		1%	*5%*	*10%*	*15%*	*20%*	*1%*	*5%*	*10%*	*15%*	*20%*
1.	First week	0.70	1.78	2.15	2.46	2.71	0.75	1.71	2.05	2.40	2.70
2.	Second week	2.34	2.69	2.62	2.52	2.53	2.38	2.70	2.59	2.65	2.65
3.	Third week	1.63	1.42	1.45	1.44	1.51	1.70	1.44	1.51	1.51	1.52
4.	Fourth week	1.32	1.25	1.28	1.30	1.32	1.35	1.30	1.35	1.35	1.30
	Total	5.99	7.14	7.50	7.72	8.07	6.18	7.15	7.50	7.91	8.17

Source : Neelakanthan, S., *et al.*, Effect of Temperature, Inoculum and Agitation on Biogas Production from Cattle Dung, Indian Jour. of Dairy Science, 29(3), 226-229 (1976).

It is to be seen that whereas increasing addition of inoculum to feed helps to improve the yield proportionately the effect of agitation on yield is marginal.

6.4 Effect of pH of Digester Contents on Biogas Yield

Biogas production is greatly influenced by pH of digester contents. It is essentially a measure of acidity and alkalinity of a solution. A pH value of 7 is regarded as neutral, less than 7 as acidic and greater than 7 as alkaline. During anaerobic fermentation, micro-organisms require, a neutral or mildly alkaline environment for efficient gas production. A too acidic or too alkaline environment is viewed as detrimental for bacterial activity. A pH between 7 and 8.5 is optimum range for increased gas yield. The pH of digester contents is mainly affected by the amount of carbon dioxide and volatile fatty acids produced in digester as intermediate products during fermentation. For a normal anaerobic fermentation process, concentration of volatile fatty acids comprising acetic acid in particular should be below 2000 parts per million. When acetic acid concentration exceeds this limit, bacterial activity is generally found to retard. Variation in pH value during a typical anaerobic fermentation process is shown in Fig. 6.1.

As per studies made in China, during the period when ambient temperature ranges between 22 to 26°C, it takes about six days for pH to acquire a stable value. Similarly during the period when temperature ranges between 18 to 20°C, it takes about 14 to 18 days for pH to attain a stable value [18].

Normally pH stabilises spontaneously as fermentation proceeds and does not require adjustment as the system automatically tends to counter level of acidity or alkalinity of substrate by natural buffering provided by ammonia and bicarbonate ions. The buffering provided by carbon dioxide and bicarbonate

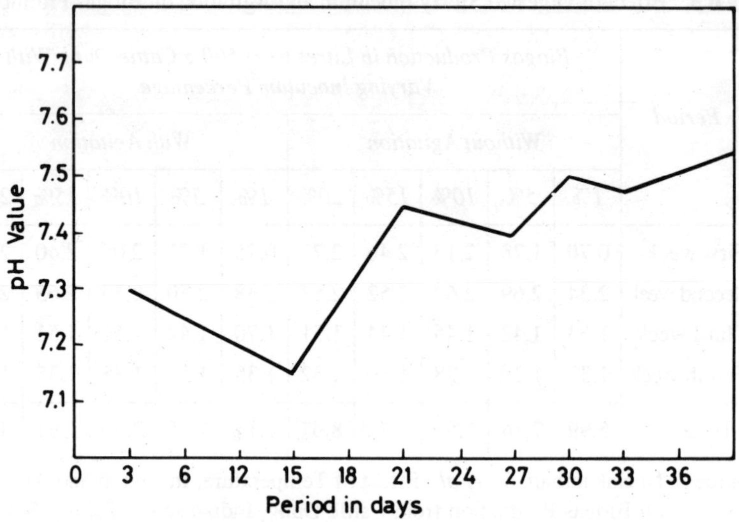

Fig. 6.1 Variation in pH value at different time intervals during fermentation

during fermentation can be described by the following mathematical relationship [1].

$$pH = 6.3 + \log [HCO_3/\text{dissolved } CO_2]$$

When biogas plant is fed with a very improper mix of feedstock, there occurs an appreciable decline in pH following excessive formation of volatile fatty acids. As a result proportion of carbon dioxide in biogas starts increasing leading to a further drop in pH value which adversely affects system's self-regulation. In this situation, it sometimes becomes necessary to bring the pH value to a desired range which can be done by introducing additives. When pH value is low due to high acidity, lime water is preferred to be added to the digester contents. Lime is generally added in the form of lime-supernatant than as it is. Care needs to be exercised that lime supernatant is added in requisite quantity only as any excess lime can start chemically reacting with carbon dioxide content of biogas thus adversely affecting biogas yield. The following chemical reaction takes place [18].

$$CO_2 + Ca(OH)_2 = CaCO_3 + H_2O$$

On the other hand when pH of digester contents is more, it can be decreased by adding requisite quantity of hydrochloric acid to it. However use of sulfuric and nitric acids for this purpose is generally not preferred.

6.5 Effect of C : N Ratio on Biogas Production

For efficient plant operation it is necessary to maintain proper composition of feedstock so that ratio of carbon to nitrogen in feed remains within the desired range. Both carbon and nitrogen provide requisite nutrients for efficient functioning of anaerobic bacteria. Carbon provides necessary energy to micro-organisms for their sustenance whereas nitrogen helps in building their cell structures. Depending upon relative richnesses in carbon or nitrogen content, feed materials can be classified as nitrogen-rich or carbon-rich materials. It is generally found that during digestion micro-organisms utilise carbon 25 to 30 times faster than nitrogen which in other words means that carbon content in feedstock should be 25 to 30 times more than nitrogen. To meet this requirement constituents of feedstock are kept in a manner so as to ensure a C : N ratio of 25 to 30 : 1, and concentration of dry matter as 7 to 10 per cent. However, in some situations, it is possible to achieve same gas yield with a low C : N ratio and higher concentration of dry matter. As per results of experiments carried out in China [8] for equivalent gas yield, when C : N ratio lies in the vicinity of 13 : 1, it is possible to keep the range of concentration of ammoniacal nitrogen between 400 to 500 ppm, when C : N ratio is 25 : 1 this range is maintained between 300 to 400 ppm, and when C : N ratio is around 30 : 1, this range is kept between 100 to 200 ppm. In plants working on animal and municipal wastes, C : N ratio lies around 30 : 1 and hence no further adjustment of this ratio is required. However, in plant substrates which utilize food-processing wastes and agricultural crop residues as feed, this ratio may be around 100 : 1 or even higher, and in such situations additional nitrogen generally in the form of ammonia becomes necessary for bringing this ratio within the desired range. C : N ratios for commonly used waste materials is given in Table 6.6. Waste materials that are low in carbon can be combined with materials high in nitrogen and vice versa to attain desired C : N ratio of 30 : 1 [1, 4, 6, 14, 16].

Table 6.6 Carbon and Nitrogen Ratio in Different Organic Wastes

Waste Material	*Nitrogen Percentage on Dry Weight Basis*	*C : N*
Animal, Human Faeces		
Cattle wastes	0.29	25
Poultry droppings	6.30	7.3
Sheep wastes	0.55	29
Pig wastes	0.60	13
Horse wastes	0.42	25
Urine	15-18	0.8
Blood	10.14	3
Slaughter house wastes	7-10	2
Night soil	0.85	29
Plant Waste		
Wheat straw	0.3	128
Rice straw	0.63	67
Corn stalks	0.75	53
Bean stalks	1.30	32

Peanut stalk and leaves	0.59	19
Rotted sawdust	0.10	200-500
Purslane	4.5	8
Amaranthus	3.6	11
Cockstoot	2.6	19
Lucerne	2.8	18
Seaweed	1.9	79
Household Wastes		
Raw garbage	2.2	25
Bread	2.1	25
Potato tops	1.5	26
Cabbage	3.6	12
Tomato	2.3	128

Source :
1. Barnett, A., *et al*, Biogas Technology in the Third World: A Multidisciplinary Review, IDRC, Ottawa, Canada, 1978, p. 51.
2. Fry, L.J. and Merrill, Methane Digesters for Fuel Gas and Fertiliser, Newsletter No. 3, New Alchemy Institute, Santa Cruz, California, 1973.
3. Gotass, H.B., Composting, World Health Organisation, Geneva, Switzerland, 1956.
4. NAS, Methane Generation from Human, Animal and Agricultural Wastes, Report of an Ad. Hoc. Panel of the Advisory Committee on Technology Innovation, National Academy of Sciences, Washington, DC, 1977.
5. Singh, R.B., Biogas Plant: Generating Methane from Organic Wastes, Gobar Gas Research Station, Ajitmal, Etawah, 1974, p. 33.

A high carbon to nitrogen ratio means that nitrogen will be exhausted earlier than carbon. Conversely a low carbon to nitrogen ratio or too much nitrogen in relation to carbon results in high ammonium concentrations which may become toxic to anaerobic bacteria. It is possible to adjust the C : N ratio by suitable additives. For instance, sawdust which has a high C : N ratio could be added to poultry manure which has a slightly low C : N ratio by the micro-organisms. In nutshell, if C : N ratio is too high the process suffers from limited nitrogen availability, and if it is too low, ammonia may be found in quantities large enough to be toxic to the bacterial population. Even in situations where C : N ratio is close to 30 : 1, it will undergo efficient anaerobic fermentation only if waste materials are also biodegradable at the same time [7, 14].

There is some evidence to suggest that C : N ratio varies with temperature. For instance gas yield from a plant working on cow wastes in certain colder regions of the country sometimes drops to one third of its normal value which however can be stimulated by adding easily digestible materials like powdered leaves, kitchen wastes and powdered straw that promote multiplication of micro organisms. According to the results of a study conducted by M.A. Idnani and O.P. Chawla [1, 14], biogas production from 0.5 kg of cowdung was almost doubled from about 17.2 litres to 31.5 litres at 7°C by addition of 200 ml of urine. It was also reported that dehydrated urine-soaked materials such as powdered leaves, straw and straw dust stimulated gas production considerably. Thus, for plant owners who own cattle shed, it is advisable that they should construct a drain for collecting urine and allowing it to soak waste materials to be used as feed for biogas plants after sun-drying. In the same way, urine available from human-beings can also be gainfully utilized by making it to drain into pits that can hold soakable waste materials. Use of urine-soaked waste materials is particularly advantageous during winter months when gas production otherwise is low [14].

6.6 Effect of Loading Rate on Biogas Yield

Loadıng rate hormally expressed as amount of waste materials fed per unit volume of digester capacity is an important parameter that affects gas yield. Gas output is commonly expressed as m^3 of gas produced per kg of volatile solids destroyed. A number of studies have been carried out to analyse the impact of varying loading rate on gas yield. It is not practically possible to describe them individually and only some representative results are included herein here. Based on results of experiments carried out, usually prescribed loading rate for plants working on municipal wastes ranges from 0.03 to 0.1 lb of volatile solids per ft^3 of digester capacity (or equivalently 0.48 to 1.6 kg of volatile solids per m^3 of digester capacity) for a retention period ranging between 30 to 90 days. A summary of gas yield obtainable from a 100 m^3 capacity plant for varied feed rates and retention periods is given in Table 6.7 which gives gas yield and feed intake in both FPS and MKS units [7]. Another set of results by J.A. Gore et al. [5] describing effect of varying loading patterns on biogas yield and composition are given in Table 6.8. They analysed the impact of daily and alternate day loading on biogas yield. A 50 kg daily charge and 100 kg on alternate day basis produced 2.9043 m^3 and 2.9285 m^3 of gas respectively. For a particular size of plant, there is an optimum feed charge rate which will produce maximum gas and beyond which further quantity of charge will not proportionately produce more gas. Alternate day charging is particularly convenient for farmers who have their cattle sheds miles away from homes where their plants are generally situated. According to studies made by G.J. Mohanrao [13], a daily loading rate of 0.42 lb of volatile solids per ft^3 of digester capacity (or equivalently 16 kg of volatile solids per m^3 of digester capacity) produces 0.62 to 1.19 ft^3 of gas per lb of raw dung fed (or equivalently 0.04 to 0.074 m^3 of gas per kg of raw dung fed). L. Ke-xin *et al.* [8] carried out experimental studies in China for analysing the effect of varying batch interval in batch-fed plants on biogas yield important results of which are given in Table 6.9

Table 6.7 Gas Yields from A 100-m^3 Capacity Biogas Plant for Varied Feed Rates and Retention Periods

Volatile Solids Fed to Digester				*Gas Yield Expressed in Terms of Different Parameters*							
lbs per day	*kg per day*	*Retention Period (days)*	*Equivalent Number of Cows*	*ft^3 gas/cow -day*	*m^3 gas/cow -day*	*ft^3 gas/ft^3 digester/day*	*m^3 gas/m^3 digester/day*	*ft^3/day*	*m^3/day*	*ft^3 gas/lb. VS/day*	*m^3 gas/kg VS/day*
763	346	35	50	47.3	1.34	0.7	0.67	2,365	67	3.1	.19
1,221	554	21	80	57.0	1.61	1.3	1.29	4,560	129	3.7	.23
2,270	1030	11	150	47.5	1.35	2.0	2.0	7,130	202	3.1	.20

Source : PSU, Biogas from Manure, Special Circular 260, College of Agriculture Extension Service, Pennsylvania State University, Pennsylvania, 1979.

Table 6.8 Effect of Varying Loading Patterns on Biogas Yield and Composition

Quantity of feed (kg)	*Duration (days)*	*Tempera-ture (°C)*	*Daily Gas Yield (m^3)*	*Per cent Gas Composition*		*pH*	*Biogas Yield*	
				CH_4	*CO_2*		*m^3/kg/m^3/kg/day*	
Loading every day								
180	15	25.0	3.9600	55	45	8	0.022	0.022
150	22	28.2	4.5439	53	47	8	0.030	0.030
100	23	26.7	3.4183	54	46	8	0.034	0.034
50	23	25.3	2.9043	53	47	8	0.058	0.058
Loading on alternate days								
180	29	27.3	3.9267	52	48	8	0.021	0.042
150	23	29.1	3.4998	53	47	8	0.023	0.046
100	25	29.4	2.9285	53	47	8	0.029	0.058

Source : Gore, JA., *et al.*, More Gas From Gobar Gas Plants, Indian Farming, March, 1981.

Table 6.9 Effect of Varying Batch Interval in Batch-fed Plants on Gas Yield

Loading Pattern	*Retention Period (days)*	*Biogas Yield (m^3 of gas/m^3 of digester volume/day)*
Initial fermentation	50	0.350
First batch of 10% feed-stock	17.5	0.345
Second batch 10% feed-stock	12.5	0.310
Third batch of 10% feed-stock	7.5	0.300

Source : Ke-xin, L., *et al.*, Fermentative Technology of Chinese Rural Digester, Bioenergy Laboratory, Chengdu Institute of Biology, Academia Sinica, China, 1979.

As per the results of another study undertaken by G.J. Mohanrao [12, 13], recommended loading rate for plants working on night-soil ranges from 0.065 to 0.139 lb of volatile solids per ft^3 of digester capacity (or equivalently 1.04

to 2.23 kg of volatile solids per m^3 of digester capacity). Higher loading rates are recommended only in those regions where mean ambient temperatures are higher. Comparative volumetric loading rate varied from 1.9 to 0.89 ft^3 (0.054 to 0.025 m^3) per capita per day and were based on a per capita volatile solids contribution of 0.123 lb per day (0.56 kg per day). A night-soil loading rate of 0.1 lb volatile solids per ft^3 (1.6 kg per m^3) of digester capacity per day is recommended for temperate climates. Average gas yield was found to vary from 0.8 to 1.2 ft^3 (0.023 to 0.034 m^3) on per person basis.

Based on pilot-plant studies carried out on a digester of 35.31 ft^3 (1 m^3) capacity with a gasholder of 24.5 ft^3 (0.69 m^3) capacity, maximum gas yield was observed for a loading rate of 1.5 lb of dung/ft^3 digester volume/day (24 kg of dung/m^3 digester volume/day). At this loading rate, maximum gas yield per unit weight of volatile solids destroyed was obtained although per cent reduction of volatile solids was only two-thirds of that with smaller loading. Complete results of these experiments are summarised in Table 6.10.

Other factors remaining same, retention period (which refers to the average length of time the substrate remains in the digester) has an important bearing on gas yield. Its selection depends on plant. H.M. Lapp, *et al.* [10] have evolved norms for retention period selection as related to digester temperature which are given in Table 6.11. It is to be seen that duration of retention period decreases with rise in mean temperature at which the digester is expected to operate. Results of another study carried out by P.O. Chung *et al.*, [2] for analysing the effect of varying retention period on a set of chosen performance indicators are given in Table 6.12. It is to be seen that all the indicators keep rising upto the 12th day of retention after which some of them start declining.

V.J. Srivastava and D.P. Chynoweth [19] developed a mathematical model to describe gas yield as a function of organic loading rate corresponding to two different digester designs – the continuously stirred tank and the non-mixed vertical flow reactor. Analysis of the digester operation with the help of the model indicates that optimum gas yield can be achieved by selecting a digester design and an operating technique that will increase solids conversion through longer solids and micro-organisms retention. Pretreatment of feed was identified as one of the contributing factor for increasing the biogas yield.

6.7 Effect of Salinity on Biogas Yield

B.J. Mehta, *et al.* [11] studied the effect of salinity on biogas yield which was found. to have negative impact on gas output. Tap water, brackish water (1 per cent sulphur) and sea water (3.5 per cent sulphur) were used for the preparation of cowdung slurry. Biogas output was found to be inversely related to the salinity of cow manure slurries prepared with tap water, brackish water and seawater. Addition of sea mud was found to have no significant impact on biogas yield from saline slurries.

Table 6.10 Effect of Cowdung Loading on Gas Production in a Pilot-Scale Biogas Plant

Loading Rate of Fresh Cow Dung		*Loading Rate (1) of Volatile Solids*		*Retention Time (Days)*	*Volume of Gas Produced (2)*		*Per cent Reduction in Volatile Solids*	*Gas Production per Unit Weight of Volatile Solids Destroyed*	
lb/ft^3-day	*kg/m^3-day*	*lb/ft^3-day*	*kg/m^3-day*		*ft^3/day*	*m^3/day*		*ft^3/lb*	*m^3/kg*
0.5	8.0	0.973	1.17	50	7.8	0.22	22.8	14.1	0.88
1.0	16.0	0.150	2.40	25	15	0.42	19.5	13.2	0.82
1.5	24.0	0.234	3.76	17	18	0.52	14.5	15.6	0.97
2.0	32.0	0.330	5.29	12	17	0.47	14.2	10.1	0.63

Note : 1. Expressible as weight per unit of digester volume per day.
2. Expressible as volume of gas produced per unit digester volume.

Source : Mohanrao, G.J., Scientific Aspects of Cow Dung Digestion, Khadi Gramodyog, 20, April, 340-347 (1974).

Table 6.11 Norms for Retention Period as Linked to Digester Temperature

S. No.	*Digester Temperature (°F)*	*Retention Period (Days)*
1	60	56
2	80	30
3	100	24
4	120	16

Source : Lapp, H.M., et al., Methane Production from Animal Wastes: Fundamental Considerations, Canadian Agricultural Engineering, 17 (2), Dec, 1975, p. 98.

Table 6.12 Effect of Varying Retention Period on Different Plant Performance Parameters

Performance Indicator	*Value of Indicator for Varying Retention Periods*			
	4 days	*8 days*	*12 days*	*16 days*
1 gas/kg TS destroyed	–	803	1001	993
1 gas/kg VS destroyed	–	822	1023	1019
1 gas/kg COD destroyed	101	1318	1501	1455
1 gas/kg BOD destroyed	2944	3222	3930	3890
Per cent TS reduction	–	56.6	61.1	66.7
Per cent VS reduction	–	66.9	71.5	78.5
Per cent COD reduction	49.2	57.4	67.7	75.8
Per cent BOD reduction	62.6	73.4	80.4	86.7

Note : Loading rate 0.768 g VS/litre/day.

Source : Chung, P.O., *et al.*, Small Methane Generator for Waste Disposal, Proc. Third Symposium on Livestock Wastes, American Society of Agricultural Engineering, 238-240 (1974).

6.8 Relationship Between Loading Rate, Hydraulic Retention Time and Total Solids Concentration

Total solids concentration. All waste materials fed to a plant consist of solid matter and water. Solid matter is made of volatile organic matter and non-volatiles (fixed solids). During anaerobic fermentation process volatile solids undergo digestion and non-volatiles remain unaffected. Fresh cattle wastes, for instance, consist of around 20 per cent total solids and 80 per cent water. Total solids in turn consist of 70 per cent volatile solids and 30 per cent fixed solids. For optimum gas yield through anaerobic fermentation normally 8 to 10 per cent TS in feed is desired [20]. This is achieved by making slurry of fresh cattledung in water in ratio of 1 : 1.

Loading rate. This is usually expressed as kilograms of volatile solids (kg VS) fed to the digester per day per cubic metre of digester volume. Thus a 5 m^3 biogas plant with 50 kg of dung fed per day should have loading rate of around 1.4 kg VS/day.m^3. Temperature – controlled – mechanically stirred large biogas plants can have loading rates as high as 5 kg VS/day.m^3. It is to be mentioned that if loading rate is too high, pH of the digester content tends to fall due to its becoming acidic following mability of micro-organisms to biodegrade all feed materials [20].

Hydraulic retention time (HRT). HRT is defined as the average time spent by the input slurry inside the digester before it comes out. Rate of gas generation is initially high and then gradually declines as the digestion approaches towards completion. Thus the time required for 70-80 per cent digestion is considerably less than that needed to achieve cent per cent or complete digestion. HRT is chosen so as to achieve at least 70-80 per cent digestion. HRT varies between 20 to 120 days depending upon the design and operating temperature of the digester. HRT for digesters operating in countries of tropical region like India is usually taken as 40-50 days. In countries of colder climates like China, digesters are designed for HRT of about 100 days [20].

The total solids concentration, loading rate (LR) and HRT are related by the following equation [20].

$$\text{LR} = k\frac{\text{TS}}{\text{HRT}}$$

For cattledung, LR is expressed in units of kg. VS/day. m^3, TS in percentage and HRT in days, and k is found to have an approximate value of 7.

6.9 Effect of Temperature on Biogas Production

Temperature of digester contents has a major effect on biogas yield. There are different temperature ranges during which mesophilic and thermophilic bacterias are most active causing maximum gas yield. It is generally found that former are most active between the range 35° to 40°C and the latter between 50° to 60°C. Choice between mesophilic and thermophilic fermentation is governed by the natural climatic conditions in which the plant is located. However, it is possible to create conditions for thermophilic fermentation by external heat but such a course is generally uneconomic. Length of fermentation period is linked with digester temperature. For same yield, when digester temperature is high, fermentation period is usually kept shorter. On the other hand when digester temperature is low, fermentation period is generally enhanced following reduced bacterial activity. To be specific, as per the studies made in China, when digester operates at a temperature of 15°C it takes nearly a year for the digestion cycle to complete whereas if the temperature is around 35°C, cycle can be easily completed in less than a month's time or so. When digester temperature is, say, 25°C it takes about 50 days for digestion of cattle wastes to complete whereas if the temperature, ranges between 32 to 38°C, digestion is generally complete within 28 days.

For increased gas yield, it is preferred to construct biogas plants sun-facing and build in a manner so as to protect them from cold winds. For minimising heat losses from the plants, as per the practice followed in China, plant domes can be covered with locally available crop residues. Crop residues can also be composting in an annular ring around digester to enable heat generated during composting to be gainfully utilised for warming up digester contents. For maintaining higher digester temperature, wherever possible sun-warmed water should be used for slurry-preparation. For stimulating microbial activity during winters, apart from measures which minimise heat loss from digester, powdered leaves or wheat straw or animal urine are also sometimes recommended to be added to animal wastes slurry.

Several studies have been carried out to study effect of temperature on biogas yield. Results of a study carried out by R.B. Singh [16, 17] are given in Tables 6.13 to 6.16. Effect of seasonal variation on biogas yield as observed in North India is given in Tables 6.13 and 6.14. Effect of temperature on gas yield for varying retention periods is given in Table 6.15. Based on studies conducted in the USA, effect of temperature on biogas composition for varying lengths of retention periods is given in Table 6.16. Effect of seasonal variation on biogas yield in South India on the basis of studies carried out at Gandhigram Institute at Madurai is given in Tables 6.17, 6.18 and 6.19. Monthwise mean ambient temperature in celcius in important Indian cities which affects gas yield is given in Table 6.20. A pictorial representation showing effect of temperature on biogas yield as observed in European countries is given in Fig. 6.2.

Table 6.13 Effect of Seasonal Variation in Northern India on Biogas Yield from a 100 ft^3 Capacity Plant

Period	*Biogas Yield (ft^3/day)*
Jan. 1 to 15	63.30
Jan. 16 to 31	64.58
Feb. 1 to 15	65.37
Feb. 16 to 28	66.12
Mar. 1 to 15	67.11
Mar. 16 to 31	68.32
Apr. 1 to 15	76.20
Apr. 16 to 30	88.12
May. 1 to 15	102.31
May. 16 to 31	110.15
Jun. 1 to 15	106.71
Jun. 16 to 30	101.21
Aug. 1 to 15	100.11
Aug. 16 to 31	99.13
Sep. 1 to 15	98.72
Sep. 16 to 30	86.52
Oct. 1 to 15	88.22
Oct. 16 to 31	78.27
Nov. 1 to 15	68.10
Nov. 16 to 30	64.10
Dec. 1 to 15	62.31
Dec. 16 to 31	62.12

Source : Singh, R.B., Gobar Gas: A Fuel for the Future (In Hindi) Gobar Gas Research Station, Ajitmal, Etawah, 1975, p. 66.

Table 6.14 Effect of Seasonal Variation in Northern India on Biogas Yield from a Plant Utilising Cow Wastes as Feedstock

Indicator	*Unit*	*Value of Indicators in Different Months*											
		Jan	*Feb*	*Mar*	*Apr*	*May*	*Jun*	*Jul*	*Aug*	*Sep*	*Oct*	*Nov*	*Dec*
Average temperature	°F	36.8	61.4	71.9	76.3	82.4	86.6	83.2	82.1	80.6	78.5	66.7	62.6
Wastes quantity	lb	4.99	4.44	4.85	3.66	4.74	4.58	4.91	4.84	4.59	4.72	4.57	4.90
Water quantity	lb	4.99	4.44	4.85	3.66	4.74	4.58	4.91	4.84	4.59	4.72	4.57	4.90
Biogas yield	ft^3	2.67	2.95	3.74	4.96	5.83	6.51	6.39	6.11	5.54	5.29	4.02	3.53
Biogas yield	ft^3/lb	0.53	0.66	0.77	1.06	1.22	1.42	1.33	1.26	1.20	1.12	0.88	0.11

Source : Singh, R.B., Gobar Gas: A Fuel for the Future (In Hindi), Gobar Gas Research Station, Ajitmal, Etawah, 1975, p. 71 (Based on a study report of IARI, New Delhi).

Table 6.15 Effect of Temperature and Retention Periods on Biogas Yield

Temperature °F	*Biogas Yield in ft^3 per lb of cattle feed at Increasing Retention Periods in Days*					
	4th	*8th*	*10th*	*15th*	*20th*	*30th*
95.0	1.40	2.30	2.58	2.86	2.95	3.05
104.0	2.33	2.98	3.20	3.67	3.85	4.10
105.8	2.49	3.09	3.25	3.71	3.95	4.20
107.8	2.59	3.14	3.33	3.74	3.95	4.20
109.4	2.58	3.09	3.26	3.67	3.82	4.09
113.0	2.08	2.55	2.70	3.05	3.14	3.36
118.4	2.64	3.14	3.30	3.57	3.82	4.06
127.4	3.33	3.88	4.07	4.29	4.51	4.70
132.8	3.74	4.20	4.35	4.54	4.74	4.88
.0	4.23	4.49	4.57	4.79	4.93	4.86

ırce : Singh, R.B., Gobar Gas: Fuel of the Future, Gobar Gas Research Station, Ajitmal, Etawah, 1975, p. 47.

Table 6.16 Effect of Temperature on Biogas Composition

Retention Period (days)	*Biogas Composition in Per cent at Two Different Temperatures*			
	At 95°F		*At 140°F*	
	CH_4	*CO_2*	*CH_4*	*CO_2*
4	69.7	30.3	53.5	46.5
4	69.8	30.2	55.0	45.0
6	64.3	35.7	51.9	48.1
8	58.6	41.4	50.2	49.8
10	57.2	42.8	52.3	47.7
15	53.8	46.2	53.5	46.5
20	53.4	46.6	49.1	50.9
30	53.8	46.2	53.7	46.3

Source : Singh, R.B., Gobar Gas: A Fuel of the Future, Gobar Gas Research Station, Ajitmal, Etawah, 1975, p. 48. (Based on findings of a study made by Prof. J. Fotter of the Illinois University, USA).

Table 6.17 Effect of Monthly Variation on Biogas Yield from a Set of Different Digesters

Type	*Average Daily Biogas Yield in Litres per kg of Feed*												*Mean Daily Yield (l/kg of feed)*
	Jan	*Feb*	*Mar*	*Apr*	*May*	*Jun*	*Jul*	*Aug*	*Sep*	*Oct*	*Nov*	*Dec*	
A	30.06	35.71	33.74	37.86	41.18	38.83	42.00	42.13	40.69	41.40	36.68	32.11	37.67
B	30.35	35.62	39.45	40.28	46.00	40.55	41.04	42.54	49.69	41.01	36.19	35.05	39.03
C	31.27	33.50	31.17	31.60	34.55	33.70	33.43	40.72	17.70				
D	39.18	43.16	42.57	41.73	48.90	47.03	49.60	6.27	43.56	42.11	40.61	36.08	43.32
E	24.42	30.52	36.00	37.24	43.66	41.79	40.66	38.51	38.19	33.96	33.51	31.16	35.79
F	42.04	43.15	41.47	40.65	49.75	47.63	49.72	49.96	46.09	46.49	42.94	40.72	45.05

Table 6.18 Effect of Monthly Variation on Daily Biogas Yield (m^3) from a Set of Different Digesters

Type	*Average Daily Gas Yield in m^3*												*Mean Daily Yield (m^3)*
	Jan	*Feb*	*Mar*	*Apr*	*May*	*Jun*	*Jul*	*Aug*	*Sep*	*Oct*	*Nov*	*Dec*	
A	2.40	2.85	2.69	3.02	3.29	3.07	3.36	3.33	3.45	3.73	3.30	2.89	3.12
B	2.46	2.85	3.05	3.22	3.68	3.24	3.28	3.40	3.37	3.69	3.25	3.15	3.21
C	1.87	2.01	1.87	1.89	2.07	2.02	2.67	3.25	1.50				
D	3.17	3.45	3.40	3.34	3.91	3.76	3.96	3.70	3.70	3.79	3.65	3.24	3.59
E	1.95	2.47	2.78	2.97	3.49	3.34	3.25	3.08	3.24	3.05	3.01	2.80	2.98
F	3.36	3.35	3.31	3.25	3.98	3.81	3.97	3.99	3.91	4.18	3.86	3.66	3.80

Table 6.19 Effect of Monthly Variation in Daily Biogas Yield Per Unit of Digester Volume from Different Capacity Digesters

Type	*Average Daily Gas Yield per Unit of Digester Volume (m^3/m^3)*												*Mean Daily Yield (m^3/m^3)*
	Jan	*Feb*	*Mar*	*Apr*	*May*	*Jun*	*Jul*	*Aug*	*Sep*	*Oct*	*Nov*	*Dec*	
A	0.33	0.38	0.36	0.41	0.44	0.41	0.45	0.45	0.47	0.50	0.45	0.39	0.42
B	0.36	0.42	0.45	0.47	0.54	0.48	0.48	0.50	0.50	0.54	0.48	0.46	0.47
C	0.25	0.26	0.25	0.25	0.27	0.26	0.35	0.43	0.20	NA	NA	NA	
D	0.42	0.46	0.45	0.44	0.52	0.50	0.53	0.49	0.49	0.51	0.49	0.34	0.47
E	0.21	0.27	0.30	0.32	0.38	0.36	0.35	0.33	0.35	0.33	0.33	0.30	0.32

Note: Digester volume A: 7.34 m^3, B: 6.72 m^3, C: 7.48 m^3, D: 7.43 m^3, E: 9.11 m^3, F: 8.04 m^3

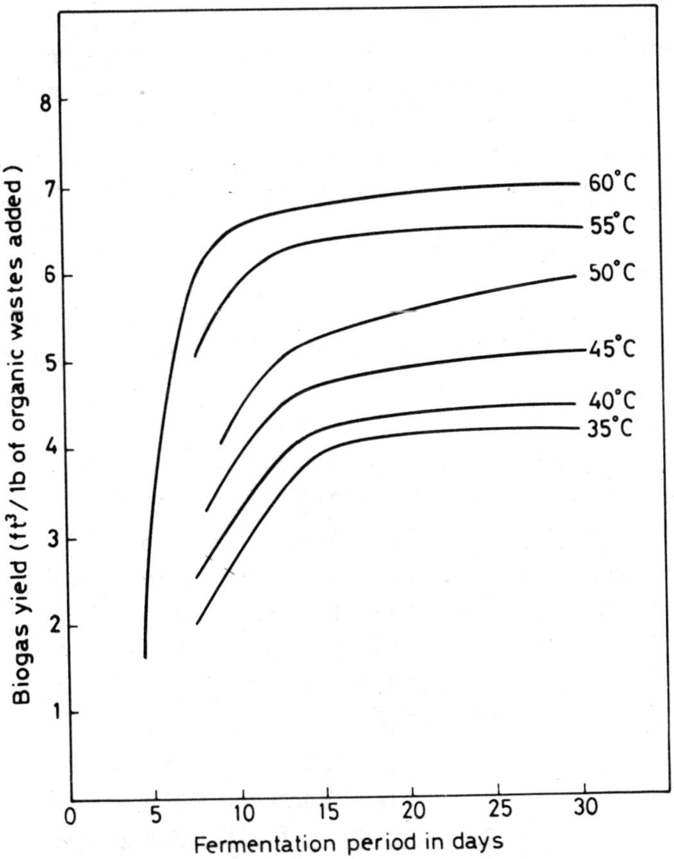

Fig. 6.2 Effect of digester temperature on biogas yield

Table 6.20 Monthwise Mean Ambient Temperature (°C) in Important Indian Cities that Affect Biogas Yield

S. No.	City	Jan	Feb	Mar	Apr	May	Jun	Jul	Aug	Sep	Oct	Nov	Dec
1.	Agra	14.8	18.0	23.8	29.7	34.5	35.0	30.9	29.3	28.9	26.2	20.6	16.1
2.	Ahmedabad	20.3	22.7	27.1	31.3	33.5	32.7	29.5	28.2	28.7	28.4	24.5	21.1
3.	Ajmer	15.0	17.2	22.8	27.8	32.1	31.9	30.4	28.7	27.8	23.9	19.4	15.7
4.	Allahabad	16.4	19.1	25.1	30.7	34.7	34.3	30.1	29.1	29.0	26.5	21.1	17.1
5.	Bangalore	20.9	23.1	25.7	27.3	26.9	24.3	23.2	23.3	23.3	23.2	21.7	20.5
6.	Baroda	20.5	22.5	26.6	30.8	33.4	32.1	28.9	28.1	28.3	27.5	23.9	21.2
7.	Bombay	24.3	24.9	26.9	28.7	30.1	29.1	27.5	27.1	27.4	28.3	27.5	25.9
8.	Calcutta	20.2	23.0	27.9	30.7	31.1	30.4	29.1	29.1	29.2	27.9	23.9	20.6
9.	Darjeeling	6.1	7.7	11.3	14.4	15.7	17.0	17.6	17.6	17.3	15.1	11.3	8.1
10.	Dehradun	12.6	14.8	19.4	24.5	28.9	29.5	26.7	26.1	25.5	22.1	17.5	13.9
11.	Gwahati	17.5	23.9	27.4	30.3	32.5	29.1	26.1	25.8	25.7	25.1	22.3	29.6
12.	Hyderabad	17.5	19.5	23.3	25.9	26.9	28.1	28.9	29.0	28.7	26.3	22.3	18.7
13.	Indore	17.9	19.9	24.5	29.1	32.3	30.1	26.1	25.1	25.1	24.1	20.5	18.3
14.	Jaipur	15.1	18.1	23.2	28.7	33.2	33.3	29.9	28.1	28.1	25.7	20.5	16.7
15.	Jammu	13.3	15.8	20.6	26.8	32.4	34.1	30.7	29.1	28.6	25.4	19.8	15.2
16.	Jamshedpur	19.3	21.7	26.7	31.2	33.4	31.7	28.9	28.7	28.7	26.9	22.3	19.2
17.	Jodhpur	17.0	19.8	25.0	29.8	34.0	33.9	32.4	30.5	30.3	27.0	22.4	18.4
18.	Kanpur	16.2	18.8	25.0	30.2	33.8	34.3	31.5	29.8	29.4	25.5	20.6	16.8
19.	Kodaikanal	12.5	13.3	14.6	15.7	16.5	15.3	14.3	14.5	14.5	13.9	12.9	12.5
20.	Lucknow	16.1	18.9	24.6	30.1	33.9	33.7	30.1	29.3	29.1	26.3	21.0	16.9
21.	Madras	24.5	25.9	27.9	30.5	32.7	32.5	30.7	30.1	29.7	28.1	25.9	24.6
22.	Mahabaleshwar	18.9	20.5	23.1	24.1	23.5	19.4	17.9	17.6	18.0	20.1	19.5	18.7
23.	Mussorie	6.3	7.8	11.7	16.5	19.9	20.3	18.3	17.9	17.1	14.9	11.6	8.5
24.	Mysore	22.3	24.7	26.9	27.7	23.5	24.5	23.5	23.7	24.0	24.0	22.9	21.7
25.	Nagpur	19.6	22.8	27.3	31.3	34.5	33.2	28.9	27.6	24.2	24.2	22.2	19.6
26.	New Delhi	14.3	16.9	22.7	28.6	32.5	34.3	31.3	29.9	29.3	25.9	20.3	15.7
27.	Ootacamund	12.0	12.9	14.4	15.9	16.7	15.5	14.6	12.4	14.3	13.5	12.4	12.2
28.	Patna	17.3	19.9	25.7	30.5	29.9	31.9	29.8	29.3	29.3	27.5	22.5	18.3
29.	Poona	21.3	23.1	26.5	29.3	19.6	27.5	24.9	24.6	25.0	25.5	22.9	21.1
30.	Shillong	9.5	11.7	16.0	18.9	17.9	20.5	21.1	20.9	20.1	17.3	13.3	10.5
31.	Shimla	15.5	6.7	10.1	14.2	18.2	19.4	18.8	18.3	16.7	14.1	11.1	8.3
32.	Srinagar	1.1	3.5	8.5	13.3	28.3	21.7	24.6	23.9	20.5	14.1	7.7	3.5
33.	Trivandrum	26.8	27.3	28.3	28.7	26.9	26.5	26.1	26.3	26.6	26.7	26.6	26.7
34.	Udaipur	16.0	18.7	23.7	28.1	31.7	30.6	27.3	26.1	26.5	25.5	20.1	17.3

Optimum gas production occurs in two temperature ranges. Mesophilic bacteria thrive in temperatures around 95°F (35°C), and thermophilic bacteria in the range 120°F to 140°F (49°C to 60°C). Fig. 6.3 shows that gas production decreases when the bacteria are subjected to temperature variations beyond these ranges [7]. While thermophilic bacteria produce somewhat more gas, often the extra gas is not worth the energy needed to raise the digester temperature from 95°F (35°C) to 120°F (49°C). In China average biogas yield during summer lies somewhere between 0.2 to 0.3 m^3 per m^3 of digester volume. As far as yield during winters is concerned, it is significantly different in northern and southern parts of China. During winters in southern areas where ambient temperature ranges between 12 to 13°C, daily gas yield varies from 0.07 to 0.08 m^3 per m^3 of digester capacity. However, during winters in northern parts where atmospheric temperature lies between 8 to 10°C which are much colder than southern China, biogas yield generally varies between 0.03 to 0.05 m^3/m^3 of digester volume. On an average for a period of four months in northern parts, due to exceedingly low temperatures, biogas yield from family units is scarcely enough to meet the fuel needs of the households.

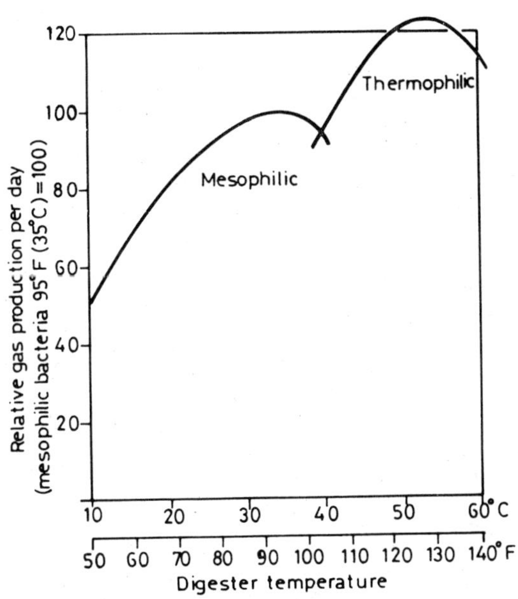

Fig. 6.3 Effect of temperature on gas production rate. (Roediger, H. Die anaerobe alkalische Schlammfaulung. *Wasser-Abwasser*, H.1, Verlag R. Oldenbourg, Muchen u. Wien. 1967.)

S. Neelakantan, *et al.* [15], carried out experiments at the National Dairy Research Institute, Karnal to analyse the effect of temperature variation on anaerobic fermentation of cattle wastes. For this purpose, 200 g of dried dung powder, 1 kg of dung, 100 ml of digested slurry inoculum were mixed with 1.8 litres of water to prepare input slurry for fermentation. Volume of gas produced was measured by water displacement method. Gas yield was studied during four blocks of two months each, namely, Oct-Nov, Dec-Jan, Mar-Apr and May-Jun when ambient temperature ranged between 20-25°C, 13-18°C, 21-20°C and 32-35°C in the city of Karnal. During these periods, total gas yield was measured as 13.81, 7.84, 14.96 and 17.97 litres; with per cent digestion in first month of each block as 64.4, 64.4, 71.4 and 71.4, respectively.

6.10 Effect of Inhibitory Factors and Materials Affecting Microbial Activity

There are several factors and class of materials which act as inhibitor to bacterial activity during anaerobic fermentation. When these chemicals and factors reach certain level, bacterial activity almost stops thereby severely affecting gas yield. For example, when volatile acid concentration reaches a value of 200 ppm, or ammoniacal nitrogen concentration exceeds a value of 1500 ppm, microbial activity is greatly retarded. A retarded microbial activity ultimately results in low quality gas with low methane content in relation to other constituents. There are several factors which contribute to excessive acid formation such as higher proportion of acid forming over methanogenic bacteria in substrate, high acid contents in feed, delays in addition of fresh feed and delayed withdrawal of sludge and improper feed composition. Problem of excessive fatty acid formation can be minimised by having a proper composition of feedstock along with adequate mix of some sludge seeding bacteria. Based on studies carried out in China [18], commonly experienced proportions of biogas constituents and biochemical indices of intermediate products that serve to indicate faulty digester operation are given in Table 6.21.

Table 6.21 Commonly Experienced Proportions of Biogas Constituents and Biochemical Indices of Intermediate Products that Serve to Indicate Faulty Digester Operation

Digester	*Gas Constitution in Per Cent*				*Volatile Fatty Acids*				*Ammoniacal Nitrogen*	*Acetic Ether (ppm)*	*Acetone (ppm)*	*pH*
	CH_4	CO_2	N_2	O_2	*Acetic Acid*	*Propoinic Acid*	*Butyric Acid*	*Total Acids*				
1	50.1	29.8	4.1	0.8	2727.0	441.0	100.0	3268.0	418.6	9.1	74.8	7.48
2	43.6	11.0	26.7	4.4	2640.0	630.4	163.4	3433.8	542.6	52.0	23.9	7.50
3	68.6	12.3	1.7	0.3	3970.0	721.0	106.0	4797.0	612.6	23.0	78.9	7.60
4	60.5	18.8	4.5	1.0	3959.2	304.0	459.0	5722.0	531.6	8.8	29.8	7.30
5	43.6	26.9	12.9	1.1	2871.0	800.7	84.7'	3756 1	1048.6	26.0	50.6	7.40

Source : SPOBD, Biogas Technology and Utilisation, Chengdu Seminar, Sichuan Provincial Office of Biogas Development, Sichuan, China, 1979.

Table 6.22 Moderately Inhibitory Range of Concentration of Harmful Materials that Tend to Retard Anaerobic Fermentation Process

Inhibitor	*Concentration Range*	
ABS (Detergent Compound)	20-40	ppm/l
Sulphates (SO_4)	5000	ppm
Sodium Chloride (NaCl)	40000	ppm
Nitrates (NO_3)	0.05	mg/ml
Copper (Cu)	100	mg/ml
Chromium (Cr)	200	mg/ml
Nickel (Ni)	200-500	mg/ml
Sodium (Na)	3500-5500	mg/l
Potassium (K)	2500-4500	mg/l
Calcium (Ca)	2500-4500	mg/l
Magnesium (Mg)	1000-1500	mg/l

ppm : parts per million

Table 6.23 Stimulatory and Inhibitory Concentrations of Sodium, Potassium, Calcium and Magnesium

Element	*Concentration in mg/l*		
	Stimulatory	*Moderately Inhibitory*	*Strongly Inhibitory*
Sodium	100-200	3,500-5,500	8,000
Potassium	200-400	2,500-4,500	12,000
Calcium	100-200	2,500-4,500	8,000
Magnesium	75-150	1,000-1,500	3,000

Table 6.24 Dilution Norms to Minimise Effect of Toxicity Due to Excessive Ammonia and Volatile Fatty Acid Formation

Source of Manure	*Dilution Expressible as Ratio of Manure (Manure + Water)*
Swine	1 : 2.9
Dairy	Undiluted
Beef, 700 lb (320 kg) cow	1 : 2.5
Poultry (layer)	1 : 2.5
Poultry (broiler)	1 : 10.2

A variety of materials act as inhibitory to the fermentation process. For instance, presence of certain metals such as copper in waste materials act toxic during anaerobic digestion. Some of the commonly recognised toxic materials include common alkali and alkaline-earth cations such as sodium, potassium, calcium and magnesium. Traces of pesticides such as DDT and chlorinated hydrocarbons which usually accompany crop residues as feed tend to produce toxic effect during anaerobic fermentation. For minimising toxicity, dilution with water or addition of certain non-toxic materials to substrate is recommended. In this context it should be noted that harmful effect of certain toxic materials is generally influenced by their concentration. Normally when a material is present in low concentration, its presence may be stimulatory but when this concentration exceeds certain value, its effect becomes toxic. When poultry manure is used as feed, toxicity on account of excessive ammonia formed can be minimised by adjusting C : N ratio by adding shredded straw or bagasse which have a high C : N ratio, or simply by diluting the slurry with water. Moderately inhibitory range of concentration of toxic materials is given in Table 6.22. Stimulatory, moderately inhibitory and strongly inhibitory concentrations of elements like sodium, potassium, calcium and magnesium is given in Table 6.23. When inhibitory materials are present in excess, dilution of digester contents by water is generally found helpful in limiting their toxic effect. A set of prescribed dilution norms in respect of some animal wastes to minimise effect of toxicity due to undesirable concentration of ammonia and fatty acids are summarised in Table 6.24 [18].

6.11 A Case Study of Plant that Works at Low Temperature

G. Kopiske and H. Eggersgliib [9] studied performance of low temperature biogas plants. According to their findings at low temperature biogas plants with some design modifications can also function quite effectively as in warm climate. The apparently long growth phase can be reduced considerably by intensive initial inoculation. Thus the inert biological process will not take a long time to reach an optimal steady state. However, individual planning and effort is necessary to make low temperature systems work economically in practice. Their experience is based on a plant built by the UTEC GmbH in erstwhile Federal Republic of Germany. The plant consisted of a sublevel concrete tank of 10 m diameter and 4 m depth. It also had facility for insulation, gas tightness, heating and agitation. The digester was fed by natural slope as well as by pumping slurry from a small distant tank. Gas storage was realised with a 20 m^3 CPE bag inside a converted steel tank. Biogas was burnt in a central atmospheric gas burner for heating and hot water supply. Fig. 6.4 shows a simple sketch of the system. Insulation was provided by 12 cm polystyrol at roof and wall segment. Heating was provided through 120 m PPH-pipes mounted at the wall. Agitation was provided by 11 kW, 281 rpm. propeller mixer.

Figure 6.5 shows impact of ambient temperature in different months on yield and composition of biogas. It was attempted to initiate the growth phase without inoculation. After three months, gas yield stagnated at 15 l/kg of VS at an average temperature of 18°C. The plant was then inoculated with inoculum from another digester. After initial intensive mixing the agitation system was worked for 60 minutes/day. At that time heating system was not used. In spite of sinking process temperature when it fell down to 10°C, the gas yield rose upto 100 lit/kg of VS. The digester was then heated upto 15°C to support the multiplication of the bacteria. Entire quantity of gas was needed to warm up the farm house during winter months. Accordingly no heat was available for the digester and the process temperatures again fell to 12°C. Gas production did not increase further. The pH level of the substrate stayed constant at 7.1. Laboratory tests showed a good fermentation ability of the substrate in mesophilic temperature range.

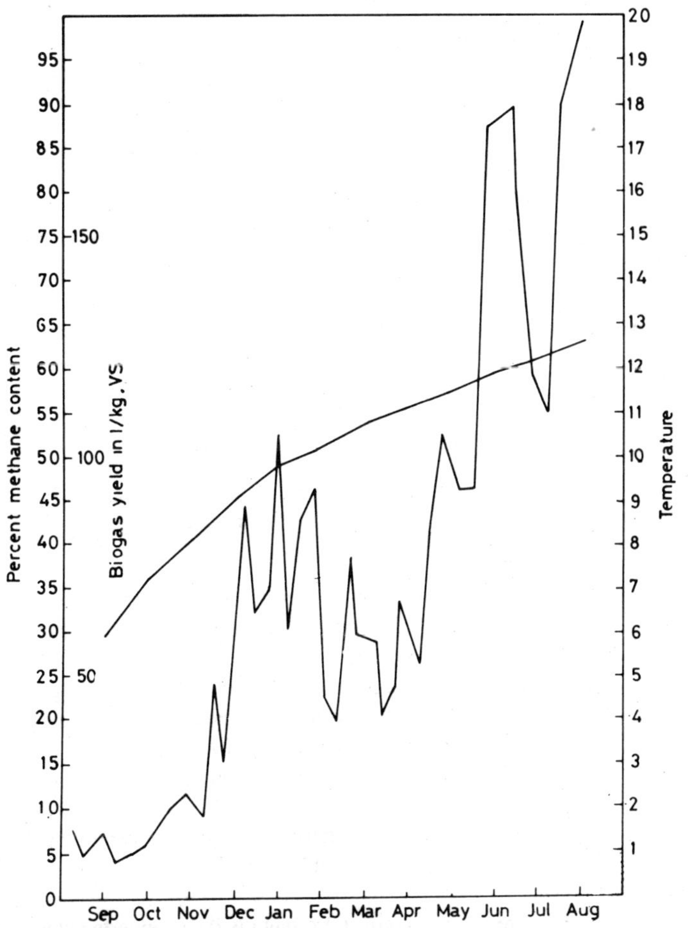

Fig. 6.5 Variation in temperature, methane content and biogas yield during different months

For analysing scum-formation, it the plant was operated without agitation. Considerable scum was found after 2 months with a thickness of 1.5 m. The effective retention time was thus reduced greatly. To overcome this problem the scum was mixed up again with the installed agitator which proved to be too ineffective to mix the complete digester volume. Effective agitation is required to achieve satisfactory inoculation of the fresh slurry. This is necessary to accelerate the start of gas production. Due to this reason, a new agitator was installed. Following improved agitation, gas yield was found to rise upto 208 l/kg of VS.

As a result of experimentation with a low temperature biogas plant in Wohlenhausen, following conclusions were derived [9].

- Fermentation process is sensitive to temperature variations and various additives.
- Simple regulation systems are adequate for achieving normal gas yield.
- Multiplication of bacteria population is a slow process requiring longer gas phase to reach an optimal operation.
- Process energy demand is low inspite of high agitation rate.
- Short term peak energy demand can be fulfilled with 100 per cent gas production.
- Big digester volume normally requires a fair amount of initial inoculum.

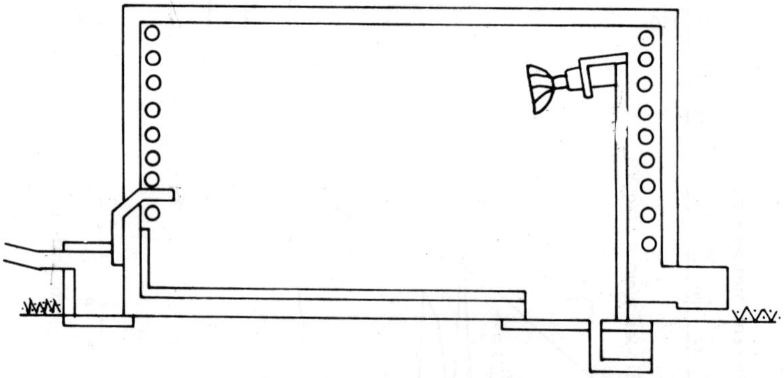

Fig. 6.6 Main component of digester and technical data

Digester	:	Converted concrete tank (10 m diameter, 4 m depth)
Insulation	:	12 cm polystyrol at roof and wall segment.
Heating	:	120 m PPH pipes mounted at the wall
Agitation	:	Propeller mixer, 11 kw, 281 rpm
Feeding	:	Natural slope and pumping from a distant tank
Gas Consumer	:	Atmospheric gas burner for central heating and hot water supply, 34 kw
Gas Storage	:	20 m^3 CPE bag inside a converted steel tank.

6.12 Concluding Remarks

Efficiency of anaerobic fermentation essentially depends on intensity of bacterial activity. For efficient performance of a biogas plant, it is therefore necessary to regulate factors which affect bacterial activity. For instance, there should be continuous availability of nutrients to sustain the growth of microorganisms. Similarly, formation of volatile fatty acids beyond a particular range should not be allowed to take place. Improper dispersal of nutrients within the digester or excessive gas pressure which tend to lower fermentation efficiency, should be properly controlled. In the same way, temperature, and pH of digester contents should also be made to lie between the desired range of values. Loading rate of feedstock and withdrawal rate of digested slurry should be properly balanced and continuously monitored. Based on Chinese studies addition of 4 per cent of fresh waste materials and withdrawal of 4 per cent of digested slurry every fifth day helps to maintain efficient gas production. During winters, relatively more quantity of waste materials are required to be added. Furthermore, to counter the effect of low temperatures, waste materials with more nutrients should be fed during winter periods [18]. As gas yield is found to drop down during winter months, it is often advantageous to cover the digester of fixed-dome type biogas plants with straw and carry out composting on digester surface wherever possible. If digester conditions are not well-buffered, acidity level within the digester tends to rise which in turn causes higher carbon dioxide content in biogas with the result that it burns with a yellow flame. To overcome this difficulty, it is often advisable to control digester acidity by adding lime water or ashes.

REFERENCES

1. Barnett, A., *et al.*, *Biogas Technology in the Third World: A Multidisciplinary Review*, IDRC, Ottawa, Canada, 1978, p. 51.
2. Chung, P.O., *et al.*, *Small Methane Generator for Waste Disposal,* Proc. Third Symp. on Livestock Wastes, American Society of Agricultural Engineers, 238-240 1974.
3. Coppinger, E.R., *The Operation of a 50,000 Gallon Anaerobic Digester at the Monroe State Dairy Farm,* Ecotope Group, 2332, East Madison, Seattle, WA, June, 1979.
4. Fry, L.J. and Merrill, Methane Digesters for Fuel Gas and Fertiliser, *Newsletter* No **3**, New Alchemy Institute, Santa Cruz, California, 1973.
5. Gore, J.A., *et al.,* More Gas from Gobar Gas Plants, *Indian Farming,* March, 1981.
6. Gotass, H.B., *Composting*, W.H.O., Geneva, Switzerland, 1956.
7. Homan, E., *Biogas from Manure,* Special Circular 260, College of Agriculture Extension Service, Pennsylvania State University, Pennsylvania, 1979.
8. Ke-xin, L., *et al., Fermentative Technology of Chinese Rural Digester,* Bio-energy Laboratory, Chengdu Institute of Biology, Academia Sinica, P.R. China, 1979.

9. Kopiske, G. and Eggersgliib, in 'Strub, A., Charter, P. and Schleser, G. (eds), *Energy from Biomass,* Proc. E.C. Conference, Applied Science Publishers, London, 1982,' 591-594.
10. Lapp, H.M., *et al.*, Methane Production from Animal Wastes: Fundamental Considerations, *Canadian Agricultural Engineering,* **17(2)**, Dec., 1975, p. 98.
11. Mehta, B.J., *et al.*, Effect of Salinity on Biogas Production, *Research and Industry,* **33(1)**, 16-18 1988.
12. Mohanrao, G.J., Scientific Aspects of Cow Dung Digestion, *Khadi Gramodyog,* **20(7)**, 340-347 1974.
13. Mohanrao, G.J., *Aspects of Night Soil Digestion, Sewage Farming and Fish Culture,* A Working Paper, All India Institute of Hygiene and Public Health, 1975.
14. NAS, *Methane Generation from Human, Animal and Agricultural Wastes,* Report of an Ad. Hoc. Panel of the Advisory Committee on Technology Innovation, National Academy of Sciences, Washington, DC, 1977.
15. Neelakanthan, S., *et al.*, Effect of Temperature, Inoculum and Agitation on Biogas Production from Cattle Dung, *Indian Journal of Dairy Science,* **29(3)**, 226-229 1976.
16. Singh, R.B., *Biogas Plant: Generating Methane from Organic Wastes,* Gobar Gas Research Station, Ajitmal, Etawah, 1974, p. 33.
17. Singh, R.B., *Gobar Gas: A Fuel for the Future* (in Hindi), Gobar Gas Research Station, Ajitmal, Etawah, 1975, p. 66.
18. SPOBD, *Biogas Technology and Utilisation,* Chengdu Seminar, Sichuan Provincial Office of Biogas Development, Sichuan, P.R. China, 1979.
19. Srivastava, V.J. and Chynoweth, *D.P., Kinetic Analysis of Biogasification of Biomass, Waste Blend and its Engineering, Significant Energy from Biomass and Wastes,* Elesevier, 1021-1034 1987.
20. TERI, *Fixed Dome Biogas Plants: A Design, Construction and Operation Manual,* TERI, New Delhi, 1987, p. 3.
21. UN., *Guidebook on Biogas Development,* United Nation is, New York, 1980, p. 99.

Chapter 7

BIOGAS YIELD FROM DIFFERENT ORGANIC WASTES

7.1 Introduction

A number of studies have been carried out world over concerning biogas production from different animal wastes and agricultural crop residues using different additives under varying operating conditions. Plant materials generate considerably higher biogas yield per kilogram of total solids than several animal wastes. Biogas yield depends on several factors such as digestion temperature, C : N ratio, pH of digester contents, loading concentration, characteristics of manurial solids and operating conditions. Wastes generated by animals differ on account of feed ration and accordingly due to non-availability of requisite feed, cattle in Indian subcontinent generate one third manure as compared to those in several European countries. Vegetable matter from young plants generates more gas than from old plants, and dry vegetable matter generates more gas than green vegetable matter [1]. Pretreatment by grinding of plant materials markedly improves both the rate of biogas production and the total gas yield. Different plant materials and manures produce biogas at different rates [25].

7.2 Comparative Yield from Different Organic Wastes

The more complex a substrate the longer it takes to be biodegraded. J.L. Burford and F.T. Varani [5] reported composition of biogas and from various components of agricultural plant matter and manures (Table 7.1). Batch digestion is quick and inexpensive method of testing comparative yield from different waste materials in terms of concentrations and digestion characteristics [32]. In seventies, U.P. Sharma and T.M. Paul [29] carried out studies at the National Dairy Research Institute, Aarey Milk Colony, Bombay on comparative biogas yield and composition from different animal manures using Orsat method (Tables 7.2 and 7.3). In case of poultry droppings, gas production was very quick which started within three hours of supplying feed whereas in case of cow and buffalo wastes it took from one to two weeks for biogas generation to begin.

Table 7.1 Comparative Biogas Yield and Composition from Carbohydrates, Fat and Proteins

Component	*Chemical Formula*	*Digestion Products*	*Methane Content (Per Cent)*	*Biogas Yield (l/kg VS destroyed)*
Carbohydrate	$(C_6H_{10}O_5)_n$	CH_4+CO_2	50	886
Fat	$C_{50}H_{90}O_6$	CH_4+CO_2	70	1535
Protein	$6C{\cdot}2NH_3{\cdot}3H_2O$	$CH_4+CO_2+(NH_4)_2CO_3$	84	587

Source : 1. Burford, J.L. and Varani, F.T., Energy Potential Through Bioconversion of Agricultural Wastes, A Report, Colorado Energy Research Institute, Colorado.

2. Trevelyan, W.E., The Methane Fermentation: A Discussion Paper, Tropical Science, 17 (4), 193-209 (1975).

Table 7.2 Composition of Different Animal Wastes used in Analysing Their Effect on Biogas Yield and Composition

Organic Waste	*Dry Matter (per Cent)*	*Volatile Solids on Dry Weight Basis (Per Cent)*	*Fat (Per Cent)*	*Crude Protein (Kjeldahl Per Cent)*	*Crude Fibre (Per Cent)*	*Ash (Per Cent)*
1. Cow wastes	18	84	2.4	9.2	21	16
2. Buffalo wastes	19	83	3.0	9.9	19	17
3. Horse wastes	32	82	3.0	10.1	25	18
4. Sheep droppings	32	82	3.8	17.1	13	19
5. Camel wastes	32	84	3.3	10.8	16	16
6. Goat droppings	32	78	3.2	12.5	13	22
7. Pig wastes	45	72	4.5	19.4	26	28
8. Poultry droppings	47	65	1.5	26.9	19	34

Table 7.3 Comparative Biogas Yield and Gas Composition from Different Kinds of Animal Wastes

Organic Waste	*Biogas Yield*			*Biogas Composition*	
	l/kg Wet Solids	*l/kg Dry Solids*	*l/kg VS*	*Methane (Per Cent)*	*Carbon Dioxide (Per Cent)*
1. Cow wastes	12	60	72	50	36
2. Buffalo wastes	12	65	79	50	35
3. Horse wastes	31	91	111	55	30
4. Sheep droppings	31	91	113	54	28
5. Camel wastes	34	100	119	53	30
6. Goat droppings	37	106	135	53	23
7. Pig wastes	47	107	149	56	26
8. Poultry droppings	74	126	191	56	25

7.3 Biomethanation as part of an Integrated Farming System

M. Mahadevaswamy and L.V. Venakataram [19] examined the feasibility of an integrated farming system for bioconversion of poultry droppings to produce biogas and utilisation of the effluent to produce blue green algae *spirulina platensis*. Poultry droppings when used as feed yielded 8.64 ft^3/lb of TS (0.54 m^3/kg of TS). An integrated bioenergy system using 2 per cent total solids as the sole nutrient medium for *spirulina platensis* produced 0.15-0.17 lb (7-8 g) dry blue algae per day. Algae was harvested by filtration and the sun-dried algal biomass used as a poultry feed component. M. Mahadevaswamy and L.V. Venkataraman [2C] also described an integrated system for biogas production from rabbit droppings and utilisation of the effluent feeding major carp (*Rahu* and *Labeo rohita*) and common carp (*Cyprinus carpio*). Rabbit droppings produced 0.24 m^3 of biogas/kg of total solids. Biogas effluent when used at the rate of 30 kg/100 m^2 in ponds as an exclusive source of feed for carp yielded 8.9 kg/100 m^3/120 days of fish with a desired combination of colour, flavour and taste.

7.4 Effect of Additives on Biogas Yield

D.B. Madamwar and B.M. Mithal [18] carried out two sets of experiments, one at controlled temperature of 100°F (38°C) and the other at ambient temperatures 60-104°F (15-14°C) to study the impact of adding pectin to cattle dung slurry as feed on biogas yield. Pectin not only enhanced gas yield but also imparted process stability during periods of fluctuating temperature. It did not enter into the digestion process but as an absorbent to accumulate substrate for creating areas of high localised concentration for greater bacterial growth and consequent higher gas yield. S. Kumar *et al.* [15] analysed the effect of adding commercial charcoal to cattle-dung slurry as feed on biogas yield. Addition of 5 per cent commercial charcoal to cow dung slurry on dry weight basis raised the yield by 17 and 35 per cent in batch and semi-continuous process respectively. S.S. Geeta *et al.* [9] studied the impact of adding inert materials like vermiculite, charcoal and lignite to bovine excreta as feed on biogas yield. These additives were found to increase biogas yield by 15 to 30 per cent. Pebbles, glass marbles and plastic mesh when suspended in digester slurry reportedly led to increase in gas yield by 10-20 per cent.

C.R. Prasad [23] studied the effect of adding bagasse, Gulmohar leaves (*Delomix ragia*) wheat straw, groundnut shells and leguminous plant leaves as additives to cow dung on biogas yield, gas composition and extent of biodegradation. These additives were separately mixed with cow dung in the ratio of 1 part (oven dry) to 10 parts of fresh dung containing 19.2 per cent of total solids on weight basis. Anaerobic fermentation was carried out under batch process in bottles in laboratory at ambient temperatures between 30-32°C for 9 weeks. Volumes of biogas generated in 24 hours was measured every day and gas composition analysed periodically. Biogas yield, per cent decomposition, and methane and carbon dioxide content in gas from cow dung and individual additives comprising bagasse, Gulmohar leaves, wheat straw, groundnut shells are given in Table 7.4. It indicates that it is advantageous to use these materials as additives for obtaining higher gas yield.

Table 7.4 Effect of Additives on Feed Decomposition, Biogas Yield and Gas Composition

Feed	*Per Cent Decomposition*	*Biogas Yield (litre/g of feed)*	*Composition*	
			CH_4	CO_2
Cow dung	26.1	196.9	57.6	40.2
Cow dung + bagasse	44.4	260.4	60.0	39.2
Cow dung + Gulmohar leaves	45.1	240.9	60.0	39.2
Cow dung + wheat straw	44.5	265.4	51.2	48.0
Cow dung + groundnut shells	25.1	137.0	60.0	40.0

Based on experimental studies, K.V. Lo, *et al.* [17] reported that digestion of a mixture of cheese whey and dairy manure under mesophilic conditions in a rotating biological contact reactor with a hydraulic retention time of less than 5 days yielded 3.74 litres of biogas per day. The resulting effluent was found to be high in COD content. O. Odeyemi [22] studied biogas yield from poultry droppings, sheep manure, sewage and cassava leaves. In a batch loaded laboratory digester biogas yield in a 21-day cubation period was reported to be 93 litres/kg from poultry droppings, 30 from sheep manure and 60 from sewage. Cassava leaves generated thrice the volume of methane from sheep manure and more than three times that obtainable from cow dung; hence it is a valuable feed for biomethanation in humid tropics which supports cassava cultivation.

7.5 Biogas Yield from Crop Residues

A.C. Varshney, *et al.* [35] studied biogas potential of certain crop residues such as castor shells and leaves, pearl millet husk, mustard straw and potato haulm. M.S. Kalra and J.S. Pawar [14] carried out experimental studies to investigate biodegradability of rice-straw which caused biogas yield of 4 ft^3/lb (200 litre/kg) under batch digestion from experimental digesters of 50 gallon (190 litres) capacity. Rice-husk did not undergo complete digestion following its high lignin content and unfavourable non-lignin carbon-to-nitrogen ratio. A mixture of rice-straw and cattle dung in equal proportion on dry weight basis yielded 9.1 per cent more gas than rice-straw alone. A. Rahman, *et al.* [24] reported yield in batch loaded laboratory digesters from rice-straw, sawdust and water-hyacinth inoculated with cowdung. Under identical conditions, rice-straw produced 310 litre biogas with 77.8 per cent methane content.

sawdust 144.8 litre and 62.28 per cent and water-hyacinth 368 litre with 77.13 per cent (methane) respectively. The pH of the digester contents was initially low and found to be decreasing as the digestion picked up. Total microbial populations of bacteria, fungi, *actinomycetes, cellulolytia* organisms and *coliform* bacteria were found to decrease with time.

D.M. Badger, *et al.* [3] carried out laboratory-scale studies for anaerobic fermentation of crops and organic wastes at loading concentrations of 3-10 per cent total solids. Concentration of nutrient elements in feed and resulting residue (sludge) following fermentation are summarised in Table 7.5. In general the concentration of an element as a percentage of the dry weight was higher in the residue than in the undigested material. This occurs on account of conservation of nutrient elements during digestion and gasification of some of the organic material which makes it rich fertiliser. Biogas yield decreased with increasing loading concentrations. Above 5 per cent total solids, digestion of crops with high soluble carbohydrate content needed continual additions of alkali to maintain a pH of 7 ± 0.5. Volatile solid contents, volatile solids destruction, yield and composition of biogas from anaerobic digestion of various crops and organic wastes at 5 per cent total solids loading concentration and 37°C after 17-36 days of digestion are given in Table 7.6. The composition of biogas varied with the material from 50 per cent methane from newsprint to 68 per cent from cattle manure. The balance was carbon dioxide in all cases with less than 0.001 per cent hydrogen sulphide. In another similar study carried out at the Kasetsart University in Bangkok, S. Tumwasron [33] reported biogas yield (Table 7.7) and gas composition (Table 7.8) from agricultural crop residues. Comparative nutritive values of residues before and after anaerobic fermentation are given in Table 7.9. S.A. Waksman [36] presented chemical composition of select agricultural crop residues for assessing their biodegradability (Table 7.10). With growing age of agricultural plants, sugars, amino acids, proteins, minerals etc. all of which are soluble in water tend to decline whereas non-water-soluble components like cellulose, hemicellulose and lignin tend to grow. Actual nutrient content with possible exception of nitrogen is though seldom low enough to limit the rate of anaerobic fermentation, biological availability of organic matter is usually a rate-limiting factor. The higher is proportion of lignin in substrate the less cellulose is available for bacterial action with the result that anaerobic fermentation becomes less efficient. In other words, presence of lignin largely controls the rate of anaerobic fermentation. Lignin is considered to be the main nonbiodegradable component of the substrate. Lignin is insoluble in water, ether, alcohol, dilute alkali and is insusceptible to bacterial attack. According to A.M. Buswell and W.D. Hatfield [6], low availability of nitrogen is usually a rate-limiting factor for anaerobic fermentation of nonleguminous plants. Nitrogen content in agricultural crop residues depends on type of species, age of plant, environmental conditions and availability of nitrogen-based soil nutrients. Amount of crop residues available can be estimated by multiplying expected or available harvest weights by what is called residue coefficients. If only a portion of available crop residues can be harnessed, only that part need be multiplied with residue coefficients. These coefficients (Table 7.11) are ratios of weight of dry matter of crop residues to weight of harvest measured at field moisture level [21]. For grains and straw, field moisture level can be assumed as 15 per cent.

Table 7.5 Concentration of Nutritive Elements in Agricultural Plant Materials Before and after Anaerobic Fermentation

Item	*Status*	*Concentration of Nutritive Elements (Per Cent of Total Solids or Dry Matter)*				
		Nitrogen	*Phosphorus*	*Potassium*	*Calcium*	*Magnesium*
Grass	Before	1.6	0.3	3.3	0.7	0.2
Sugarbeet	After	2.0	0.5	7.1	2.4	0.8
leaves	Before	3.7	0.3	4.4	1.4	0.5
	After	4.3	1.0	6.1	3.9	1.3
Kale	Before	2.5	0.2	3.1	2.4	0.4
	After	4.8	0.2	8.6	6.0	1.0
Maize	Before	1.8	0.2	2.0	0.5	0.3
	After	2.0	0.2	–	4.4	0.6
Oats	Before	1.4	0.2	2.8	0.6	0.2
	After	2.5	0.7	–	2.6	0.6
Wheat straw	Before	1.0	0.1	1.6	0.2	0.1
	After	1.2	1.0	2.1	0.6	0.3
Lakeweed	Before	3.1	0.6	2.1	1.8	0.3
	After	3.8	0.6	5.7	2.4	0.8
Lucerne	Before	3.4	0.2	3.4	0.8	0.2
	After	6.1	0.3	5.9	2.5	0.5

Table 7.6 Comparative Biogas Yield from a Set of Agricultural Plant Materials

Agricultural Plant Material	*Weight of TS (kg)*	*Weight of VS (kg)*	*Digestion Period (days)*	*VS Destroyed*		*Biogas Yield (l/kg of)*			*Methane Content (Per cent)*
				kg	*Per Cent*	*TS added*	*VS added*	*VS destroyed*	
1. Grass hay	1.00	0.93	30	0.61	73	462	497	587	54
2. Grass hay (ground)	1.00	0.93	18	0.59	70	507	546	678	54
3. Kale	1.00	0.89	36	0.60	73	440	493	733	60

Contd...

4.	Sugar beet leaves	1.00	0.77	21	0.55	79	380	450	691	66
5.	Maize	1.00	0.95	33	0.69	81	500	526	725	65
6.	Oats	1.00	0.86	18	0.50	66	470	546	940	54
7.	Wheat straw	1.00	0.94	31	0.68	79	412	440	620	58
8.	Wheat straw (ground)	1.00	0.94	18	0.78	91	530	564	780	58
9.	Lakeweed	1.00	0.79	17	0.44	62	380	481	864	56
10.	Cow wastes	1.00	0.83	36	0.29	39	216	260	745	68
11.	Sheep wastes	1.00	0.86	24	0.44	56	415	482	943	67
12.	Newsprint	1.00	0.99	25	–	–	235	237	–	50
13.	Lucerne hay	1.00	0.91	17	0.61	69	346	380	567	65

Note : Ambient temperature (37°C), total solids concentration (5 per cent)

Source : Badger, D.M., *et al.*, Biogas Production from Crops and Organic Wastes, New Zealand Journal of Science, 22, 11-20 (1970) [For both Tables 7.5 and 7.6]

Table 7.7 Biogas Yield Obtainable from Set of Agricultural Crops/Residues as Feedstock

Crop Residue	*Biogas Yield (m^3/kg DM)*	*pH*	*Temperature Ranges (°C)*
Rice straw	0.0056	6.0-6.8	33-44
Leucaena	0.0075	6.7-7.5	30-48
Paragrass	0.0073	6.5-7.8	32-44
Cucumber leaves	0.0052	6.0-7.1	31-45
Duck weed	0.0054	6.8-7.7	32-44
Corn top	0.0054	6.3-7.0	31-43
Water hyacinth	0.0203	6.8-8.0	31-44

Table 7.8 Composition of Biogas Obtainable from Set of Agricultural Crops/Residues as Feedstock

Crop Residue	*Composition in Per Cent*					
	Methane	*Carbon Dioxide*	*Oxygen*	*Hydrogen*	*Nitrogen*	*Hydrogen Sulfide*
Rice straw	23	25	1.2	0.4	50.4	–
Leucaena	25	19	6.4	0.7	48.9	–
Paragrass	4	23	2.6	0.6	69.8	–
Cucumber leaves	–	14	1.7	1.3	82.0	–
Duck weed	11	32	1.6	0.2	55.2	–
Corn top	7	28	0.8	0.2	64.0	–
Water hyacinth	8	16	2.1	0.5	73.4	–
Sugarcane top	9	22	2.0	0.4	66.6	–

Table 7.9 Comparative Nutritive Characteristics of Agricultural Crops/Residues Before and After Anaerobic Fermentation

Crop Residue	*Before Fermentation (Per cent of Dry Matter)*					*After Fermentation (Per cent of Dry Matter)*			
	DM	*N*	*P_2O_5*	*K_2O*	*C/N*	*DM*	*N*	*P_2O_5*	*K_2O*
Rice straw	46	1.2	0.4 ·	0.5	75	19	0.9	0.2	0.9
Leucaena	28	4.5	0.3	2.0	22	17	0.4	0.2	1.6
Paragrass	30	1.6	0.3	4.3	58	18	1.2	0.2	0.5
Sugarcane top	20	1.2	0.3	1.4	46	15	0.9	0.2	1.2
Cucumber leaves	25	2.1	0.3	3.2	32	20	1.8	0.3	2.9
Duck weed	23	2.5	1.5	1.0	24	17	2.2	1.4	0.8
Corn top	19	1.3	0.3	2.0	78	14	0.6	0.1	0.3
Water hyacinth	12	2.5	0.5	5.5	33	10	2.3	0.8	3.7

Source : Tumwasron, S., Biogas Technology and its Applications at Different Farm Levels in Central Thai Village, A Report, Animal Science Department, Kasetsart University, Bangkok, Thailand, March, 1981, p. 6-7. [For Tables 7.7, 7.8 and 7.9]

Table 7.10 Chemical Composition of a Select Agricultural Crop Residues for Assessing their Biodegradability

Crop Residues	*Per Cent of Dry Matter*						
	Fats & Waxes	*Water Solubles*	*Hemi Celluloses*	*Cellulose*	*Lignin*	*Protein*	*Ash*
Young rye plants	2.35	29.54	12.67	17.84	10.61	12.26	12.55
Mature wheat straw	1.10	5.57	26.35	39.10	21.60	2.10	3.53
Soyabean tops	3.80	22.09	11.08	28.53	13.84	11.04	9.14
Alfalfa tops	10.41	17.24	13.14	23.65	8.95	12.81	10.30
Young corn stalks	3.42	28.27	20.38	23.05	9.68	2.61	7.40
Matured corn stalks	5.94	14.14	21.91	28.67	9.46	2.44	7.54
Young pine needles	7.65	13.02	14.68	18.26	27.63	8.53	3.08
Old pine needles	23.92	7.29	18.98	16.43	22.68	2.19	2.51
Oak leaves green	7.75	22.02	12.50	15.92	20.67	9.18	6.40
Oak leaves brown	4.01	15.32	15.60	17.18	29.66	3.47	4.68

Table 7.11 Residue Coefficients for Estimating Availability of Crop Residues

Crop	*Residue Coefficient Range*
Soyabean	0.55-2.60
Corn	0.55-1.20
Cotton	1.20-3.00
Wheat	0.47-1.75
Barley	0.82-1.50
Paddy (unhusked rice)	0.38-1.25
Rye	1.20-1.95
Oats	0.95-1.75
Grain sorghum	0.50-0.85
Sugarbeet	0.07-0.20
Sugarcane	0.13-0.25

Gliricidia macullata, a tree introduced in India from the West Indies, is grown for shade and green leaf manuring. It can be easily grown on wastelands and along roadsides. Based on experimental studies, V.N. Gunaseelam [12] established the potential of *Gliricidia* leaves as feed for biogas production and the use of digested effluent as a rich fertiliser which is superior to fresh *Gliricidia* leaves. Experiments were carried out in 3 litre batch digesters at room temperature (32 ± 3°C) and gas yield of 165-180 ml methane per g. VS added, and VS reduction of 37-39 per cent was reported. S.K. Sharma *et al.* [28] studied the potential of *Mirabilis* leaves as feed for biomethanation. They reported that under identical experimental conditions *Mirabilis* leaves produced nearly 400 litres of gas/kg of dry matter which is double the amount produced by cattle dung (200 litre/kg of dry matter). Methane content in biogas obtained from *Mirabilis* leaves was 69 per cent as against 62 per cent normally obtained from cattle dung. Total reduction of volatile matter in *Mirabilis* leaves was 42.6 per cent as compared to 22 per cent in cattle dung.

A.S. Traore [31] carried out studies to analyse prospects of biogas production from *Calotropis procera,* a latex plant found in West Africa. This plant, which contains hydrocarbons as well, is found in great abundance in Burkino Faso with the highest density occurring in Sahelian and semi-desert zones. *C. procera* is a bush latex plant, 1-3 m high, its leaves are 10-13 cm wide and 17-19 cm long. This plant is fast-growing, draught-resistant and grows throughout the Sahelian countries, notably in Burkino Faso. A high speed blade grinder was used to obtain crushed leaf powder of the plant. The fine particles thus obtained were suspended in water and used for fermentation test runs at 30°C in batch mode. It provided a valuable substrate for biogas production. The highest yield obtained varied form 2.9 to 3.6 litres/day litre at initial pH of 7.5

with digester loading rate being 4 per cent (w/v) suspension of dry leaves. The acidogenic phase of the fermentation process was found to be very fast with 66 per cent of dry material loaded being degraded during the first two days of incubation at 30°C. The resulting biogas was found to contain 56-59 per cent methane by volume.

Studies have been carried out recently to examine the prospects of coir industry wastes, namely, coir pith for anaerobic fermentation. K.S. Jagadeesh [13] carried out studies at the University of Agricultural Sciences, Dharwad, Karnataka by using coir pith as additive to cow dung in the ratio of 1 : 4 as feedstock. In his experiment coir pith was soaked in calcium hydroxide for 24 hours and washed with water to remove the traces of alkali. This was followed by mixing with cowdung and water to obtain slurry with a total solid content of 9.2 per cent. The slurry was then fed to a 4 m^3 capacity KVIC biogas plant. After a lapse of 40 days, the plant was fed with slurry once in 24 hours and digestion was continued for 80 days. In the first week there was no gas production but later it averaged around 2.5 m^3. During the 50-55th day interval it went upto 2.74 m^3/day which was equivalent to 0.152 m^3 of gas/kg of dry matter/day. When compared the corresponding yield with cow dung alone was 0.20-0.25 m^3/kg dry matter/day.

K.G. Gollakota and K.K. Meher [11] studied the effect of particle size, temperature, loading rate and stirring on biogas production from oil-expelled caster cake. Experiments were carried out in 5 litre capacity single-stage digesters protected from light at 30 and 37°C. Four particle sizes in the range less than 0.5 to 2 mm were used. Volumetric loading rates ranged from 4 to 12 g TS/1-day. Stirring of cultures was continuous with a magnetic stirrer or intermittent with a platform shaker. The HRT was 15 days in all cases and the fermenters were fed once a day. Biogas yield was measured in terms of litre of biogas/g of TS. Both rate and yield were higher with particles 2-1.4 mm and 0.5 mm and less with particles of intermediate size, i.e. 1.4-1.0 mm and 1 to 0.5 mm. Both rate and yield were higher at 37°C than at 30°C. W.J. Ng *et al.* [37] carried out experiments for using Palm Oil Mill Effluent (POME) as feed for anaerobic fermentation which was claimed to be an effective means for POME treatment. At hydraulic retention time of 25 days and more, biogas yield during mesophilic stage was found to be more than during thermophilic stage.

7.6 Common Biogas Yield Norms Based on Several Studies

Based on several studies carried out, United Nations [34] attempted to standardise norms of biogas yield from cattle, pig and poultry manure, night-soil, pretreated crop wastes and water-hyacinth which are given in Table 7.12. It also prescribed optimum proportions of night-soil, pig manure and crop residues in several mixed substrates for achieving maximum biogas yield (Table 7.13). National Academy of Sciences [21] compiled biogas yield data of different waste materials (Table 7.14). T.K. Ghose [10] studied biogas from cattle manure, water-hyacinth, algae, bagasse, rice-husk and their different combinations (Table 7.15). Comparative estimates of biogas yield from different animal wastes as experienced in China [30] are given in Tables 7.16 and 7.17. Comparative range of biogas yield from cattle, poultry, swine and sheep manure [2, 7, 8 and 27] as reported by different scientists in different units of measurement are given in Tables 7.18 and 7.19. Biomass Energy Institute at Winnipeg in Canada [4] provided estimates of equivalent energy from different animal wastes (Table 7.20). Biogas yield from these wastes is given in Table 7.21. Biogas yields as reported by R.D. Laura and M.A. Idnani [16] from variety of mixed substrates comprising agricultural plant materials at ambient temperatures are given in Table 7.22. The information included herein provides fairly detailed account of biogas yields from different organic wastes.

Table 7.12 Norms of Biogas Yield from Different Types of Feedstocks

Source Providing Feedstock	*Biogas Yield per kg of Animal Wastes*
Cattle (cows and buffaloes)	22 to 40 litres (0.8 to 1.4 ft^3)
Pig	40 to 60 litres (1.4 to 2.1 ft^3)
Poultry (cnicken)	65.5 to 115 litres (2.3 to 4.1 ft^3)
Human	20 to 28 litres per person using toilet (0.7 to 1 ft^3)
Pretreated crop waste	34 to 40 litres
Water-hyacinth	40 to 50 litres

Note : From the above table it is to be seen that under average conditions of 40 to 60 days of retention period which is necessary to produce 1 m^3 (35 ft^3) of gas per day, following quantities of animal wastes in different categories is generally required; Cattle wastes : 32 kg per day, Pig wastes : 20 kg per day, Poultry droppings : 12 kg per day.

Source : UN, Guidebook on Biogas Development, United Nations, New York, 1980, p. 8.

Table 7.13 Optimum Proportions of Night-Soil, Pig wastes and Crop Residues in Mixed Substrate for Achieving Maximum Gas Yield

S. No.	*Substrate*	*Optimum Mix*
1.	Night-soil and wheat-stalks	1.75 : 1
2.	Night-soil and rice-husk	4.40 : 1
3.	Night-soil and corn-stalks	1.13 : 1
4.	Night-soil and fallen leaves	0.85 : 1
5.	Night-soil and soyabean stalks	0.45 : 1
6.	Night-soil, wild green grass/waterweeds	1.00 : 10
7.	Night-soil, pig wastes and rice-husk	1 : 1 : 1
8.	Night-soil, pig wastes and wheat-stalks	1 : 2 : 1
9.	Night-soil, pig wastes and corn-stalks	3 : 4 : 4

Contd...

10.	Night-soil, pig wastes, wild green grass and waterweeds	1 : 10 : 180
11.	Pig wastes and wheat-stalks	4.55 : 1
12.	Pig wastes and rice-husks	3.65 : 1
13.	Pig wastes and corn-stalks	2.95 : 1
14.	Pig wastes and fallen leaves	2.22 : 1
15.	Pig wastes and soyabean stalks	1.10 : 1
16.	Pig wastes wild green grass/ waterweeds	1.00 : 25

Note : Since C : N. ratio in cow, sheep and horse manure which can be utilised as feedstock normally lies within the desirable range of 20 : 1 to 30 : 1, for obtaining higher yields it is preferable to mix them if available with the above substrates.

Source : UN, Guidebook on Biogas Development United Nations, New York, 1980, p. 69. [Based on experiences of a study tour to China during the period Sept 1 to 21, 1979].

Table 7.14 Biogas Yield from Different Waste Materials

Raw Materials	*Biogas Yield per Unit Weight of Dry Solids*		*Temperature*		*Methane Content Per cent*	*Fermentation Period (Days)*
	ft^3 lb	*m^3 kg*	*°F*	*°C*		
1. Cow dung	5.3	0.33	–	–	–	–
2. Cattle wastes	5	0.31	–	–	–	–
3. Cattle wastes (Indian)	3.6-8.0	0.23-0.50	52-88	11.1-31.1		
4. Cattle wastes (German)	3.1-4.7	0.20-0.29	60-63	15.5-17.3	–	–
5. Beef wastes	13.7	0.86	95	34.6	58	10
6. Beef wastes	17.7	1.11	95	34.6	57	10
7. Chicken droppings	5.0	0.31	99	37.3	60	30
8. Poultry droppings	7.3-8.6	0.46-0.54	90.5	32.6	58	10-15
9. Poultry droppings (3) (4)	8.9	0.56	123	50.6	69	9
10. Swine wastes (3) (4)	11.1-12.2	0.69-0.76	90.5	32.6	58-60	10-15
11. Swine wastes	7.9	0.49	91	32.9	61	10
12. Swine wastes(3)(4)	16.3	1.02	95	34.6	68	20
13. Sheep wastes(3)	5.9-9.7	0.37-0.61	–	–	64	20
14. Super beet leaves	8	0.50	–	–	–	14
15. Algae	5.1	0.32	113-122	45-50	–	11-20
16. Night-soil	6	0.38	68-79	20-26.2	–	–

Note : 1. Based on total solids
2. Based on volatile solids fed
3. Includes both faeces and urine
4. Animals on growing and finishing rations.

Source : NAS, Methane Generation from Human, Animal, and Agricultural Wastes, National Academy of Sciences, Washington, DC, 1977, p. 66.

Table 7.15 Comparative Biogas Yield and Other Characteristics from a Group of Individual and Mixed Substrates

Individual and Mixed Substrates	*Per Cent VS*	*Per Cent VS Digestible*	*Biogas Yield*		*Methane (Per Cent)*	*Methane Yield (l/kg of Waste)*	*Energy Equivalent (kcal/kg of Waste)*
			l/kg Waste Added	*l/kg VS Digested*			
1. Gobar (G)	80	67	214	400	66	141	1128
2. Water-hyacinth (WH)	75	69	225	500	70	157	1460
3. Algae (A)	75	72	251	650	72	252	2016
4. Bagasse (B)	65	60	152	400	65	98	784
5. Rice-husk (RH)	70	65	182	400	67	122	975
6. G+WH (1 : 1)	78	67	364	700	72	262	2096
7. G+A (1 : 1)	78	70	407	750	73	297	2377
8. G+RH (1 : 1)	75	67	225	450	67	151	1206
9. G+WH+A (1 : 1 : 1)	77	70	458	850	75	343	2744
10. G+B+RH (1 : 1 : 1)	70	64	200	450	70	140	1120

Note : Calorific value of Biogas: 540 Btu/ft^3 = 4.81 kcal/1
Calorific value of CH_4 : 900 Btu/ft^3 = 8.0 kcal/1

Source : Ghose, T.K., Fuels from Biomass, Biochemical Engineering Research Centre, IIT, New Delhi, 1978.

Table 7.16 Comparative Estimates of Biogas Yield from Different Animal Wastes as Experienced in China

Measure	*Unit*	*Cattle*	*Horses*	*Pigs*	*Chicken*	*Men*
1. Fresh faeces generated (wet)	kg/day/head	16	16	2.9	0.14	0.20
2. Urine generated	kg/day/head	6	3.5	1.8	–	1.00
3. Total solids produced	kg/day/head	3.2	3.2	0.6	0.05	0.14
4. Volatile solids produced	kg/day/head	2.6	2.6	0.5	0.02	0.11
5. Biogas yield	m^3/day/head	1.3	1.3	0.3	0.01	0.06
6. Equivalent energy produced	kcal/day /head	7540	7540	1450	58	320

Note : Total solids are estimated at a norm of 20 per cent of faeces. Volatile solids are estimated at a norm of 80 per cent of total solids. Biogas is assumed to be produced at a norm of 0.5 m^3/kg VS/day. Calorific value of biogas is assumed as 5800 kcal/m^3.

Source : Smil, V., China's Energy: Achievements, Problems, Prospects, Praeger Publishers, NY, 1976, p. 106.

Table 7.17 Biogas Yield Obtainable from a Group of Agricultural and Animal Wastes

Source of Feedstock	*Biogas Yield (m^3/ton of dry solids)*	*Methane Content (Per Cent)*
Livestock waste:		
1. Cattle wastes	260-80	50-60
2. Pig wastes	561	
3. Horse wastes	200-300	
Agricultural waste:		
4. Fresh weeds	600	70
5. Flax or hemp stalks	369	
6. Wheat straw	432	59
7. Green leaves	210-294	58
8. Rice husks	615	
9. Sewage waste	640	50
10. Liquid waste from wine or spirit making factory	300-600	58
Compounds:		
11. Carbohydrates	750	49
12. Lipids	1440	72
13. Proteins	980	50

Source : Sichuan Provincial Office of Biogas Development, Biogas Technology and Utilisation, Chengdu Seminar, China, 1979. Quoted in 'UN, Guidebook on Biogas Development, United Nations, New York, 1980, p. 8.'

Table 7.18 Comparative Range of Biogas Yield from Different Animal Wastes

Waste	*Range of Biogas Yield*		
	1/kg of TS added	*1/kg of VS added*	*1/kg of VS destroyed*
Cattle waste	50-117 (74)	65-700 (276)	260-1770 (596)
Poultry waste	41-292 (182)	67-930 (257)	318-1067 (340)
Swine waste	170-452 (291)	254-760 (495)	492-1920 (1912)
Sheep waste	91	113-232 (168)	370-610 (477)
Grass	131-419 (282)	163-497 (352)	346-786 (461)

Note : Values given in brackets refer to average biogas yield

Sources :
1. Anthonisen, A. and Cassel, E.A., Studies on Chicken Manure Disposal, Report from Clarkson College of Technology, USA.
2. Burford, J.L. and Varani, P.T., Energy Potential Through Biconversion of Agricultural Wastes, Colorado Energy Research Institute, Colorado.
3. Converse, J.C., *et al.*, Dairy Manure Degradation Under Mesophilic & Thermophilic Temperatures, paper presented at the Winter Meeting of

American Society of Agricultural Engineers, Chicago, Dec. 15-18, 1975.

4. Fischer, J.R., et al., Producing Methane Gas from Swine in a Pilot Size Digester, paper presented at the Mid-Central Regional Meeting of the American Society of Agricultural Engineers, St. Joseph, Missouri, Mar. 25-26, 1955.

Table 7.19 Biogas Yield and Methane Content Obtained from Different Animal Wastes Under Varying Loading Rates

Source Providing Feedstock	*Loading Rate added (kg VS/m³/day)*	*VS Added per cent*	*Methane Content (Per Cent)*	*Temperature (°C)*	*Detention Period (days)*	*Biogas Yield (m³/kg VS Added)*
1. Swine	0.32-3.20	41-54	59	35	10-50	0.49-0.64
2. Swine	2.42-3.05	53-62	68	35	20	0.37-0.54
3. Swine	2.41-4.01	26-76	57-60	32-52	10-15	0.48-1.05
4. Swine	1.92-3.85	44-61	58-61	33	10-15	0.26-0.45
5. Dairy	2.56-3.53	42-53	65	35	17-20	0.12-0.19
6. Dairy	2.41-4.01	16-29	57-68	32-52	10-15	0.26-0.90
7. Dairy	2.08-3.53	11-16	52-64	23-35	15-26	0.07-0.16
8. Dairy	1.92-3.85	18-27	61-66	33	10-15	0.06-0.10
9. Dairy	1.60-2.89	38-53	74-79	35	12-20	0.04-0.06
10. Poultry	2.72-4.97	20-45	11-49	35	23-26	0.10-0.27
11. Poultry	1.92-3.85	57-68	58	33	10-15	0.28-0.36

Source : Schulte, D.D., *et al.,* Methane Production Through Bioconversion of Agricultural Residues, paper presented at the International Solar Energy Society Meeting, Winnipeg, Manitoba, Aug. 11-15, 1976.

Table 7.20 Equivalent Varied Energy of Feed Materials from Different Animals with Their Total Liveweight as 1000 kg On Uniform Basis

Source of Feedstock	*Propane (litres/day)*	*Petrol (litres/day)*	*Fuel Oil (litres/day)*	*Electricity (kwh/day)*
1. Growing swine	1.5	1.0	1.0	2.6
2. Dairy	1.9	1.2	1.2	3.4
3. Poultry	3.6	2.3	2.4	6.2
4. Beef	1.5	1.0	1.0	2.6

Note : While making the above estimates it is assumed that 35 per cent of gross energy is used up in operating the digester.

Table 7.21 Biogas Yield in Different Measures from Various Animals

Measure	*Unit*	*Growing Swine*	*Dairy Cows*	*Laying Hens*	*Beef Feeders*
1. Fresh faeces plus urine	1/day/1000 kg	65	82	53	60
2. Volatile solids generated per day	kg VS/day/1000· kg liveweight	4.80	8.60	9.50	4.90
3. Fraction of VS convertible to biogas	–	0.50	0.35	0.60	0.45
4. Biogas yield	m^3/day/1000 kg Liveweight	2.62	3.28	6.21	2.66
5. Biogas yield	m^3/day/m^3 of digester volume.	1.10	1.10	1.30	1.30

Note : The above estimates are based on a norm of 1.09 m^3 of biogas (comprising 65 per cent methane and 35 per cent carbon dioxide) per kg of volatile solids destroyed.

Example : For estimating volume of undiluted faeces plus urine generated per day by one of the four types of animals given above, corresponding faeces generation rate is needed to be multiplied by average weight and numbers of concerned animals and resulting product divided by 1000.

As an example, volume of undiluted faeces plus urine generated per day by 4000 laying hens = 53 (1/day/1000 kg) × 4000 (layers) × 2 (kg layer) × (1/1000 kg) = 924 litres /day

Source : BEI, Biogas Production from Animal Manure, Biomass Energy Institute, Winnipeg, Manitoba, 1978, p. 10.
[For Tables 7.20 and 7.21].

7.7 Biogas Production from Dry Dung Cakes

In view of the fact that dung cakes are more convenient to tramsport as compared to fresh manure and do not stink, D.R. Ranade, *et al.* [26] examined the possibility of producing biogas from them. Experiments were carried out in two sets. In the first set, wet dung was diluted with an equal quantity of water to obtain slurry with 10 per cent total solids concentration. The same slurry with 10 per cent TS concentration was also used for preparing dung-cake samples. These dung-cakes were later fermented by preparing their slurry at (i) two months and, (ii) eight months intervals after they were formed. In another set of experiment about eight months old dung cakes purchased from market were used. Anaerobic fermentation was carried out separately with wet manure, two and eight month old cake samples prepared in laboratory and readymade dung cakes purchased from market, and the results of biomethanation compared.

Average gas yield with wet dung as feed was 3.68 litres/day, total gas generation over a period of 54 days being 204 litres with methane content as 55 and carbon dioxide content as 42 per cent. In the experiment with dry dung

Table 7.22 Biogas Yield from Mixed Substrates Comprising Agricultural Plant Materials and Cow Wastes at Ambient Temperatures

Raw Material	*Biogas Yield per Unit Weight Dry Solids*				*Gas Composition After 21 days (2)*		
	After 24 days		*After 80 days*		*CH_4*	*H_2*	*CO_2*
	ft^3/lb	*m^3/kg*	*ft^3/lb*	*m^3/kg*			
1. Cow dung	1.0	0.063	3.3	0.21	60.0	1.1	34.2
2. Cow dung + 0.4% cane sugar	1.1	0.070	3.3	0.21	57.6	2.1	38.4
3. Cow dung + 1% ashes	0.98	0.061	3.0	0.19	60.4	2.9	34.4
4. Cow dung + 2.4% fresh leguminous leaves (25% dry matter, 2.31% N)	1.0	0.063	3.2	0.20	61.6	4.0	32.0
5. Cow dung + 1.2% sarson oil cake (94% dry matter, 4.74% N)	1.0	0.063	3.3	0.20	67.7	–	30.4
6. Cow dung + 1% cellulose	1.3	0.084	3.3	0.21	52.8	–	44.0
7. Cow dung + 0.4% casein (126% N)	1.4	0.087	3.5	0.22	64.0	2.4	32.0
8. Cow dung + 1% cane sugar + 1% urea (44.5% N)	1.4	0.087	4.2	0.26	68.0		30.6
9. Cow dung + 1% cane sugar + 1% $CaCO_3$	1.5	0.091	3.0	0.24	70.0	–	28.0
10. Cow dung + urine (4% solids) at 20 ml/100 g	1.4	0.087	3.9	2.24	67.0	–	32.0
11. Cow dung + 0.4% charcoal	1.0	0.65	2.6	0.16	65.6	–	32.0
12. Cow dung + 20% dry non leguminous leaves (1.71% N)	1.3	0.081	3.5	0.22	68.0	0.6	28.0

Note : Ambient temperature taken as 70°F and 21°C.

Source : Laura, R.D. and Idnani, M.A., Increased Production of Biogas from Cow dung by Adding Other Agricultural Waste Materials, Journal of the Science of Food and Agriculture, 164-167 (1971).

cakes purchased from the market, the average gas production was 2.78 litres/day and total gas produced over a period of 54 days was 150 litres with methane content as 47 and carbon dioxide also 47 per cent. This means in relation to wet dung case, gas yield declined by 24.3 per cent and methane content less by eight per cent. In the experiment with two month old dry dung prepared in the laboratory, mean gas production was 3.2 litres/per day and total gas yield over a period of 54 days was 172.3 litres with methane and carbon dioxide content being 67 and 28.5 per cent, respectively. When compared with wet dung case, this shows that mean gas production declined by 13 per cent and total yield was less by 32.3 litres but methane content went up from 55 to 66 per cent. In the experiment with eight months old dry dung cakes prepared in the laboratory, average gas yield was 2.72 litres/day with total gas production over a period of 54 days as 147.2 litres and methane content as 58.7 per cent. This shows that in relation to the wet dung case, daily gas yield declined by 26 per cent and total yield for the same period was less by 57 litres. Comparative results of these experiments shows that one to two months old dung cakes can be harmlessly used for biomethanation without any appreciable decline in gas yield and also with some improvement in calorific value following somewhat higher methane content.

7.8 Concluding Remarks

Results summarised in this chapter indicate that gas yield varies from substrate to substrate and it depends on several factors which need to be controlled. In case of animals wastes, gas yield also depends on quality of feed which the animals receive. In case of crop residues, gas yield also depends on maturity of agricultural plants as the composition of plant matter changes with growing age. For optimising plant performance, biogas plant owner should maintain desired loading concentration, digester temperature, C : N ratio, pH of digester contents and other controllable parameters.

REFERENCES

1. Agarwal, P.K., Prabhu, K.A. and Vishwanathan, L., *Microbial Fermentation of Agricultural Wastes into Biogas Biomanure by a Semi-wet Process,* A Report, Biochemistry Division, National Sugar Institute, Kanpur, 1977.
2. Anthonisen, A. and Caseel, E.A., *Studies on Chicken Manure Disposal,* Report from Clarkson College of Technology, USA.
3. Badger, D.M., *et al.*, Biogas Production from Crops and Organic Wastes, *New Zealand Journal of Science*, **22**, 11-20 1979.
4. BEI, *Biogas Production from Animal Manure*, Biomass Energy Institute, Winnipeg, Manitoba, 1978.
5. Burford, J.L. and Varani, F.T., *Energy Potential Through Bioconversion of Agricultural Wastes,* A Report, Colorado Energy Research Institute, Colorado.
6. Buswell, A.M. and Hatfield, W.D., *Anaerobic Fermentation*, Bull. No. 32, Illinois State Water Survey, Urbana, Illinois, 1939.

7. Converse, J.C., *et al., Dairy Manure Degradation Under Mesophilic and Thermophilic Temperatures*, Proc. Winter Meeting of American Society of Agricultural Engineers, Chicago, Dec. 15-18, 1975.
8. Fischer, J.R., *et al., Producing Methane Gas from Swine Manure in a Pilot Size Digester*, Proc. Mid-Central Regional Meeting of the American Society of Agricultural Engineers, St. Joseph, Missouri, Mar. 25-26, 1955.
9. Geeta, G.S., Raghavendra, S. and Reddy, T.K.R., Increase of Biogas Production from Bovine Excreta by Addition of Various Inert Materials, *Agricultural Wastes*, **17(2)**, 153-156 1986.
10. Ghose, T.K., *Fuels from Biomass*, Biochemical Engineering Research Centre, IIT, New Delhi, 1978.
11. Gollakota, K.G. and Meher, K.K., Effect of Particle Size, Temperature Loading Rate and Stirring on Biogas Production from Castor Cake (Oil Expelled), *Biological Wastes*, **24(4)**, 243-249 1988.
12. Gunaseelan, V.N., Anaerobic Digestion of *Gliricidia* Leaves for Biogas and Organic Manure, *Biomass*, 1988 (0144).
13. Jagadeesh, K.S., Anaerobic Fermentation of Coir Pith, *Current Research*, **19**, 189 1990.
14. Kalra, M.S. and Panwar, J.S., Anaerobic Digestion of Rice Crop Residue, *Agricultural Wastes,* **17(4)**, 263-269 1986.
15. Kumar, S., *et al.*, Note on Stimulation of Biogas Production from Cattle Dung by Addition of Charcoal, *Biological Wastes*, **20(3)**, 209-215 1987.
16. Laura, R.D. and Idnani, M.A., Increased Production of Biogas from Cow dung by Adding other Agricultural Waste Materials, *Journal of the Science of Food and Agriculture*, 164-167 1971.
17. Lo, K.V., *et al.*, Mesophilic Anaerobic Digestion of A Mixture of Cheese Whey and Dairy Manure, *Biomass*, **15(1)**, 45-53 1988.
18. Madamwar, D.B. and Mithal, B.M., Effect of Pectin on Anaerobic Digestion of Cattle Dung, *Biotech. Bioeng.*, April, 1986.
19. Mahadevaswamy, M. and Venkataraman, L.V., Bioconversion of Poultry Droppings for Biogas and Algal Production, *Agricultural Wastes*, **18(2)**, 93-101 1987.
20. Mahadevaswamy, M. and Vekataraman, L.V., Integrated Utilisation of Rabbit Droppings for Biogas and Fish Production, *Biological Wastes*, **25(4)**, 249-256 1988.
21. N.A.S., *Methane Generation from Human, Animal and Agricultural Wastes*, Report of Ad. Hoc. Panel of the Advisory Committee on Technology Innovation, National Academy of Sciences, Washington, DC, 1977.
22. Odeyemi, O., *Biogas Generation from Cassava Leaves Compared with Two Animal Manures and Sewage Sludge,* Proc. Second Inter. Seminar on Energy Conservation and the Use of Renewable Energies in Bio-industries, Oxford, Sep. 6-10, 554-558 1982.
23. Prasad, C.R., Utilisation of Organic Wastes in Biogas Plants, *Khadi Gramodyog*, Sep., 514-518 1985.
24. Rahman, A., *et al.*, Microbial Production of Biogas from Organic Wastes, *Jour, of Fermentation Technology,* **64(1)**, 45-49 1986.
25. Rajasekaran, P., *et al.*, Bio-Processing of Certain Wastes in Relation to Energy, *Madras Agri. Jour.*, **73(8)**, 421-424 1986.

26. Ranade, D.R., *et al.*, Use of Dry Dungcakes for Biogas Production, *Khadi Gramodyog*, 502-506 1979.
27. Schulte, D.D., *et al., Methane Production Through Bioconversion of Agricultural Residues,* Proc. Inter. Solar Energy Society Meeting, Winnipeg, Manitoba, Aug. 11-15, 1976.
28. Sharma, S.K., *et al., Mirabilis* Leaves: A Potential Source of Methane, *Biomass,* **13(1)**, 13-24 1987.
29. Sharma, U.P. and Paul, T.M., Studies on Biogas Sources, *Current Science,* **45(10)**, May 20, 376-377 1976.
30. Smil, V., *China's Energy: Achievements, Problems, Prospects*, Praeger Publishers, New York, 1976, p. 106.
31. Traore, A.S., Biogas Production from *Calotropis procera*: A Latex Plant Found in West Africa, *Bioresource Technology*, **41**, 105-109 1992.
32. Trevelyan, W.E., The Methane Fermentation: A Discussion Paper, *Tropical Science*, **17(4)**, 193-209 1975.
33. Tumwasron, S., *Biogas Technology and its Application at Different Farm Levels in Central Thai Village,* A Report, Animal Science Department, Kasetart University, Thailand, 1981.
34. UN, *Guidebook on Biogas Development*, United Nations, New York, 1980.
35. Varshney, A.C., *et al.*, Evaluation of Crop Residues in North Gujarat, *Biological Wastes*, **19(3)**, 227-231 1987.
36. Waksman, S.A., *Soil Microbiology,* John Wiley and Sons, New York, 1952.
37. W.J., Ng, *et al.*, Energy Yields from Anaerobic Digestion of Palm Oil Mill Effluent, *Biological Wastes*, **19(4)**, 257-266 1987.

Chapter 8

BIOGAS YIELD FROM WATERWEEDS

8.1 Introduction

Waterweeds provide widely available feedstock for biogas generation. Water-hyacinth and algae are two major categories of waterweeds. A number of experimental studies have been carried out in past both in India and abroad to estimate potential, and develop the related technology. Water-hyacinth regarded as neglected weed for many years is at present being increasingly utilised. It is nowadays being used as a cleaning agent to purify industrial effluents and municipal refuse, as a biomass source to produce paper and board, and as an energy-resource to produce biofuels. Water-hyacinth is one of the fastest growing plants which grows on clean as well as polluted waters. It is a warm-weather plant which flourishes in tropical and sub-tropical regions alike. It grows in deltas, swamps and irrigation canals which often results in overrunning and clogging of waterways. If not frequently removed and put to use, water-hyacinth poses difficulties in shipping, navigation and hydropower generation besides offering breeding ground for insects.

Based on studies carried out in the USA and India and other parts of the world, seaweeds (algae) have emerged as valuable input material to generate biogas. Algae is one of the simplest and primitive waterweeds having no true roots, stems, leaves, flowers and seeds. These are essentially photosynthetic plants which can be both aquatic or subaerial. Aquatic algae grows in water of low salinity whereas subaerial grows while being exposed to air. Algae is also known by other names such as pond scum, water-moss or simply as seaweed. Algae is believed to be known to mankind as a useful resource since around 3000 BC. Industrial use of seaweeds began during the seventeenth century in Western Europe when they were used to produce soda and potash from ash of burnt brown seaweeds (kelp). Ash was also known to be used for producing glass and soap.

8.2 An Overview of Biogas Generation from Water-hyacinth

Water-hyacinth regarded as neglected weed for many years is a source of biogas and other products. Water-hyacinth (*Eichhornia crassipes*) consists of a fleshy vertical stem called rhizome from which roots, leaves and flowers sprout. The rhizome floats just below the water surface protected by shields of folded leaves. Being a warm weather plant, except during the period when water is in frozen state, temperature of around 10°C supports its growth. Anaerobic fermentation of water-hyacinth generates adequate biogas yield. One metric ton of dry water-hyacinth following anaerobic digestion can generate 26,500 ft^3 of biogas with a heat value of 600 BTU/ft^3. There are, however, no fixed norms available about the rate of gas generation per unit weight of water-hyacinth fed as composition of feedstock varies from plant to plant [11].

A hectare of water surface yields about 50 tonnes of water-hyacinth which is currently growing over several million hectares of water surface in India. Except for Gujarat, Punjab and Kashmir, it is abundantly available in several parts of the country particularly in water logged areas of West Bengal, Assam, Kerala and Andhra Pradesh. Growth of water-hyacinth in India is believed to have enormously increased from 1905 onwards. In 1945, it covered 1.5 million hectares of water surface in West Bengal and was occasionally described as a 'Bengal Terror'. Out of 1,23,000 hectares of water area in Assam 1,11,000 hectares are estimated to be infested with water-hyacinth. About 10,360 hectares in Manipur and 3,850 hectares in Andhra Pradesh are infested with this weed. It is widely available in wastelands and neglected ponds in Kerala and most tank and rice fields in Madhya Pradesh. Fourteen out of 26 districts in Rajasthan find water-hyacinth as a growing menace [10].

8.3 Composition and Characteristics of Water-hyacinth

A typical water-hyacinth plant contains 24.8 per cent root, 41.9 per cent stalk and 33.3 per cent leaf. Weights and proportions of water-hyacinth however vary from plant to plant and from region to region. According to the results of a study carried at the Central Food Technological Research Institute, Mysore, fresh hyacinth contains 95.5 per cent water, 3.5 per cent organic matter, 0.04 per cent nitrogen, 1 per cent ash, 0.06 per cent phosphorus (P_2O_5) and 0.2 per cent potassium (K_2O). Dried plant normally contains 75.8 per cent organic matter, 1.5 per cent nitrogen and 24.2 per cent ash. This ash in turn contains 28.7 per cent K_2O, 1.8 per cent Na_2O, 12.8 per cent CaO, 21 per cent chlorine and 7 per cent P_2O_5. Water hyacinth contains 17-22 per cent crude protein, 15-18 per cent fibre, and 16-20 per cent ash [19]. Composition of sewage-grown water-hyacinth as reported by B.Wolverton and R.C.Mc Donald [19] is summarised in Table 8.1.

As per the studies carried out at the National Technology Laboratory (NSTL) in Mississippi (USA), growth of plant in warm enriched domestic sewage generates about 18 tonnes of wet biomass per hectare per day. These plants grow at a rate of 215 tonnes of dry plant material per hectare. Water hyacinth is also appropriate in treating effluents mixed with toxic metals. The NSTL studies revealed that these plants rapidly absorb gold, silver, cobalt,

Table 8.1 Composition of Sewage-grown Water-Hyacinth

S. No.	*Constituent*	*Per Cent on Dry Weight Basis*
1.	Carbon	32-35
2.	Hydrogen	5.4-5.8
3.	Nitrogen	2.8-3.5
4.	Potassium	2.0-3.5
5.	Sodium	1.5-2.5
6.	Calcium	0.6-1.3
7.	Phosphorus	0.1-1.0
8.	Sulphur	0.3-0.4
9.	Magnesium	0.2-0.3
10.	Iron	0.03-0.05
11.	Zinc	0.005-0.05
12.	Manganese	0.005-0.008

strontium, cadmium, nickel, lead and mercury. They also absorb or metabolise phenols and other organic substances which are often present in water supply systems of large cities. Results of experiments carried out at the NSTL to evaluate the effectiveness of water-hyacinth in purifying the canal waters are given in Table 8.2. Results of studies carried out by the NSTL to determine the pollution-abatement capabilities of effluent by water-hyacinth grown in domestic sewage lagoons are given in Table 8.3. It was found that half hectare lagoon filled with water-hyacinth can purify daily night-soil of 1000 heads to acceptable levels with a minimum sewage retention time of two weeks. Water-hyacinth can reduce pollution levels by 75 to 80 per cent.

Table 8.2 Effect of Treating Laboratory Effluents in 0.2 Hectare Canal Covered with Water-Hyacinth for Reducing Organic Substances

Flow	*Unit*	*Organic Carbon*	*BOD*	*Total Dissolved Solids*	*Nitrogen*	*Phosphorus*	*Silver*
Inflow	mg/1	75	33	380	2.36	0.48	0.99
Outflow	mg/1	13	3.6	212	0.43	0.08	0.001

Table 8.3 Effect of Treating Secondary Sewage Effluent with Water-Hyacinth after Primary Filtration in an NSTL 0.28 Hectare Area and 6.8 million litre Capacity Lagoon

Period	*Total Suspended Solids (mg/l)*		*Biochemical Oxygen Demand (mg/l)*		*Flow Rate (litres/day)*
	Inflow	*Outflow*	*Inflow*	*Outflow*	
1. Before adding water-hyacinth (May 1975)	31	49	22	32	437000
2. After adding water-hyacinth					
Jul 1975	33	7	27	11	437000
Aug 1975	35	13	26	5	437000
Sep 1975	43	6	22	7	437000
Oct 1975	48	9	25	7	1059000
Nov 1975	50	10	29	15	1059000
Dec 1975	52	24	32	24	1892500

Source : Wolverton, B. and Mc Donald, R.C., Don't Waste Waterweeds, New Scientist, Aug 12, 318-320 (1976).

As per the NSTL studies, a hectare of water-hyacinth fed on sewage nutrients can yield 0.9-1.8 tonnes of dry plant material per day. This biomass can produce 220-440 m^3 of biogas having heat value of 7-14 million BTU. In addition, the sludge that remains after fermentation is a rich fertiliser, following its retaining most of nitrogen, phosphorus and other mineral contents.

The Regional Research Laboratory at Jorhat in Assam analysed composition and characteristics of water-hyacinth (Tables 8.4 and 8.5) which were obtained by following standard procedures prescribed by the US Technical Association of Pulp and Paper Industry [13].

H. Siregar, *et al.* [14] reported that water-hyacinth when fed in batch system yielded more gas than semi-continuous system. Furthermore, with the batch system, the gas yield was maximum when water-hyacinth was submerged to a depth of one-third of the digester height. In contrast in continuous system the highest gas yield was obtained when water-hyacinth was submerged to two-thirds of the digester height.

8.4 Studies on Biogas Yield from Water-hyacinth at CMERI (Durgapur)

According to studies carried out at the Central Mechanical Engineering Research Institute in Durgapur [3, 4], 1 kg of dry water-hyacinth can produce 4 to 6 litres of gas following fermentation at temperature between 30-35°C. About 10 cm long chopped stem and evenly crushed material with moisture-content ranging between 50 to 60 per cent is appropriate for biodegradation. Chopping or crushing of lower parts of stems is helpful in quick drying to the desired level of moisture-content. Gas yield is maximum after 13 to 15 days of anaerobic fermentation during summers and after 18 to 20 days during winters. Maximum yield occurs from 15th to 20th day of fermentation. For 3000 litre capacity biogas plant installed at the CMERI, 18 quintals of coarsely chopped semi-dried water-hyacinth is required to be fed. CMERI's laboratory-scale units consisted of 200 litres capacity empty diesel drums which were able to give a yield of 50 litres of gas per day during summers when temperature ranged between 31 to 35°C. In the full-sized plant for obtaining the yield of 3000 litres of gas per day, 40-45 kg of semi-dried and coarsely chopped water-hyacinth was needed to be fed daily from 20th day onward after initial charging [3, 4].

Table 8.4 Characteristics of Water-hyacinth

Parameters	*Per Cent on Oven Dry Basis*
1. Cold water solubility	15.7
2. Hot water solubility	17.0
3. One per cent caustic soda solubility	46.8
4. Alcohol: Benzene (1 : 2) solubility	10.6
5. Lignin-content	6.7
6. Pentosan-content	14.2
7. Cellulose	29.0
8. Ash content	18.4
9. Moisture-content	13.8

Table 8.5 Comparative Characteristics of Water-hyacinth and other Agricultural Plant Materials

Raw Material	*Per Cent Content*			*Average Size*	
	Ash	*Lignin*	*Cellulose*	*Length (mm)*	*Diameter (micron)*
1. Water-hyacinth	12-18	6-10	28-35	1.6	5.5
2. Straw	15-18	12	32	1.1	16
3. Bagasse	2-6	18-21	32-41	1.4	18
4. Bamboo	2-6	20-32	30-40	1.4-3.8	9-22
5. Cotton			92-97	18	20
6. Jute		12	57	2	22
7. Indian soft woods	1	28-30	39-46	27-3.6	27-52
8. Indian hard woods	0.402	21-28	35-45	0.7-1.8	9-44

Source : RRL, Jorhat, Paper from Water-Hyacinth, A brochure, Regional Research Laboratory, Jorhat, June 31, 1980, p.13-14.

As for the availability of enriched fertiliser from water-hyacinth fed plants, the CMERI's 3000 litre capacity plant produced 5 tonnes of compost on dry basis having 2.05 per cent nitrogen, 2.5 per cent K_2O and 1.1 per cent P_2O_5. In other words 1 tonne of dry compost should contain 20.5 kg of nitrogen equivalent to about 105 kg of ammonium sulphate, 25 kg of K_2O equivalent to about 50 kg of muriate of potash, and 11 kg P_2O_5 equivalent to about 62 kg of super-phosphate. This resulting compost is also found to be a superior soil-conditioner due to its having fibrous characteristics and moisture retention property [3, 4].

8.5 Power Generation from Water-hyacinth Based Plant

K.V. Gopalakrishnan and B.S. Murthy [7] carried out studies at IIT, Madras for analysing the economics of power generation from water-hyacinth based biogas plants. Efficiency of the system can be improved by recovering part of heat generated during the process, and gainfully utilising it for warming up digester-contents for obtaining higher gas yield. Complete plant economics is worked out as follows:

Assumed heat value of gas = 4900 kcal/m^3

Biogas yield per tonne of dry water-hyacinth = 375 m^3

Assumed thermal efficiency of gas engines working on efficient heat-recovery cycle = 50 per cent

Assumed efficiency of electric generator = 95 per cent

Electric power generated per tonne of dry water-hyacinth

$$= \frac{375 \times 4900 \times 0.5 \times 0.95}{860}$$

$$= 1010 \text{ kwh}$$

From the waste heat available in the exhaust, refrigeration capacity that can be supported with an assumed a COP of 50 per cent

$$= 375 \times 4900 \times 0.5 \times 0.5$$

$$= 0.46 \times 10^6 \text{ kcal/tonne of dry plant}$$

For supporting a generating set of 100 kw to produce 1200 kwh/day, the quantity of water-hyacinth needed,

$$= \frac{1200}{1010} = 1.2 \text{ tonnes/day}$$

Growth rate of water-hyacinth

$$= 150 \text{ tonnes/hectare/year}$$

$$= \frac{150}{365} = 0.406 \text{ tonnes/hectare/day}$$

Hence, the area needed to grow water-hyacinth for generating 1200 kwh/day

$$= \frac{1.2}{0.406} = 3 \text{ hectares}$$

The refrigeration capacity which can be supported by the same area,

$$= \frac{0.46 \times 10 \times 1.2}{3000 \times 24} = 7.66 \text{ tons}$$

In other words this is equivalent to saying that one hectare of water-hyacinth plantation can generate 400 kwh/day and support a refrigeration capacity of 2.55 tons. The proposed system is shown in Fig. 8.1. Water-hyacinth can be grown in ponds, tanks or natural lakes. The inputs to the growth area for water-hyacinth cultivation would include raw sewage and carbon dioxide derived from biogas plant digesters of the system or the power plants used in the system. Water hyacinth is regularly harvested by a suitable mechanical means, slurried and fed to biogas digesters.

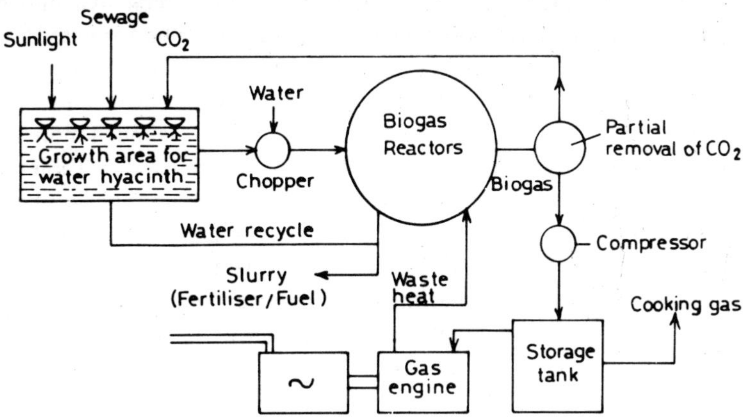

Fig. 8.1 Schematic representation for water-hyacinth based decentralised power generation

8.6 Findings of a JNTU (Hyderabad) Study

A pioneering study with water-hyacinth as feed was carried by N. Sreeramulu and B.N. Bhargava [15] at the Jawaharlal Nehru Technological University which was sponsored by the Department of Science and Technology. Water hyacinth used in this study was collected from Hussain Sagar, an open lake situated between the twin cities of Secunderabad and Hyderabad in Andhra Pradesh. Leaves used in the experiments were first cut to 2 to 3 cm long pieces for feeding into the digester. In the beginning leaves were kept under mud slurry in a container for a period of three months which provided necessary anaerobic conditions for fermentation to start. Initially about 60 g of starter-seed thus prepared was blended with 1000 ml of distilled water. The sediment was allowed to settle for 30 minutes and the supernatant was then decanted into digesters. Digester was sealed to the atmosphere with a single-holed rubber stopper fitted with a glass tube. Digester was connected with a rubber tube to another sealed container fitted with double-holed rubber stopper. The container was filled with water which was acidified with sulphuric acid. Digester was covered with aluminium foil to prevent exposure to light. Additional feedstock was added to digesters only when earlier feed practically ceased to yield further gas. Displacement of water in the second container by resulting biogas generated in first one provided a convenient method for monitoring gas yield.

Biogas yield was analysed in wet and partially dried conditions by carrying out batch type experiments at ambient temperature (26 ± 4)°C. It was found that green leaves caused higher yield as compared to dry or partially dried leaves. With green leaves, the yield varied from 51 to 64 litres per kg of feed whereas in case of partially dried leaves it varied from 47 to 58 litres. Higher temperatures caused more significant increase in the yield in case of semi-dried leaves than wet or fully green leaves. In contrast as per the NSTL studies [19], biogas yield was merely 11 to 18 litres per kg of green hyacinth plant. Comparative higher yield reported vide JNTU studies can be attributed to much higher ambient temperatures prevalent at Hyderabad than Mississippi. If we compare the gas yield from water-hyacinth and cowdung which are 50 to 64 litres per kg and 37 litres per kg respectively, we find it is much more in case of former. Complete results of these studies are summarised in Tables 8.6 to 8.9. Volatile-solid content of water-hyacinth which is a main contributing factor for gas yield compares very closely with that of cowdung and night-soil. Heat value of gas obtained from water-hyacinth and cowdung are more or less same. Sludge obtained from water-hyacinth based digester compares well in richness with that from cowdung and night-soil based plants.

Table 8.6 Comparative Characteristics of Water-hyacinth, Cow Dung and Night-Soil

Feedstock	*Moisture Content (Per cent)*	*Total Solids (Per cent)*	*Volatile Solids (Per cent of TS)*
Water-hyacinth	82	18	82.5
Cow dung	14-82	18-26	10-15
Night-soil	85-90	10-15	80-90

Table 8.7 Comparative Gas Yield and Heat Values Obtained from Anaerobic Fermentation of Water-hyacinth and Cow Dung

Feedstock	*Biogas Yield (m^3/kg of wet mass)*	*Heat Value (kcal/m^3)*	*Net Heat Value (kcal/kg of wet mass)*
Cow dung	0.037	2660	90.4
Water-hyacinth	0.053	2600	137.8

Table 8.8 Effect of Varying Loading Rate on Biogas Yield with Water-hyacinth as Feedstock

Feedstock	*Feed (VS. kg/m³.d)*	*Biogas Yield (m³/kg. VS.d)*	*Biogas Yield (m³/m³ of digester vol/day)*
Water-hyacinth	0.29	0.61	0.18
Digester – 1	0.58	0.40	0.24
Digester – 2	0.44	0.54	0.24
Cow dung		0.09-0.2	0.25-0.45

Table 8.9 Comparative Characteristics of Sludge Obtained from Anaerobic Fermentation of Water-hyacinth, Cow Dung and Night-Soil

Characteristic	*Per cent on Dry Weight Basis*		
	Water-hyacinth	*Cow dung*	*Night-soil*
Nitrogen (N)	2.48	1.4-1.8	3-5.0
Phosphorous (P_2O_5)	1.20	1.1-2.0	2.5-4.4
Potassium (K_2O)	2.70	0.8-1.2	0.7-1.0

8.7 Findings of an RRL (Jorhat) Study

B.G. Unni, *et al.* [17] working at the Biochemistry Division of the Regional Research Laboratory in Jorhat carried out a study to analyse the kinetics of biogas yield from water-hyacinth digested in unstirred conditions. Agitation of water-hyacinth based slurry is particularly desirable to prevent stratification which if not checked creates problems in digestion. Initially the fermentation was started with animal dung slurry which was subsequently mixed with increasing quantity of water-hyacinth in powdered form and finally with freshly chopped water-hyacinth. Laboratory scale digester consisted of 5 litre capacity digester bottle. Digestion was carried out for a period of 65 days with ambient temperature varying from 28° to 32°C and pH value between 6.5 to 6.9. Experiments were carried out for varying retention periods and volatile solid concentration of influent. Total solids, volatile solids and volatile acids were measured using standardised procedure prescribed by the American Public Health Association vide its manual 'Standard Methods for Examination of Water and Water Wastes, 1971.' Cellulose, hemicellulose and lignin were analysed by following the method described by H.K. Goering and R.J. van Soest in their book 'Forage Fibre Analysis, US Dept of Agriculture, Washington, D.C.' Carbohydrates were analysed by acetone reaction and proteins by the Kjeldahl nitrogen estimation procedure.

Composition and other characteristics of water-hyacinth as per cent of dry solids are given in Table 8.10. Effect of influent VS concentration and retention period on per cent VS digestion is given in Table 8.11. Volatile solids digestion rises with increase in retention period. For a given retention period, digestion increases with increase in influent VS concentration. Biogas yield corresponding to varying retention period and for different volatile solid concentrations in influent is given in Table 8.12. Gas yield was analysed to increase with increase in influent VS concentration and decrease in retention period. Proportion of methane in biogas was found to increase with rise in retention period. Gas productivity was analysed to be linearly related to the VS loading rate. [17].

Table 8.10 Composition and Characteristics of Water Hyacinth Used in Experiments

Parameter	*Values as Per Cent of Dry Solids*
1. Dry solids as % fresh weight	4-5
2. Volatile solids	70-80
3. Total liquids	2.1
4. Total Proteins (Kjeldahl N × 6.25)	27.5
5. Cellulose	24.7
6. Hemicellulose	27.7
7. Lignin	9
8. Water soluble organic matter (measured by dichromate method)	8.60
9. Water soluble carbodydrate (measured by anthrone method)	1.10
10. Water extractable proteins (Kjeldahl N × 6.25)	0.50
11. Water extractable (RNA)	0.02
12. Water extractable (DNA)	0.03

Table 8.11 Per Cent Digestion of Volatile Solids with Varying Retention Time and for Different VS Concentrations in Influent

VS in Influent g/l	*Per Cent Digestion of VS with Varying Retention Time*			
	10 days	*20 days*	*30 days*	*40 days*
16	30.0	40.0	41.8	54.4
28	32.0	48.4	51.4	58.6
40	38.0	53.0	55.6	58.5

Note : Per cent digestion of volatile solids

$$= \frac{\text{matrix(Influent VS concentration–steady state effluent VS concentration)} \times 100}{\text{Influent VS concentration}}$$

Table 8.12 Biogas Yield with Varying Retention Time and for Different VS Concentrations in Influent

VS in Influent g/l	*Biogas Yield in ml/g of VS added/day with Varying Retention Time*			
	10 days	*20 days*	*30 days*	*40 days*
16	147	194	175	194
28	127	164	190	212
40	136	155	161	178

8.8 A Weed Harvester Developed by CIFT (Cochin)

Central Institute of Fisheries Technology, Cochin, developed a harvester which is manufactured by a Maharashtra firm, Sri Vardhman Industrial Engineering Pvt. Ltd., Kolhapur [2]. Application of the machine makes weed removal faster. It is particularly appropriate for harvesting floating weeds like water-hyacinth (*Eichhornia sp.*), salvinia (*Salvinia sp.*), water lattuce (*Pitia sp.*), submerged weeds like *Hydrilla, Vallisneria* and *Naias* and rooted weeds like loture (*Nymphea*) and lilies (*Nelumbium*). The CIFT developed two kinds of machines, one meant for removal of only submerged weeds and the other for removal of both submerged and floating weeds. It is designed to clear about 1.5 hectares of area in 8-hours shift operation. The machine consists of a shallow draft and self-propelled barge made up of mild steel tanks. It has a cutter to chop the weeds as they pass through the cutter. A releaser releases the weeds from the carrier rake which then fall into the hopper of a roller crusher wherein weeds are crushed to smaller sizes. Crushed weeds are kept in tilting type trays. Barge is driven by paddle wheels with independent clutch control which is also used for steering the barge. All machine controls are located at one place so that one person can handle them conveniently [For further details reader may contact the manufacturer].

8.9 An Overview of Biogas Generation from Algae

Algae is another major waterweed which can be utilised for biomethanation. Several studies have been carried out both in India and abroad on use of algae as feedstock. A study was carried out at the University of California. Slurry consisted of 5 per cent algae and 95 per cent water. Algae was cultivated in large shallow ponds and stirred occasionally. Harvesting was done either by centrifuging or by adding aluminium sulphate which helped in settling it. Results of studies showed that composition and volume of gas produced per kg of digested algae was more or less the same as from sewage sludge. A period of 11 days was taken as digestion period. One acre pond with algae cultivation was estimated to generate 11 kW of electric power.

Growth of algae in water depends on many factors including salinity and alkalinity of water, concentration of dissolved oxygen and carbon dioxide, presence of pollutants and contaminants in air and water, ambient and water temperature, wind velocity and intensity of light penetration in water. Several species of algae form chains or filaments that aggregate to generate colonies. However, the common species known as *phytoplankton* grow as single cells that are suspended in water.

8.10 Common Algae Varieties

There are several ways in which algae can be categorised. One common way is to classify them as micro or macro-algae depending upon size. Micro-algae comprises several microscopic cells which range in size from less than 2 μm to 1 mm. Some microalgae varieties which have high protein content provide valuable food ingredients for both livestock and human-beings. Mass cultivation of micro-algae as a food ingredient is commercially practised in Mexico and several far east countries, *chlorella* and *spirulina* are the species which are commonly grown for this purpose. *Spirulina* has about 74 per cent protein, 14 per cent insoluble acids and 8.5 per cent carbohydrates. Algae can also be classified into five groups (Table 8.13) on the basis of their pigment content and certain biochemical characteristics. These groups are blue green algae (*myxophyceae*), green algae (*chlorephyceae*), red algae (*rhodophyceae*), brown algae (*phacophyceae*) and golden algae [8]. Aquatic weeds have higher moisture which sometimes can be as high as 90-95 per cent. Submerged plants have higher proportions of crude-protein emergent plants. Chemical composition of select aquatic weeds with high energy potential is given in Table 8.14 [8].

Table 8.13 Different Types of Algae

Category	*Description*
1. Blue-green algae	Common species are filamentous or colonial. Some forms such as *spriulina* are very rich in protein.
2. Green algae	Includes both microalgae and macroalgae species. *Chlorella* is a common unicellular green algae. Ulva, is a macroalgae belongs to this category.
3. Red algae	These are macroscopic seaweeds having reddish-brown or bright red colour. Some of their common species are *Chondrus Crispus* (Irish moss) and *Gracilaria.* This group is commercially important and yields hydro-colloids, carrageenan, and agar.
4. Brown acid algae	These are macroscopic seaweeds. *Sargassum notans* and *Macrocystis pyrifera* (or giant kelp). The latter species yield hydrocolloid alginic acid.
5. Golden algae	These are mostly unicellular or colonial planktonic species.

Source : Kinderman, E.M., *et al.*, Technical and Economic Evaluations of Biomass Utilization Processes, Tech. Report No. 1, SRI International, Menlo Park, California, Sep., 1980.

Table 8.14 Composition of Select Aquatic Weeds With High Energy Potential Expressed in per cent Dry Weight

Description	*Ash*	*Crude Protein*	*Fat*	*Fibre or Cellulose*
Submerged				
1. Partially submerged species	2.8	17.5	3.5	32.6
2. *Hydrilla*	27.1	18.0	3.5	32.1
3. *Eichhornia Crassipes*	1.7	16.0	3.4	28.1
4. *Lemna Emergent*	12.3	24.4	2.8	10.4
5. *Typha Latifolia*	6.9	10.3	3.9	33.2

Source : Kinderman, E.M., *et al.*, Technical and Economical Evaluations of Biomass Utilisation Processes, Tech. Report No. 1, SRI International, Menlo Park, California, Sep., 1980.

Aquatic algae found in suspension are called planktonic and those living on bottom as benthic. Planktonic algae generally consists of flora, fauna, and fungi of organisms in suspension. Benthic algae on the other hand comprise attached and bottom-dwelling organisms and are known by different names depending on the place of location. Algae living on stones are called epilithic, those attached to mud or sand are called epipetic, those living on plants and vegetations are called epiphytic, and those attached to animals are called epizoic.

Harvesting of micro-algae is somewhat difficult. Single-cell species offer more difficulty in harvesting than multi-cell ones. It is more convenient to cultivate and harvest filamentous blue-green algae like *Spirulina, Ocillatoria* or colonial green algae than single cell species. Planktonic algae which are found in suspension can be harvested by drawing a plankton net through water. This net is usually made of silk of fine mesh of about 180 pores per square inch. Silk net acts as a strainer which helps in filteration and accumulation of planktonic algae and other fine species

8.11 Estimates of Regionwise Algae Availability in India

Algae is abundantly available in different parts of India. Y.A. Doshi and A.V Rao [6] made estimates of seaweeds (algae) availability in different coastal areas of India (Table 8.15). Although coastline of India extends over 5700 km, seaweeds are mostly limited to narrow littoral and sublittoral belts of country's marine environment. Seaweed resources are limited to rocky coral formations in Gujarat, Andhra and Tamil Nadu states, in Karwar, Ratnagiri, Varkala, Visakhapatnam, Chilka and Pullicat lakes; and islands of Laccadives, Andaman and Nicobar. *Agarophytes* and *alginophytes* are found in commercially viable quantities on east and west coasts (Table 8.16). Seaweeds can be used for

Table 8.15 Seaweed Availability in Different Coasts of India

Region	*Species*	*Estimates of Availability (Tonnes)*
1. Gujarat Coast		
(a) Gulf of Kutch	*Alginophytes*	18705
(b) Okha to Dua	*Alginophytes*	5.41
	Agarophytes	29
	Ulva	39
2. Maharashtra Coast	*Alginophytes*	3.10
	Ulva	5
3. Tamil Nadu Coast	*Alginophytes*	3.10
	Agarophytes	1.048
	Hypnea	7.98
4. Lakshadweep Islands	*Alginophytes*	9-15
	Agarophytes	961-2074
	Other seaweeds	2675-5509

Source : Doshi, Y.A. and Rao, A.V., Chemicals from Indian Seaweeds, Journal of Scientific and Industrial Research, 139 (7), 1980, p. 364.

producing variety of chemicals, minerals, vitamins, amino acids, and proteins. They contain a variety of polysaccharides which constitute an industrially important group of phycocolloids. This group forms 10-65 per ccnt of dry weight of harvested seaweeds. Among polysaccharides, agar-agar and carrageenam obtained from red algae, and algin from brown algae, have considerable economic importance [9].

Table 8.16 Yield and Physical Properties of Agar-agar (Agarophytes Variety) Available on Tamil Nadu Coast

Seaweed	*Monthwise Yield*		*Gelation Strength (g/cm^2)*	*Gelation Temperature (°C)*	*Melting Temperature (°C)*
	Per cent	*Month*			
1. *Gelidiella Acerosa*	50	Jan	325	38-52	61-86
2. *Gracilaria Edulis*	45-51	Jan	139	37.5-51	48-75
3. *Gracilaria Foliifera*	40.4	Jan	56	37.5-51	48-70
4. *Gracilaria Corticata*	50.4	Jan	22	40-49	50-60

Source : Krishnamurthy, V.S. *et al.*, Proc. of the Symposium on Marine Algae of Indian Ocean Origin, Central Salt and Marine Chemicals Research Institute, Bhavnagar, 1979.

8.12 Findings of a CSMCRI (Bhavnagar) Study

S.J. Tarawadi and V.D. Chauhan [16] carried out experimental studies at the Central Salt and Marine Chemicals Research Institute, Bhavnagar, for producing biogas from *Sargassum tenessium*, a brown seaweed that was collected from the coastal Gujarat and stored in dried conditions. Before drying, the collected brown algae was cleared off extraneous materials such as sand and salt by soaking in water. Measured characteristics of dry algal powder used in the experiments are summarised in Table 8.17.

Table 8.17 Properties of Dry Algal Powder used as Feed Material

Characteristic	*Per cent*
1. Volatile solids	77.5
2. Non-volatile solids	22.5
3. Chemical oxygen demand	1.0
4. Total nitrogen	1.9
5. Salts soluble in water	22.0
6. Salts soluble in ether	2.2
7. Cellulose	7.0
8. Lignin	2.0
9. Proteins	12.3
10. Phosphorus (P_2O_5)	0.1
11. Potassium (K_2O)	1.2
12. Carbon nitrogen ratio (C : N)	23 : 1

Two strains obtainable by decomposition of Sargassum were identified which could cause about 28 per cent breakdown of seaweeds in about 45 days fermentation period. These strains were used to cause primary biodegradation of seaweed for yielding biogas. In a separate experiment, effect of adding bacterial strains to cow dung slurry on gas yield was also studied. In a batch type laboratory digester, 250 g of cow dung was added to 1 kg of Sargassum used as a feed for anaerobic fermentation. With gas production having started from 13th day onwards, 49.3 litres of gas yield was obtained out of a total of 60.6 litres of gas generated. In the second experiment in which no bacterial strain was added, gas yield was 21.7 litres out of a total of 28.7 litres of gas produced with gas generation having started from the 21st day onwards.

Effect of variable feed rate on gas yield was also analysed. For this purpose two laboratory-scale digesters of 6-litre capacity with facilities for feeding, discharging and gas collection were set up. Varying feed rates of 1, 2 and 2.5 kg VS/m^3 day were used to determine the optimal rates for biogas yield. It was observed that a feed rate of 1 kg VS/m^3 day caused maximum gas yield with gas formation having started from the seventh day onwards, and accordingly further investigations were carried out by varying other parameters but keeping feed rates fixed at this value. Effect of daily feed versus on alternate on gas yield was also analysed. In case of daily feeding, a feedstock comprising 6 g algal powder, 850 g seaweed, 50 g ulva (green seaweed), 100 g cowdung and 100 ml of bacterial strain produced 3.7 litres of biogas per day [5]. In case of alternate day feeding but keeping all other parameters same, resulting biogas yield was 3.58 litres per day (Table 8.18).

Table 8.18 Effect of Daily Versus Alternate Day Feed on Biogas Yield from Dry Algae

Loading Pattern	*Digestion Period (days)*	*Biogas Commenced on day*	*Biogas Yield (litres)*	*Biogas Yield (litres/day)*
Loaded daily				
1 kg VS/day	133	7th day	494.1	3.70
Loaded on alternate days				
1 kg VS/day	56	11th day	200.6	3.58

Under experimental conditions, maximum yield of biogas was found in the range of 400-500 litres per kg of dry algal powder. Production of gas in semi-continuous mode was greater than in batch-operation. Semi-continuous digester operation resulted in biogas yield in range of 470-500 litres per kg of seaweed. In batch-type digester at room temperature 26-31°C, biogas yield was 49 litres per kg of seaweed. Carbon : nitrogen ratio was kept at 23 : 1 and pH was maintained in the range of 6.8-7.4. Comparative performance of seaweed digester and cow dung digester showed that biogas yield per kg of volatile matter feed was 0.083-0.36 m^3/day for marine algae whereas for cow dung based digester it ranged between 0.09-0.2 m^3/day. Comparison of heat value of gas obtained from different feedstocks showed that it was quite high in case of marine algae (4900 kcal/m^3) being only marginally less than that of pure methane (5320 kcal/m^3). Salinity of water more than 2 per cent if used in slurry preparation had detrimental effect on gas yield. Sludge obtained from seaweed digestion could be used as an enriched fertiliser and soil-conditioner [5, 16].

8.13 Findings of an IARI (New Delhi) Study

G.S. Venkataraman and B.D. Kaushik [1, 18] carried out studies to analyse the effect of adding algae in feed on gas yield from a biogas plant primarily running on cowdung slurry as feedstock. Algae used in these experiments was collected from Pirana sewage oxidation ponds in Ahmedabad city which included several species like *oscillatoria chalybea, euglena, scenedesmus, spirulina,* and *merismopedia* in which however *oscillatoria chalybea* formed the largest proportion. Laboratory-scale digesters used in the experiments consisted of a battery of 10 litre aspirator bottles each with slurry containing 4 kg of fresh cowdung and four litres of water. Experiments were carried out at ambient temperatures of 16-18°C, 20-22°C and 30-31°C. The resulting biogas was collected over saturated brine solution and measured by liquid displacement method. For analysing the effect of algae, in one experiment, feed comprising cowdung slurry was suplemented with 3 per cent dried algae.

Results of experiments were analysed for two different variants, namely, with and without algal supplementation at different ambient temperatures which are graphically shown in Fig. 8.2. It was found that with increasing fermentation period, gas yield increased initially but slowed down later. Addition of algal powder was reported to stimulate yield which was about 1.64 times more (3114 cm^3/kg of feed) compared to the yield of 1900 cm^3/kg obtained from cowdung slurry alone. Addition of algae powder however did not significantly affect composition and combustibility of gas produced. Addition of algal powder is thus an effective way of improving gas yield which in small amounts can be grown in sewage oxidation ponds.

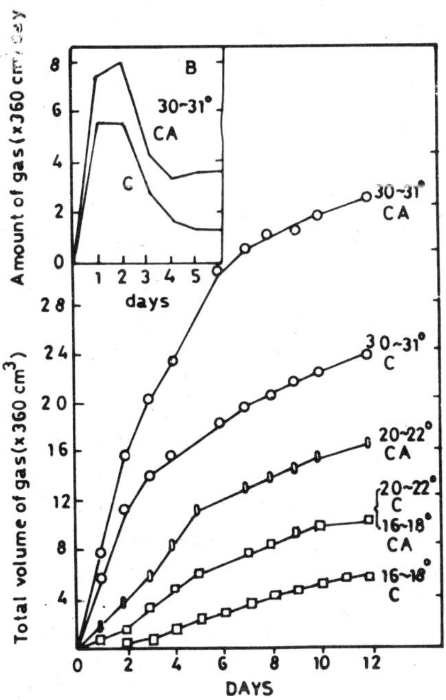

Fig. 8.2 Gas production by cowdung with (CA) and without (C) algal supplementation at different ambient temperatures; B, rate or gas production per day by cowdung with (CA) and without (C) algae at 30-31°C

8.14 Concluding Remarks

Experimental studies carried out in India and abroad have clearly established the appropriateness of certain waterweeds like water-hyacinth and algae as a feed material for biomethanation. Whereas results of experimental studies have been encouraging, not many worthwhile attempts have been made to harness these sources on commercial scale. There is a need to utilize large quantity of vegetable organic biomass generated in tropical and sub-tropical regions for biomethanation. The main problem is not of assessing their potential but deals with practical aspects of handling large volume of biomass and mixing it with appropriate agricultural waste materials [12]. Utilisation of waterweeds as feed material can be particularly advantageous for community plants whose economies-of-scale warrant their extensive utilisation. However, if for any biogas plant, waterweeds are not available in adequate quantities, these can be used in combination with other organic wastes as supplementary feed.

REFERENCES

1. Chawla, O.P., *Advances in Biogas Technology,* Indian Council of Agricultural Research. New Delhi, p. 84-95.
2. CIFT, *Aquatic Weed Harvester,* A Report, Central Institute of Fisheries Technology, Cochin, Kerala, 1976.
3. CMERI, *Biogas Research Activities at CMERI,* A Report, Central Mechanical Engineering Research Institute, Durgapur, 1980.
4. CMERI, *Water-hyacinth Fed Biogas Plant as Designed and Developed at CMERI,* A Report, Central Mechanical Engineering Research Institute, Durgapur, 1980.
5. CSMCRI, *Biogas from Seaweeds,* A Working Paper, Central Salt and Marine Chemicals Research Institute, Bhavnagar, 1980.
6. Doshi, Y.A. and Rao, A.V., Chemicals from Indian Seaweeds, *Journal of Scientific and Industrial Research,* **39(7)**, 364-371 1980.
7. Gopalakrishnan, K.V. and Murthy, B.S., Potentialty of Water Hyacinth for Decentralised Power Generation in Developing Countries, *Regional Journal of Energy, Heat and Mass Transfer,* **1(4)**, 349-357 1979.
8. Kinderman, E.M., *et al., Technical and Economic Evaluation of Biomass Utilisation Processes,* Tech. Report No. 1, SRI International, Menlo Park, California, Sep., 1980.
9. Krishnamurthy, V.S., *et al., Proc. of Symp. on Marine Algae of Indian Ocean Origin,* Central Salt and Marine Chemicals Research Institute, Bhavnagar, 1979.
10. Murthy, S.R., Biogas from Water Hyacinth, *Decan Chronicle,* Secunderabad, April 8, 1980, p. 7.
11. Potty, V.H., *Utilisation of Water Hyacinth,* A Report, Industrial Development and Consultancy Services, Central Food Technological Research Institute, Mysore, Oct. 23, 1981.
12. Rao, S.V.R., Anaerobic Digestion of Two Aquatic Weeds for Energy Generation, *Resource Management and Optimisation,* **3(4)**, 343-352 1986.
13. RRL, *Production of Paper from Water Hyacinth,* A brochure, Regional Research Laboratory, Jorhat, June 31, 1981, p. 1.

14. Siregar, H., *et al., Effect of Water Hyacinth Holder in the Digester Liquid on the Biogas Production,* Proc. of the ASEAN Conf. on Energy from Biomass, Penang, Malaysia, Oct. 13-15, 245-253 1986.
15. Sriramulu, N. and Bhargava, B.N., *Biogas from Water Hyacinth – Part I and II,* Jawahar Lal Nehru Technological University, Hyderabad, 1980.
16. Tarwadi, S.J. and Chauhan, V.D., Seaweed Biomass as a Source of Energy, *Energy,* **12(5)**, 275-378 1987.
17. Unni, B.G., *et al., Steady State Kinetics of Biogas Production from Water Hyacinth in Unstirred Reactor,* Proc. National Seminar on Biogas Technology, Ludhiana, Jul 19-11, 1981.
18. Venkataraman, G.S. and Kaushik, B.D., Stimulation of Gas Production from Cowdung by Algal Supplementation, *Current Science,* **44(22)**, 807-808 1975.
19. Wolverton, B. and Mc.Donald, R.C. Don't Waste Waterweeds, *New Scientist,* Aug. 12, 318-320 1976.

Chapter 9

BIOGAS GENERATION FROM INDUSTRIAL WASTES

9.1 Introduction

Biogas can be generated not only by anaerobic fermentation of animal wastes and agricultural plant materials but also host of industrial wastes. Wastes from sugar factories, textile-mills and food-processing industry are particularly appropriate for fermentation. Sugar-mill wastes like bagasse and pressmud can be gainfully utilised for biomethanation. Bagasse is traditionally used for steam generation, preparing animal feed, biomanure, furfural, paper and fibre-boards which recently has also been used as feed material for biomethanation. Pressmud which is another sugar plant waste has also considerable potential as feedstock for biomethanation. Pressmud, in normal course, is used for extraction of wax and as fertiliser in agricultural field. Distilleries cause environmental pollution by their effluents. Anaerobic digestion of distillery-wastes not only helps to recover energy but also keeps the environment clean. Biogas can be recovered by digestion of distillery wastes, spent grams, beet pulp, root and other residues. Cotton-mill wastes which are used as main or supplementary boiler fuel are also valuable feedstock for biomethanation. Willow-dust is one of the solid cellulosic textile-mill wastes which can be anaerobically fermented. Biogas can also be obtained from canteen and confectionery wastes. Wastes from food-processing industry are yet another valuable resource whose availability is going up from year to year with changing lifestyle. In this chapter it has been endeavoured to comprehensively review important studies carried out in past on biogas generation from industrial wastes.

9.2 Biogas Generation from Bagasse

Bagasse a fibrous ligno-cellulosic residue resulting from sugar-cane following extraction of juice is used for variety of applications including biomethanation. Its composition varies from place to place and fibre size depends on cane variety and processing technology used. Bagasse in general comprises 45 per cent cellulose, 28 per cent hemicellulose (pentosans), 20 per cent lignin, and 2 per cent as sugar wax, minerals and other impurities. Pentosans include products like araban, galactan and xylan. Due to variable composition, it is infeasible to estimate carbon, oxygen and hydrogen content in bagasse in precise terms. It comprises 5 to 7 per cent of hydrogen with balance equally divided between carbon and oxygen on dry weight basis. Cellulosic content of bagasse is responsible for its fuel value. Pre-drying of bagasse helps to raise its calorific value which depends on cane variety, fibre size and external area. On dry and wet basis, it is estimated to be somewhere close to 8350 BTU/lb and 3250 BTU/lb, respectively.

Moisture content in bagasse varies from region to region. For example, in northern India, it is found to vary from 33 to 36 per cent whereas in southern region it varies from 20 to 30 per cent [20].

Chemical composition of bagasse as generated in different world-regions is given in Table 9.1. In view of the fact that bagasse is primarily a fibrous material, its properties compare well with those of hard and softwood. Comparative characteristics of hardwood, softwood and bagasse are given in Table 9.2 [13]. Bagasse undergoes change in characteristics with time [17]. Comparative characteristics of fresh and one year old bagasse are summarised in Table 9.3. In past, studies were carried out at the National Sugar Institute in Kanpur [12] to study biogas potential of bagasse. It was reported that 1 tonne of dry organic matter generated 200-300 m^3 of biogas and 0.5 tonne of biomanure. Biomanure was found to be rich in nutrients like potassium, calcium, phosphorus, nitrogen and humus and superior to ordinary farm manure in several respects. Biomanure contained 75 per cent nitrogen, 1-1.2 per cent phosphorus and 0.6-0.8 per cent potassium. Based on experimental studies, G.S. Narsimhamurty and A. Purushothaman [11] reported that it is possible to obtain higher biogas yield from bagasse if instead of one digester two digesters connected in series were used. By this arrangement corresponding to a total residence time of 20 days it was possible to convert 71.4 per cent of bagasse to biogas.

Table 9.1 Comparative Properties of Bagasse Available in Different Parts of the World

Area	*Composition in Per Cent on Dry Weight Basis*			
	Cellulose	*Hemicellulose*	*Lignin*	*Ash*
India	57-63	12.59-22.80	16.9-30	1.0-3.47
Mauritius	44.4-48	27.8-29.7	13.0-14.3	1.8-2.4
South Africa	41.7-45.3	24.1-26.9	20.2-22.1	1.6-1.9
Hawaii, USA	36.5-38.8	25.8-27.3	19.4-21.1	1.3-2.0
Louisiana, USA	58.4	29.4	21.3	2.9
Florida, USA	48-49	26.6 -28.7	16.4-18.6	1.0-2.2
Puerto Rico, USA	50.9	29.6	48.1	3.9
Philippines	56.8	31.8	22.3	2.3

Note : Hemicellulose cover pentosans which include araban, galactan, xylan etc.
Source : Upadhyay, A.R., Handbook of Sugar Factory Management, Rajiv Sadan, Ratlam, 1969, p. 468.

Table 9.2 Comparative Composition of Bagasse and Wood

Characteristics	*Unit*	*Hard wood*	*Soft wood*	*Bagasse*		
				Whole	*Fibre*	*Pitch*
Cellulose	Per cent	38-48	40-45	30-39	38-43	26-36
Pentosans	–do–	20-25	10-13	24-30	27-32	28-33
Lignin	–do–	20-25	26-34	18-22	20-23	18-22
Ash (Silica, iron etc.).	–do–	0.3-1.2	0.2-0.8	1-4	0.6-1.2	1.8
Average fibre length	mm	0.7-1.6	2.7-3.6		1-2	0.25-0.40

Source : Paturau, A., By-Products of the Sugar Cane Industry, Elsevier, New York, 1969.

Table 9.3 Comparative Composition of Fresh and Stored Bagasse

Characteristics	*Composition in per cent*	
	Fresh Bagasse	*One year Old Bagasse*
Cellulose	54.2	52.0
Pentosans	23.4	23.6
Lignin	21.3	19.1
Ash	0.7	2.8

Source : Singh, M. and Paul, B.B., Role of New Technology, Economic Times, Feb. 4, 1981, p. 4.

9.3 Biogas Generation from Pressmud

Pressmud is a sugar factory waste which is rich in organic matter (20.3 per cent), nitrogen (0.4 per cent), phosphate (0.4 per cent), potash (0.02 per cent) and sterols (0.33 per cent) has considerable commercial potential as feedstock for biomethanation. In over 325 sugar mills in the country around 60 million tonnes of sugar-cane is crushed annually which generates about 2.5 million tonnes of pressmud. One tonne of pressmud following anaerobic fermentation yields 50 m^3 of biogas. If the entire pressmud available in the country is used for biomethnation, it can generate 120 million m^3 of gas valued at 180 million rupees at a rate of Rs. 1.50 per m^3 of biogas [10]. B.G. Unni, *et al.* [19] carried out laboratory-scale and pilot studies for producing biogas from pressmud separately during winter, spring and summer seasons the results of which are summarised in Table 9.4. In view of the fact that pressmud normally has high soil, sand and grits content, formulation is somewhat less smooth on account of stratification inside the digester. The digester was fed at a rate of 2.1 kg VS/m^3. d during winter season and at a rate of 1.05 kg VS/m^3. d during summer period causing biogas yield of 0.48 m^3/m^3. d and 0.59 m^3/m^3d, respectively with methane content ranging between 70 to 72 per cent. P. Rajasekran, *et al.* [14] studied the effect of adding pressmud, groundnut shell and paddy-husk as additive to cowdung slurry as feedstock. Gas output and methanogenic activity were maximum in pressmud mixed with cowdung (12, 173 ml and 222.3 $\times 10^3$/g) followed by groundnut mixed with cowdung (11,962 ml and 189.66 $\times 10^3$/g). Studies carried out, thus confirm the utility of mixing pressmud and groundnut shell as supplementary feedstock for added biogas yield.

Table 9.4 Biogas Yield and Operating Parameters from Sugar Cane Pressmud

Parameters	*Winter and Spring period (Dec 1983 to Apr 1984)*	*Summer period (May to July 1989)*
1. Feed rate (kg/d)	100	50
2. Retention time (days)	30	300
3. Digester temperature (°C)	17-24	21-26
4. Ambient temperature:		
Maximum (°C)	19-36	30-42
Minimum (°C)	7-25	18-23
5. Loading rate (kg VS/m^3d)	2.1	1.05
6. Methane content (%)	72	70
7. Biogas yield:		
(m^3/d)	4.8	5.9
(m^3/kg feed added)	0.048	0.118
(m^3/kg TS added)	0.160	0.393
m^3/kg VS added)	0.228	0.562

Source : Unni, B.G., Pillai, K.R., and Singh, H.D., Production of Biogas from Sugarcane Pressmud: Laboratory Scale and Pilot Plant Studies, Energy Management, 11 (1), 23-32 (1987).

As with other feedstocks, pressmud based plant sludge is a rich fertiliser but on comparative terms somewhat less in view of its low pH and acidic nature. The sludge can be used for making other products as well. It can be used for extraction of phytosterols (beta-sito-sterol) 69 per cent, stigmasterol 18.1 per cent, compesterol and brassicosterol 12.8 per cent which can be biodegraded with a particular bacterium to form precursors of steroidal drugs [14]. The slurry after extraction of phytosterols can be used as a biomanure on account of its high NPK values.

According to MNES estimates, 4.2 m.t. of pressmud available from 104 m.t. of cane crushed during 1992-93 could generate 3.5 $\times 10^5$ cu.m of biogas equivalent to about 700 mw of power. Existing 425 sugar mills in the country produce about 5 m.t. of pressmud annually. If the entire pressmud is used for biomethanation, it could yield 3215 kilolitres of biogas equivalent to 1.1 m.t. of wood, 0.2 million kilolitre of kerosene, 014 m.t. of LPG, or 386 million kwh of electricity [24].

9.4 Biogas Production from Canteen Wastes

Canteen wastes particularly those from industrial canteens provide valuable feedstock for biomethanation. Canteen wastes mainly comprise vegetable and fruit wastes, food left-overs bread pieces, and cooked rice. Y.Y. Yeole, *et al.* [23] carried out experiments for analysing digestibility of canteen wastes by fermenting them in 25 and 200 litre capacity laboratory-scale digesters of the KVIC design. The plants were separately fed with canteen wastes and cattle dung, and performance analysed. Chemical composition of fresh canteen wastes and cattle dung as used in the experiments is summarised in Table 9.5 below:

Table 9.5 Composition of Fresh Canteen Waste and Cattle Dung

Constituents	*Kitchen Waste*	*Cattle Dung*
	(Values in g/kg of fresh wastes)	
Moisture	811	826
Total solids	189	174
Total volatile solids	173	153
Cellulose	5	28
Hemicellulose	Traces	22
Lignin	Not estimated	17
Starch	44	23
Lipid	43	28

Biogas obtained from 25 litre experimental plant for a 35 day retention period was 31.4 litre/kg per day with 71 per cent methane content as against a yield of 17 litre/kg per day with 62 per cent methane content from cattle dung. Biogas yield from 200 litre capacity plant while feeding kitchen wastes was analysed to be 54 litres/kg per day with 68 per cent methane content. This shows that kitchen waste produces more gas as compared to cattle dung which can be attributed to easily biodegradable matters like carbohydrates, proteins, fats etc. present in the former as against more complex polymers like cellulose, hemicellulose etc. present in the latter.

9.5 Biogas Production from Distillery Wastes

Distilleries need energy, both for the extraction of fermentable matter from cereal, sugar-cane, beet, cassava and other biomass and subsequent distillation. Distilleries also cause environmental pollution by their effluents. Biogas can be obtained through fermentation of distillery wastes, spent grams, beet pulp, root and other residues. The amount of biogas produced depends on nature of feedstock. One cubic metre of distillery wastes following anaerobic fermentation can yield 14 m^3 of biogas. S.N. Kaul, *et al.* [7] made a comprehensive review of state-of-art of industrial waste water treatment including resulting methane recovery. Biogas recovery from distillery wastes having COD concentration of 50,000 mg/l can go a long way in meeting at least part of the industrial energy needs. Rampur Distillery and Chemicals Co. Ltd. at Rampur in Uttar Pradesh set up a biogas plant in eighties for producing biogas and enriched fertiliser from distillery wastes. According to Mr M.K. Gupta, Vice-President (Rampur Distillery), the company proposes to set up a biogas-based co-generation unit in near future. The plant is at present generating 40 m^3 biogas/m^3 of spent wash at N.T.P. through thermophilic fermentation (50-52°C) of distillery effluent and achieving 85 to 90 per cent BOD reduction. The resulting biogas (calorific value 5000 kcal/m^3) contains 60 to 62 per cent methane, 2 to 3 per cent hydrogen sulphide, and balance carbon dioxide [25]. As per broad estimates made by the Union Ministry of Environment and Forests, if 90 per cent of the existing distilleries in India install biogas recovery plants for reducing the spent-wash BOD, they have the capacity to contribute an equivalent saving of about 2,80,000 tonnes of coal per annum.

9.6 Biogas Production from Textile Mill Wastes

There exists considerable scope for biomethanation from textile wastes in cities like Mumbai, Ahmedabad and Coimbatore where large number of textile mills are located. As per rough estimates about 5.28 million cubic metres of biogas can be produced from anaerobic fermentation of 30,000 to 33,000 tonnes of willow-dust produced every year. This is equivalent to three million litres of diesel or 17.4 kwh of electricity. According to the results of studies carried out by the NEERI, Nagpur, 54 per cent of cotton wastes are finer than 0.707 mm., 13 per cent range between 0.707 to 2 mm and remaining more than 2 mm size. Comparative chemical analysis of typical raw cotton dust from textile mills and municipal refuse in India is summarised in Table 9.6. C : N ratio of cotton wastes is favourable for anaerobic fermentation. Cotton dust is uniform in composition as compared to municipal refuse and free from problems of shredding and segregation. Furthermore, as textile mills are mostly located in big cities and as cotton dust is clean and homogenous waste material, it can be easily packed, retailed and carried to biogas plant site [4].

Table 9.6 Chemical Analysis of Typical Raw Cotton-dust and Municipal Refuse Generated in India

Indicator	*Value of Indicator in Per cent by Weight*	
	Cotton Dust	*Municipal Refuse*
Moisture	8.0	
Organic matter	70.0	
Carbon	41.0	
Nitrogen	1.4	0.4-0.7
Phosphorus as P_2O_5	0.6	0.3-0.5
Potash K_2O	1.2	0.3-0.7
C : N ratio	29.3	30-35
pH	6.2	

Source : NEERI, Disposal of Blowroom Cotton Dust from Textile Mills by Composting, A Report, National Environmental Engineering Research Institute, Nagpur, 1970.

In past R.H. Balasubramanya, *et al.* [2] carried out experimental studies at the Cotton Technological Research Laboratory (CTRL) in Mumbai to analyse the pattern of biogas yield from willow-dust. According to their findings dry anaerobic fermentation of willow-dust consumes less water, accommodates more fermentable material per unit space, and generates improved quality of readily usable biomanure. A pilot plant for processing willow-dust for biogas production in a battery of six batch digesters was installed in a textile mill. The depth of loading of tank during anaerobic pretreatment was found to affect the biogas yield with solid to liquid ratio varying from 1 : 6 to 1 : 1.5. About 725 m^3 of biogas/ 2 tonne of willow-dust was obtained in 90 days in three solid loading levels of 2, 4 and 6 tonnes during pretreatment when the solid to liquid ratio was 1 : 6. A gradual decline in gas production was observed during anaerobic fermentation. When the solid to liquid ratio was changed to 1 : 1.5 (dry fermentation) about 500 m^3 of biogas/tonne of willow-dust was produced in 30, 60 and 120 days with 2, 4 and 6 tonnes of initial solid loading. In this case gas was produced at more or less uniform rate during corresponding anaerobic fermentation. P. Rajasekran, *et al.* [14] studied biogas yield from cellulose rich wastes such as willow – dust before and after predigestion. Cellulose wastes were predigested using 0.15 per cent sodium hydroxide along with 2 per cent lime for 72 hours. These wastes were anaerobically fermented in batch process both with and without cowdung. Biogas yield was found to increase if willow-dusts were mixed with cattle manure after predigestion [3]. Certain oil cakes like castor cake and neem cake when used in combination with industrial wastes such as willow-dust from cotton mills provided a valuable additive for biomethanation [21].

9.7 Biogas Production from Confectionery Wastes

Biogas can also be generated from confectionery wastes. Considerable solid wastes are generated while producing confectionery items like biscuits and chocolate. Solid wastes obtained from mixing and packing units are rich in carbohydrates and easily biodegradable, and thus can be a good substrate for biogas production. Generally these wastes are mainly disposed of by incineration. These wastes mainly consist of *maida* (fine wheat flour), *ghee* (milk fat), pieces and powder of biscuits with chocolate paste. D.R. Ranade, *et al.* [15] studied biogas production from confectionery wastes generated by biscuit and chocolate manufacturing plant. These wastes were anaerobically fermented in a 180 litre capacity movable drum type biogas plant. Biogas yield was analysed corresponding to three different hydraulic retention times of 20, 30 and 40 days with 10 per cent total solids in the influent slurry. It was found that confectionery wastes are easily amenable to anaerobic fermentation. The fermentation process was more stable at 40 days HRT as compared to 30 days HRT. Biogas yield at 40 days HRT was reported to be 261 litres per day with 57 per cent methane content; and at 30 days HRT as 140 and 42, respectively in the same units.

9.8 Biogas Production from Fruit and Vegetable Wastes

Fruit and vegetable wastes provide valuable input for biomethanation. In the fruit and vegetable canning industry, part of the relevant waste can be used as feedstock for digestion. The remaining waste may be transported to a municipal landfill site. Biogas thus produced may meet significant part of a factory's energy demand [1]. Composition of some common fruits and vegetables for assessing their suitability for anaerobic fermentation is given in Tables 9.7 and 9.8. Important characteristics of processed fruit and vegetable wastes for analysing their potential for is given in Tables 9.9 and 9.10. Wieger Knol, *et al.* [8] studied anaerobic fermentation of wastes from fruit and vegetable industry at the Central Institute of Nutrition and Food Research (CINFR) at Zieist (the Netherlands) way back in 1978. Wastes were divided into three groups depending upon carbohydrate content on dry weight basis as they affect rate of biomethanation. Spinach-waste and asparagus peels were placed in group 1 with their carbohydrate content in dry weight being 0.05 and 0.06 per cent. The larger group 2 comprised apple pulp, apple slurry, carrot waste, French bean waste and strawberry slurry with a carbohydrate content varying between 0.21 and 0.29 per cent. Group 3 comprised green pea slurry only with carbohydrate content in dry weight being 0.44 per cent. Source and chemical composition of different wastes used are summarised in Table 9.11. Wastes in these groups were subjected to anaerobic fermentation individually. Information concerning operating parameters and biogas yield is given in Table 9.12. The study confirmed that anaerobic digestion could be a suitable process for the treatment of waste materials from the fruit and vegetable industry.

Table 9.7 Per cent Composition of Common Fruits for Analysing their Potential for Anaerobic Fermentation

Fruit	*Water*	*Protein*	*Fat*	*Carbohydrate*	*Ash*	*Fibre*
Apple	84.1	0.3	0.4	14.9	0.02	1.2
Apricot	85.2	1.0	0.1	12.9	0.04	–
Berries	86.2	1.0	0.8	12.0	0.60	2.1
Cherries	87.0	1.3	0.5	13.3	0.05	0.2
Citrus	87.7	0.8	0.3	10.6	0.04	0.6
Olive	75.2	1.5	13.5	4.0	0.10	–
Peach	86.9	0.5	0.1	12.0	0.03	0.6
Pear	82.7	0.7	0.4	15.8	0.03	2.7
Pineapple	85.3	0.4	0.2	12.7	0.04	0.5
Plum	85.7	0.7	0.2	12.9	0.04	0.5

Table 9.8 Per cent Composition of Common Vegetables for Analysing their Potential for Anaerobic Fermentation

Vegetable	*Water*	*Protein*	*Fat*	*Carbohydrate*	*Ash*	*Fibre*
Asparagus	93.0	2.2	0.2	3.9	0.08	0.8
Lima beans	66.5	7.5	0.8	23.5	0.20	0.8
Snap Beans	88.9	2.4	0.2	7.7	0.10	1.9
Beets	87.6	1.6	0.1	9.6	0.07	0.9
Broccoli	89.9	3.3	0.2	5.5	0.20	–
Sprouts	84.9	4.4	0.5	8.9	0.10	–
Cauliflower	91.7	2.4	0.2	4.9	0.10	–
Cabbage	92.1	1.4	0.2	5.3	0.08	–
Carrot	88.2	1.2	0.3	9.3	0.08	–
Corn	73.9	3.7	1.2	20.5	0.10	0.5
Greens	91.2	2.3	0.4	4.8	0.20	0.4
Spinach	92.7	2.3	0.3	3.2	0.10	–
Mushrooms	90.0	3.5	0.4	6.0	0.10	1.7
Potatoes	77.8	2.0	0.1	19.1	0.07	1.2
Pumpkin	90.5	1.2	0.2	7.3	0.07	1.2
Squash	90.5	0.6	0.1	3.9	0.03	–
Tomatoes	94.1	1.0	0.3	4.0	0.03	0.6

Source : Lance, N.A., Handbook of Chemistry, Handbook Publishers, Inc., Sandusky, OH 1956. Quoted by E. Ashare, *et al.*, Assessment of Secondary Residues, Final Report No. SERI/TR-981 75-2, Dynatech R&D Company, Cambridge, Massachusetts, 1981. (For both the above tables).

Table 9.9 Important Characteristics of Processed Fruit Wastes for Analysing their Potential for Anaerobic Fermentation

Fruit	*Per cent TS*	*Per cent VS*	*Ash (Per cent of TS)*	*Wet Residue (Per cent of raw product)*
Apple	15.9	99.9	0.1	27.6
Apricot	14.6	99.7	0.3	13.3
Berry	13.8	96.4	3.6	7.0
Cherry	13.0	99.7	0.3	13.7
Citrus	12.3	99.7	0.3	39.4
Olive	24.8	99.5	0.5	24.0
Peach	13.1	99.8	0.2	12.9
Peas	17.3	99.8	0.2	26.4
Pineapple	14.7	99.8	0.2	29.3
Plum/Prune	14.3	99.7	0.3	44.4

TS : Total solids, VS: Volatile solids

Table 9.10 Important Characteristics of Processed Vegetable Wastes for Analysing their Potential for Anaerobic Fermentation

Vegetable	*Per cent TS*	*Per cent VS*	*Ash (Per cent of TS)*	*Wet Residue (Per cent of raw product)*
Asparagus	7.0	98.9	1.1	35.0
Lima Beans	33.5	99.4	0.6	15.8
Snap Beans	11.1	99.4	0.6	20.6
Beets	12.4	99.2	0.8	33.3
Broccoli, Sprouts & Cauliflower	11.2	98.1	1.9	42.3
Cabbage	7.9	98.9	1.1	33.0
Carrot	11.8	99.3	0.7	50.0
Corn	26.1	99.6	0.4	65.3
Spinach Greens	8.0	97.3	2.7	13.8
Mushrooms	10.0	99.1	0.9	47.8
Peas	25.7	99.6	0.4	12.8
Potatoes	22.2	99.7	0.3	32.7
Pumpkin/squash	7.2	99.4	0.6	25.0
Tomatoes	5.9	99.5	0.5	7.5

Source : Ashare, E., *et al.*, Assessment of Secondary Residues, Final Report, No SERI/TR-98175-2, Dynatech R&D Company, Cambridge, Massachusetts, 1981. (For both the above tables.)

Table 9.11 Source and Chemical Composition of Fruit and Vegetable Wastes used in the Experiments

Waste	*Source of Waste*	*Per Cent Total Solids*	*Per Cent Ash*	*Per Cent Total Carbohydrate*	*Per Cent Crude Protein*	*Per Cent Crude Fibre*
Group 1						
Spinach Waste	Raw and blanched spinach dropp- ed along production line.	12.0	1.74	0.66 (0.05)	4.22 (0.36)	2.11
Asparagus peels	Peels removed by hand.	8.9	0.72	0.50 (0.06)	1.98 (0.22)	2.14
Group 2						
French Bean Waste	Mixture of samples collected from waste containers along production line.	13.4	1.49	4.01 (0.29)	2.37 (0.18)	2.10
Strawberry Slurry	Strawberry waste with small amounts of asparagus peas.	11.4	1.75	2.42 (0.21)	2.37 (0.18)	2.10
Apple pulp	Residues following blanching and pressing during apple sauce production.	23.2	0.46	5.57 (0.24)	1.28 (0.05)	3.74
Apple Slurry	Mainly apple particles with small amounts of asparagus.	6.7	0.29	1.41 (0.21)	0.60 (0.08)	1.29
Carrot Waste	Mixture of carrot skin sludge & rejected carrots.	5.3	0.48	1.58 (0.29)	0.38 (0.07)	0.67
Group 3						
Green Peas Slurry	Mainly whole and damaged peas.	12.8	0.45	5.70 (0.27)	3.49 (0.27)	1.68

Note : 1. Wastes are categorised under groups 1, 2, or 3 depending upon their carbohydrate content on dry weight basis.
2. Figures in brackets refer to percentages on dry weight basis.

Table 9.12 Biogas Yield and Other Operating Parameters from Varying Combination of Fruit and Vegetable Wastes

Waste	*Loading Rate kg VS/m³/day*	*Biogas Yield m³/kg VS/day*	*Methane Content Per cent*	*Per cent Total Solids*		*pH*	
				Influent	*Effluent*	*Influent*	*Effluent*
Group 1							
Spinach waste	0.83-1.18	0.40	77-80	3.1-2.0	1.3-2.0	4.9	7.2-7.5
Asparagus peels	0.74-1.06	0.30	64-82	2.5-3.6	1.7-2.4	4.3	7.0-7.2
Group 2							
French bean waste	0.96-1.15	0.47	71-75	3.9-4.0	1.5-1.9	4.5	7.0-7.2
Strawberry slurry	1.02-1.15	0.34	72-82	3.9-4.3	2.0-2.7	4.7	6.4-7.0
Apple pulp	1.02-1.60	0.49	50-75	3.5-5.1	2.6-2.8	4.1-7.0	5.3-7.1
Apple slurry	0.83-1.15	0.38	72-75	2.7-3.8	1.5-2.2	3.8-7.0	6.5-7.1
Carrot waste	0.80-0.90	0.58	71-73	2.8-3.2	1.0-1.5	4.2	6.6-7.0
Group 3							
Green pea slurry	0.87-1.25	0.42	66-81	2.8-4.6	1.2-1.6	4.2	5.6-6.3
Mixed substrates							
Pig manure and apple pulp	0.90-1.12	0.45	72-75	3.2-4.3	2.1-2.9	6.9	7.1-7.3

Note : 1. Wastes are categorised under groups 1, 2, or 3 depending upon their carbohydrate content on dry weight basis.
2. For all columns maximum and minimum values are given except for biogas yield for which average values are given.

C.M. Silverio, *et al.* [16] carried out studies at the National Institute of Science and Technology (NIST) in Philippines on biogas production from banana peels. For this purpose ripe and unripe banana peels were sun-dried for 7-14 days which were later cut into smaller pieces and ground to enable homogenous slurry preparation. Chemical composition of sun-dried ripe and unripe peels used is given in Table 9.13. After anaerobic fermentation, it was found that ground banana peels yielded more biogas following increased availability of surface area for microbial action. At low temperatures addition of 2 per cent ammonium phosphate was found to speed up anaerobic fermentation process. In view of low C : N ratio of banana peels, addition of chicken manure was found to improve biogas yield. E. Elortegw, *et al.* [6] studied prospects and potential of anaerobic fermentation of banana stem and identified critical operating parameters which help to maximise biogas yield.

Table 9.13 Chemical Composition of Sun-dried Ripe and Unripe Banana Peels

Constituent	*Unripe Peel (Per cent)*	*Ripe Peel (Per cent)*
Carbon	8.76	9.57
Nitrogen	1.04	1.28
Phosphorus	2.06	1.32
Potassium	3.21	4.85
Ash	10.50	12.50
Volatile combustible matter	60.60	56.99
Moisture	13.04	13.49

M. Tanticharoen, *et al.* [18] carried out experimental studies for producing biogas from solid pineapple wastes comprising shells and cores. Experiments were carried out using four 30 litre vessels, one 200 litre plug flow reactor without any mixing arrangement, and one 5 m^3 tank with stirring facility. The two stage process was used to determine system efficiency at high loading rates and shorter retention periods. Pretreatment of the wet solids with sludge effluent prior to loading the digester resulted in greater biodigester stability than without pretreatment. Average gas yield was found to be 0.3-0.5 litre corresponding to a loading rate of 2.5 g of dry solids added per day. Some pioneering studies were also carried out at the University of California at Berkeley on anaerobic fermentation of tomato wastes in late seventies, important results of which are summarised in Tables 9.14, 9.15 and 9.16 [8, 22]. S.S. Mahajan [9] provided a detailed account of a biogas plant designed to utilise dry and green leaves, fruit-skins, stale food and similar biodegradable matter. The new plant is designed to yield twice the amount of gas per kg of fruit and vegetable wastes than a conventional plant that utilises cowdung as feedstock.

Table 9.14 Characteristics of Fresh Tomato Wastes Used as Feed

Characteristics	*Per cent TS*
Protein and Hemicellulose	43.7
Cellulose	50.5
Lignin	3.7
Silica Ash	2.1
Volatile Solids	92.0
Total Carbon	48.9
Non-Lignin Carbon	46.6
Kjeldahl Nitrogen	2.3
Non-Lignin Carbon to Nitrogen Ratio	20.7
COD/N/P Ratio	290 : 7.6 : 1

Table 9.15 Characteristics of Influent and Effluent of a Digester Fed by Tomato Wastes

Characteristics	*Influent*		*Effluent*	
	Mean	*Range*	*Mean*	*Range*
1. Total solids (%)	4.7	2.6-6.1	2.8	1.6-5.0
2. Volatile Solids (%)	92.0	81.5-96.1	75.3	64.0-80.8
3. COD (mg/l)	68,000	54,400-85,500	41,800	19,000-88,600
4. Nitrogen (%)	3.79	3.21-5.10	4.92	0.46-5.96
5. Phosphorus (%)	0.49	0.43-0.57	0.57	0.45-0.79
6. Alkalinity			2,900	2300-4400
7. pH			6.98	6.80-7.10
8. Volatile fatty acid (g/l)			7.45	264-1079

Table 9.16 Biogas Yield and Other Performance Measures from Tomato Wastes fed Digester

Characteristics	*Unit*	*Value*
Total solids reduction	Per cent	21.0
Volatile solids reduction	Per cent	27.7
COD reduction	Per cent	21.2
Biogas yield	m^3/m^3 digester day	0.46
Methane content	Per cent	59.9
Biogas yield	m^3 gas/kg VS added	0.18
Biogas yield	m^3 gas/kg VS reduced	0.66
Methane yield	m^3 CH_4/kg COD reduced	0.33

In India several studies have been carried out in past concerning anaerobic digestion of fruit and vegetable processing wastes for biogas production. Most fruits and vegetables are processed on seasonal basis for a period of 2-3 months and accordingly the resulting wastes vary in their physical and chemical characteristics. Citrus wastes are rich in toxic constituents like limonin whereas mango and pineapple processing wastes are known to be deficient in nitrogen. Over 60×10^6 tonnes of fruits and vegetables are produced in India annually of which only a small proportion is utilised in the fruit and vegetable processing industries [22]. The solid wastes normally account for 40-50 per cent of the raw material processed and has a total solid concentration of 10-15 per cent. P. Viswanath, S. Sumitra Devi and Krishna Nand [22] carried out important studies concerning anaerobic fermentation of fruit and vegetable processing wastes at the Central Food Technological Institute, Mysore. In this study, mango, orange, pineapple and tomato processing wastes were collected from the fruit processing factories, namely, Kissan Products Ltd., Bangalore; Clear Foods, Madanpalle, A.P.; Globe Foods, Mysore, Karnataka; and Mysore Fruitin, Mysore, Karnataka. Jackfruit and banana wastes were obtained from restaurants and hotels located in Mysore city. The effect of feeding different fruit and vegetable wastes, mango, pineapple, tomato, jackfruit, banana and orange was studied in 60 litre digester by cycling each waste every fifth day. The characteristics of the anaerobically digested fluid and digester performance in terms of biogas production were determined at different loading rates and at different hydraulic retention times (HRT). The maximum biogas yield of 0.6 m^3/kg. VS added was obtained at a 20-day HRT and 40 kg TS/m^3 day loading rate. The hourly gas production was observed in the digesters operated at 16 and 24 days HRT. As much as 74.5 per cent gas was produced within 12 hours of feeding at a 16 day HRT whereas at a 24-days HRT only 59.03 per cent of the total gas could be obtained [22].

9.9 Concluding Remarks

Biogas can be obtained from anaerobic fermentation of industrial wastes available from industries like sugar factories, textile mills, sugar distilleries and food processing industry. In fact all the organic matter except mineral oil and lignin are suitable substrates for biomethanation. For obtaining higher gas yield particularly from industrial wastes it is sometimes desirable to use mixed substrates. Biomethanation is a result of interaction of various groups of microbe. Control of temperature, pH value, and other environmental parameters are important conditions for vital activity of living bacteria. Residual water from distilleries is easily amenable for anaerobic fermentation. Biogas yield is found to be higher when rotten and stale industrial wastes are used. If fresh raw materials are used without pretreatment, decomposition is slower and gas yield accordingly low. In view of their established utility and potential, all-out efforts should be made wherever applicable for tapping biogas from industrial wastes.

REFERENCES

1. Ashare, E., *et al., Assessment of Secondary Residues, Final Report* No SERI/TR-98175-2, Dynatech R&D Co, Cambridge, Massachusetts, 1981.
2. Balasubramanya, R.H., Khandeparkar, V.G. and Sundaram, V., Production of Biogas and Biomanure from the Textile Processing Residue, Willow-Dust by Dry Anaerobic Fermentation, *Agricultural Wastes*, **16(4)**, 295-302 1986.
3. Balasubramanya, R.H., *et al.*, Large Scale Digestion of Willow-Dust in Batch Digestion, *Biological Wastes*, **25(1)**, 25-32 1988.
4. CPHRI, *Disposal of Blowroom Cotton Dust from Textile Mills by Composting*, A Report, NEERI (formely Central Public Health Research Inst.), Nagpur, 1970.
5. Deerr, N.N., *Cane Sugar*, Norman Rodgers, London, 1921.
6. Elortegw, N., *et al.*, Kinetics of Banana Stem Anaerobic Digestion, *Inter. Agrophysics*, **2(2)**, 127-133 1986.
7. Kaul, S.N., *et al.*, Anaerobic Biochemical Reactor Choice for Waste-Water Treatment, *Asian Environment,* **10(1)**, 1988.
8. Knol, W., *et al.*, Biogas Production by Anaerobic Digestion of Fruit and Vegetable Waste: A Preliminary Study, *Jour. of Science for Food and Agriculture,* **29**, 822-830 1978.
9. Mahajan, S.S., Biogas from Agricultural and Kitchen Garden Waste, *Invention Intelligence*, **22(5)**, 188-192 1987.
10. Mathur, S., *et al.*, Pressmud: A Prospective Source of Biogas, Steroidal Drugs and Biomanure, *Research and Industry*, **33(2)**, 151-153 1988.
11. Narsimhamurty, G.S., and Purshothaman, A., Biogas Generation from Bagasse, *Indian Chemical Engineer*, **28(3),** 54-56 1986.
12. National Sugar Institute (Kanpur), Production of Biogas and Biofertilizer from Agricultural Wastes, *Khadi Gramodyog,* **20(7)**, 358-360 1974.
13. Paturau, A., *By-Products of the Cane Sugar Industry*, Elsevier, New York, 1969.
14. Rajasekaran, P., *et al.*, Biogas Production from Cellulosic Wastes, *Rural Technology Journal*, **4(2)**, 41-43 1987.

15. Ranade, D.R., *et al.*, Biogas from Solid Waste Originated During Biscuit and Chocolate Production: A Preliminary Study, *Biological Wastes*, **28**, 157-161 1989.
16. Silverio, C.M., *et al.*, Biogas from Banana Peels, *NSTA Technology Jour.*, Jan.-Mar., 71-80 1984.
17. Singh, M. and Paul, B.S., Role of New Technology, *Economic Times*, Feb 4, 1984, p. 4.
18. Tanticharoen, M., *et al., Biogas Production from Solid Pineapple Wastes*, Proc. Inter. Symp. — Workshop on Renewable Energy Sources, Lahore, Pakistan, Mar 18-23, 1983, T.N. Veziroglu (ed.), Elsevier, Amsterdam, 1984.
19. Unni, B.G., *et al.*, Biogas Production from Sugar Cane Pressmud, *Energy Management*, **11(1)**, 23-32 1987.
20. Upadhyay, A.R., *Handbook of Sugar Factory Management*, Rajiv Sadan, Ratlam, 1969, p. 468.
21. Anonymous, Biogas Generation Using Oil-Cakes, *Urja*, **29(3)**, 49-50 1991.
22. Viswanath, P., *et al.*, Anaerobic Digestion of Fruit and Vegetable Processing Wastes for Biogas Production, *Bioresource Technology*, **40**, 43-48 1992.
23. Yeole, Y.Y., *et al.*, Biogas from Industrial Canteen Wastes, *Khadi Gramodyog*, **30(9)**, 1984.
24. Pandey, M., Turning Liabilities into Assets: Energy from Sugar Mill Wastes, Financial Express, May 14. 1995.
25. Gupta, M.K., *Personal Communication*, Rampur, May 24, 1996.

Chapter 10

BIOGAS RECOVERY FROM SANITARY LANDFILLS, ANAEROBIC LAGOONS AND NIGHT-SOIL

10.1 Introduction

Sanitary landfills, municipal refuse, anaerobic lagoons and night-soil provide valuable source for biogas recovery. Landfills are essentially dumping grouna for municipal wastes. Anaerobic conditions generally exist in top 10 feet of a landfill resulting in production of gases and leachates. Landfill gas is directly usable as a principal fuel or a supplementary fuel after removing impurities in industrial units, fertiliser plants, utilities and gas-based power plants. It can also be liquefied for being used as a liquid fuel. Much of the work concerning biogas recovery from landfill sites has been carried out in the USA and UK and in recent years considerable headway has been made in India as well. Municipal refuse generated in most cities can also be directly used as a valuable feedstock for anaerobic fermentation without landfilling. These wastes comprise 20 to 25 per cent of kitchen and household wastes and 40 to 50 per cent of some kind of paper wastes. Biogas can also be recovered from anaerobic lagoons which exist in large numbers in USA and several European countries. Biogas from lagoons has more methane concentration than obtainable from mesophilic digesters. Night-soil digestion which is traditionally regarded as a waste treatment process is equally valuable as an energy recovery process. It provides energy and biomanure, keeps environment clean and minimises groundwater contamination by percolation and leaching. In this chapter it has been endeavoured to provide detailed overview of biogas recovery from sanitary landfills, municipal refuse, anaerobic lagoons, and night-soil.

10.2 Biogas Recovery from Sanitary Landfills

Biofuels can be recovered from sanitary landfills and gainfully utilised. Within a landfill organic components of refuse breakdown under microbial attack resulting in production of various gases like carbon dioxide and methane. Biodegradation of domestic refuse in a landfill is an anaerobic process involving fermentation of polymeric carbohydrates into carboxylic acids, carbon dioxide and methane. Landfill gases have been located as deep as 600 feet at one site. Landfill gases while being formed tend to escape by choosing the path of least resistance which may entail diffusion upwards through the surface or laterally through the sides of the site. This route depends on permeability of top surface which in turn depends on top cover used and lateral permeability of refuse and permeability of surrounding rock structure. Thus three escape routes can be identified viz. diffusion through top surface, diffusion through permeable rocks from sides, and if both the surface and sides are impermeable then through cracks which will usually form at the interface between them. Rates of their outwards movements are generally found to be inconsistent [8].

Biogas recoverable from landfill has a typical composition of 50 to 55 per cent methane with remainder as carbon dioxide, traces of hydrogen sulphide, nitrogen mercaptans and some chlorinated hydrocarbons. It has calorific value 530 BTU/ft^3 which is nearly half that of natural gas (1075 BTU/ft^3). Appreciable quantities of this gas can be obtained from large sites with theoretical norm that 1 lb of refuse can produce 2.7 ft^3 of carbon dioxide and 3.9 ft^3 of methane. D.R. Anderson, *et al.* [2] reported that a maximum of 121 ft^3 of gas with 60 per cent of methane can be obtained from a cubic yard of refuse. K. Richards [22, 23] made a detailed account of fuel supplies from landfill gas in different parts of the world. A survey of landfill gas recovery projects and utilisation facilities throughout the world showed that by 1988 there were at least 146 commercial schemes in operation in a total of 15 countries with a further 32 trials underway. At a very conservative estimate, these projects could save 8,25,000 tonnes of coal equivalent (tce) annually of non-renewabe sources.

Period of gas formation varies from site to site and depends on nature of organic material in refuse, proportion of moisture content and extent of rainfall percolating down into a site. According to F.R. Boweman, *et al.* [17], maximum and nearly half of total gas production occurs within 2 to 3 years of landfill completion. R.C. Merz and R. Stone [17] found that surface irrigation of landfill affects gas generation rate. Stages of biogas production in relation to the age of a landfill can be divided into four phases. The initial aerobic phase is short one leading into second phase of high carbon dioxide production at approximately a molar equivalent to oxygen consumed so that little nitrogen is displaced. F.R. Bowerman [7] provided estimates of carbon dioxide percentages at varying intervals such as after 11 days 70 per cent, after 23 days 50 per cent and after 40 days 90 per cent of carbon dioxide was formed. This is accompanied by increases in hydrogen and decreases in nitrogen production. In third phase, methanogenesis activity begins causing methane concentrations to increase with reduction in carbon dioxide and hydrogen levels. J.N. Ramaswamy [20] reported that this steady state phase normally occurs about 180 days after the disposal in landfill begins. F.A. Rovers, *et al.* [24] estimated this period as 250 days whereas R. Beluche [4] estimated it as high as 500 days. According to

C.E. West and E. Ashare [30], the fourth steady-state phase of methane production occurs much later producing 50 to 70 per cent methane. Microbial activity can be intensified by recycling leachate through the decomposing refuse. According to G.F. Farquhar and F.A. Rovers [24], biogas generation increases with increased moisture content although methane production is found to decrease during periods of excessive water infiltration following inhibition of methane forming bacteria. Maximum production of gas occurs at 30-35°C. Optimum value of pH for gas production was found to be around 7. Composition and quantity of gas formed was dependent on the nature of city wastes.

For determining the rates of gas production, it is necessary to know area and depth of landfill, quantity and composition of refuse dumped, local geology and hydrology, age of site and other related parameters. This information helps to determine number of wells and monitoring probes required for testing, their spacings and desired withdrawal rates during testing. After installation of test wells and monitor probes, an operating crew visits site with a test rig and instrumentation for carrying out flow measurement. Various parameters of the landfill are first measured under static conditions. Flow tests are then initiated to record variation in values of these parameters under different flow rates. Tests are conducted on a twenty-four hours-a-day and seven days-a-week basis. Though total time required for production testing varies from site to site, it has however been in general found to vary between two to four months [3].

India's first major attempt to produce biogas from landfill was made during 1986 at Timarpur in Delhi. The project was a joint undertaking by the then DNES and the Delhi Administration. Biogas was recovered by drilling a series of wells. In 1986 eight wells at a spacing of 200 ft from one another were drilled. The recovered gas was used for power generation. As per the design criterion, one well was found adequate to produce 100 kilowatts of electricity from a landfill area of one acre. In other words, landfill covering an area of 10 acres can produce 1000 kw or equivalently one megawatt of electricity. Considering that the landfill at Timarpur covers an area of 80 acres, it has the potential to produce 8 mw of electricity. In 1986, cost of boring a single well was estimated as Rs. 10 lakhs which was inclusive of the cost of installation of generators, compressor, pumps and pipes. The cost of electricity generation from landfill gas was estimated as 10 paise per unit which is nearly one tenth of the power generated by conventional sources. In Delhi alone by 1986 there were over 50 landfills in existence under control of various agencies like the Municipal Corporation of Delhi, Delhi Development Authority, New Delhi Municipal Committee and the Cantonment Board of Delhi. In India as a whole as per rough estimates, about 8000 tonnes of garbage is generated per day which is likely to be doubled by the turn of the century. If this entire garbage is available for being dumped into different landfills, it can produce about 60 mw of electricity which can meet about 8 per cent of the total peak demand of the area covered. Electricity so produced can be fed into the power grid. A single landfill is sufficient for generating electricity for a minimum period of 10 years. After the entire gas produced during this period is taken out, spent refuse can be removed and refilled with fresh urban refuse. Cycle can thus be maintained to generate electricity continuously. Generation of electricity from landfills is also eco-friendly. In the absence of any such gainful project, gas formed in landfills tends to diffuse into environment causing pollution. Electricity generation from landfill gas thus helps to meet power needs and also helps to keep the environment clean and unpolluted [10].

As in many other fields, biogas recovery from sanitary landfills too has been pioneered in the United States. A number of landfill sites in California, New Jersey, Washington D.C., New York and Chicago have been used in past to extract gas with commercial success. The U.S. interest originally began with containment of hazard posed by gas migration but later on it shifted into a full-scale gas recovery programme with its uses in several utilities, industrial units, and gas-based power stations. The world's one of the first gas recovery facilities at the Palos Verdes landfill sites in Rolling Hills Estates, California met the household energy needs of over 3500 households. The Palos Verdes facility is owned and operated by the Getty Synthetic Fuels Inc., California, a wholly owned subsidiary of the Getty Oil Company. The company is regarded as a pioneer in recovery, processing and sale of methane gas from sanitary landfills. In 1981, the plant capacity was expanded to produce some 2 million cubic feet of gas per day (2 MMCFD) with heat value of 500 BTU/ft^3 which on purification produced 1 MMCFD of gas with a heat value of 1000 BTU/ft^3. Purified gas is supplied to the Southern California Gas Company's natural gas grid. The Operating Industries landfill at Monterey Park (California) is another major landfill in the U.S.A. where gas recovery facility began operations in 1979. The gas is supplied to the southern California Gas Company which in turn meets the energy needs of more than 15,000 households. Based on American and British experience, it is to be emphasised that energy recovery from landfills is economically viable normally for large and deep sites only because in shallow sites, refuse undergoes aerobic fermentation instead of anaerobic which results in low gas yield.

10.3 Biogas Recovery from Municipal Wastes

Municipal wastes generated in most cities provide valuable feed material for anaerobic fermentation. Carbohydrates present in paper, grass, leaves and other cellulosic material provide important ingredient for anaerobic fermentation. These wastes at times on account of high lignin content pose problems for fermentation. Lignin accounts for 20 to 25 per cent of natural fibres and is non-biodegradable. When lignin content is less than 5 per cent, fermentation is greatly inhibited. The pH of the digester content tends to affect these limits. For instance, alkaline substrate allows a greater tolerance for residual lignin to achieve a given level of digestibility. Sodium hydroxide has been one of the more popular agents for delignification and has been applied to a variety of materials to improve their digestibility for use as a ruminant feed. Improvement in digestibility results from increased availability of insoluble residual cell as well as formation of soluble matter. It has been analysed that a high conversion efficiency is possible at much shorter digester retention times with pretreatment than without pretreatment. However, a more efficient conversion process needs a system to solubilize or remove lignin and break the lignin-carbohydrate complex. Among others, lignin can be removed by chemical reaction as is done in wood pulping or by physical action as is done in fine milling. An important indicator to reflect changes in organic fibre characteristics is the ratio of cellulose to lignin. This ratio in municipal wastes before digestion is around 8 which following fermentation reduces to about 3 or so.

D.L. Wise, *et al.* [31] described a low capital cost batch type digestion process which can convert all of a town waste such as garbage, sewage, food processing waste, crop residues and animal wastes into biogas. Based on landfill gas recovery process, the scheme involves an inground reactor lined with an impervious material such as clay, a liquid collector pond situated at the base of the digester, a liquid recirculation pipe, and a gas removal well and pipeline [31]. J. Mata-Alvarex, *et al.* [16] carried out laboratory simulation of municipal wastes fermentation by using five test cells operated at different temperatures under enhanced digestion conditions with leachate recycle, with water spiked with added buffer, nutrients and inoculum. They also reported that 90 per cent of digestion can be completed in 25-27 days. R.A. Kumar [14] described a pilot project undertaken by the Tata Electric Co., Mumbai on recycling of sewage, garbage and garden waste and its final conversion into biogas, organic fertiliser and distilled water through anaerobic fermentation process. In this pilot plant, 1 to 2 m^3 of sewage and 1 to 2 tonnes of processed garden waste and garbage are fed into the digester daily. It produces around 100 m^3 of gas per day, 100-200 kg of sludge (fertiliser) and 500-1000 litres distilled water per day. Laboratory-scale studies on being scaled up by about 2000 times can provide the level of output from a pilot project.

Biogas yield from a set of different sources including municipal and slaughter house wastes [15]is given in Table 10.1. Comparative biogas yield from different organic wastes including municipal refuse, night-soil, distillery and paper mill wastes [18] is summarised in Table 10.2.

French engineers combined anaerobic digestion with direct combustion for maximum recovery of energy from municipal wastes. Called the Valorga process, the scheme first sorts the organic matter from the metal and inert solids and then passes all organic wastes into the digester. Downstream of the digester, a combustion unit burns the undigested matter comprising generally wood, rags and plastics for steam and hot water production. Resulting biogas from the digester is purified by removing hydrogen sulphide and carbon dioxide, dehydrated and finally compressed prior to distribution. Digester effluent is

Table 10.1 Biogas Yield from Different Waste Materials Including Municipal and Slaughter House Wastes

Solid Waste	*Biogas Yield (m^3/kg dry solids)*		*Methane Content (Per cent)*	*Methane Yield (m^3/kg dry solids)*		*Half Period (Days)*
	TS Basis	*VS Basis*		*TS Basis*	*VS Basis*	
1. Municipal sewage sludge	0.43	0.60	78	0.34	0.47	8
2. Municipal sewage skimmings	0.57	0.63	70	0.40	0.44	–
3. Municipal garbage	0.61	0.63	62	0.38	0.39	6
4. Waste paper	0.23	0.26	63	0.14	0.16	8
5. Municipal wastes (combined but free of ash)	0.28	0.31	66	0.18	0.20	10
6. Slaughter house wastes:						
6.1 Cattle paunch contents	0.47	0.53	47	0.35	0.39	13
6.2 Intestines	0.09	0.09	42	0.04	0.04	2
6.3 Cattle blood	0.16	0.16	51	0.08	0.08	2
7. Dairy wastes (sludge)	0.98	0.03	75	0.74	0.77	4
8. Yeast wastes (sludge)	0.49	0.80	85	0.42	0.68	–
9. Paper wastes (sludge)	0.25	–	60	0.15	–	–
10. Brewery wastes (hops)	0.43	0.45	76	0.33	0.34	2
11. Stable wastes (with straw)	0.29	0.35	75	0.22	0.26	19
12. Horse wastes	0.40	0.44	76	0.30	0.33	16
13. Cattle wastes	0.24	0.32	80	0.19	0.26	20
14. Pig wastes	0.26	0.42	81	0.21	0.34	13
15. Wheat straw	0.35	0.37	78	0.27	0.29	12
16. Potato tops	0.53	0.61	75	0.40	0.46	3
17. Maize tops	0.49	0.52	83	0.41	0.43	5
18. Beet leaves	0.46	0.50	85	0.39	0.43	2
19. Grass	0.50	0.56	84	0.42	0.47	4
20. Broom (25 mm cuttings)	0.44	0.45	76	0.33	0.34	7
21. Reed (25 mm cuttings)	0.29	0.32	79	0.23	0.25	18

Source : Loll, U., Microbial Energy Conversion, Erich, Goltze, KG, Gottingen, Germany, 1976.

Table 10.2 Biogas Yield from Different Waste Materials Including Night-soil, Municipal Distillery and Paper Mill Wastes

Type of Wastes	*Wastes Supplied*				*Biogas Yield*		
	Organic Matter Per cent	*Volume ($1/m^{3}/day$)*	*Organic Loading ($kg/m^{3}/day$)*	*Total ($1/m^{3}/day$)*	*Biogas/Organic Matter (1/kg)*	*Biogas/ Organic Volume (1/1)*	*Methane content (Per cent)*
1. Night-soil	1.83	28.7	0.54	293	551	10.4	43.2
2. Alcoho! distillery waste	3-6	30-125	2-3	1000-1800	500-700	10.23	50-60
3. Paper mill waste sludge	–	–	1.8	600-1500	240-500	–	70
4. Surplus sludge (domestic sewage)	2.26	60	1.36	978	719	16.3	58.6
5. Surplus sludge	2.0	200	4.0	970	245	4.9	51.0
6. Swine faeces and urine (Mesophilic)	3.6	200	7.2	24500	340	12.3	65.2
7. Swine faeces and urine (Thermophilic)	3.6	275	9.9	3470	351	12.6	61.0
8. Urban waste (Mesophilic)	7.5-7.4	40	3.66-3.88	1220	321	–	64.1
9. Urban waste (Thermophilic)	7.5-9.4	110	9.05	3050	337	–	60.2

Source : Osada, Y., Takatani, T., Takemura, M. and Tejima, T., Studies on Fuel Gas Production from Organic Wastes with Bioconversion in Japan, Tokyo, 1979.

separated into fertiliser, and combustible material is subjected to pyrolytic combustion to produce heat, a part of which is used for obtaining added gas yield [5, 6]. Almost 50 per cent of initial organic material are converted to biogas in a retention time of 15-20 days.

10.4 Biogas Production from Anaerobic Lagoons

Anaerobic lagoons are widely used in the United States and many European countries. As with other types of anaerobic treatment, methane and carbon dioxide are the two principal by-products of biodegradation. Anaerobic lagoon biogas production rates on volumetric basis are much lower than those from mesophilic digesters. However biogas production from lagoons on volatile solids added basis holds promise of being greater than for mesophilic digesters [3]. On surface area basis of lagoon, biogas yield was found to vary from 0.02 to 0.50 m^3/m^2/day whereas on volumetric basis of lagoon, it is reported to range from 0.03 to 0.23 m^3/m^3/day. As with other types of anaerobic treatment, methane and carbon dioxide are the two principal by-products of biodegradation. Lagoon biogas is found to have higher methane concentration than obtainable from mesophilic digesters. Lagoon temperatures fluctuate depending on liquid depth and ambient temperature. Loading rates for lagoons which normally vary from 0.056-0.14 kg VS.m^3/day are low as compared to those for biogas plants [26]. By contrast, most anaerobic digesters are operated at either constant mesophilic or thermophilic temperatures and loading rates varying from 2.7-17.7 kg. VS/m^3 of digester volume per day [26]. It is found that biogas yield from normally loaded lagoon (less than 0.06 kg VS/m^3/day) is not of sufficient quantity nor is production rate consistent enough throughout the year to be considered as a significant source of energy.

Biogas production from lagoon has been studied by several researchers. For a swine manure loading rate of 0.36 kg VS/m^3 day on a pilot-scale lagoon, F.J. Humenik and M.R. Overcash [12] measured biogas yield as 0.21 m^3/m^2 day (0.113 m^3/m^3.day). J.B. Allen and B. Lowery [1] studied biogas production from a full-scale swine lagoon and dairy and poultry pilot-scale lagoons. For the swine lagoon, they reported a mean biogas production rate of 0.006 m^3/m^2.day for a mean lagoon temperature of 14°C. However, the biogas yield was 0.55 m^3/m^2.day for a 3-day period when the lagoon temperature was 27.5°C. Biogas production from the poultry and dairy pilot-lagoons ranged from 0.001 to 0.02 m^3/m^2.day for a 3-day period when the lagoon temperature was 27.5°C. Biogas production from the poultry and dairy pilot-lagoons ranged from 0.001 to 0.02 m^3/m^2.day. J.A. Chandler, *et al.* [9] studied biogas production from a California swine lagoon where a 1070 m^2 floating cover was used to collect biogas. Gas yield ranged from 0.66 to 0.92 m^3/m^2.day (0.11-0.15 m^3/m^3.day). The lagoon was 6.1 m deep and the liquid temperature varied from 11 to 22°C throughout the year. L.M. Safely and P.W. Westerman [25] reported

Table 10.3 Biogas Yield from Anaerobic Lagoons Reported in Different Studies

Waste Type	*Lagoon Loading Rate (kg.VS/ m^3 day)*	*Biogas Production*		*Biogas Quality*		*Source*
		m^3/m^2.day$^{(1)}$	*m^3/m^3.day$^{(2)}$*	*CH_4(%)*	*CO_2 (%)*	
Swine	0.36	0.21	0.113	70	25	Humenik and Overcash (1976)
Swine	–	0.006-0.55	–	65-77	–	Allen and Lowery (1976)
Swine	0.11	0.66-0.92	0.11-0.15	69	–	Chandler, *et al.*, (1983)
Swine	0.05	0.04	0.04	–	5-8	Safely and Westerman (1988)
Swine	0.04	0.10	0.03	–	8-13	Safely and Westerman (1988)
Swine	0.04	0.09	0.03	–	10-14	Safely and Westerman (1988)
Poultry Digester Effluent	0.16	0.10	0.03	–	1-35	Safely and Westerman (1988)
Poultry	0.02	0.10	0.03	–	10-15	Safely and Westerman (1988)
Dairy	0.02	0.07-0.35	0.03-0.23	–	10-23	Safely and Westerman (1988)

Note : 1. m^3 biogas per m^2 of lagoon surface per day
2. m^3 biogas per m^3 of lagoon volume per day

Source : Safely, L.M and Westerman, P.W., Anaerobic Lagoon Biogas Recovery Systems, Biological Wastes, 27, 43-62 (1989).

lagoon biogas production rates ranging from 0.02 to 0.50 m^3/m^2.day (0.03 to 0.23 m^3/m^3.day). These rates were determined during summer months from swine, poultry and dairy lagoons with varying loading rates. Lagoon loading rates, lagoon depth, lagoon temperature and position of the collection unit on the lagoon surface were found to affect gas yield. A summary of biogas yield from anaerobic lagoon is given in Table 10.3. Biogas production dropped appreciably when lagoon liquid temperatures dropped below 15°C but again increased once the lagoon liquid temperature increased. Methane content in biogas from lagoons was found to range from 62 to 65 per cent for lagoon temperatures above 15°C [26].

10.5 Biogas Generation from Night-Soil

Night-soil digestion is traditionally regarded as a waste treatment process with recovery of bioenergy and biomanure viewed only as by-product. In most cities night-soil is disposed off through water carriage system (sewerage) but most rural and semi-urban areas in the country even do not have this basic facility. Night-soil scattered in fields is disposed off either by dumping or trenching. In many cases it remains unremoved making it breeding ground for flies and insects. Within this backdrop anaerobic fermentation of night-soil is an attractive proposition which not only keeps the environment clean but also provides energy and biomanure. H.B. Gotass [11] reported that per capita production of human faeces and urine lies within range 35-70 g and 50-70 g per day, respectively. Characteristics and norms for night-soil generation on dry weight basis are summarised in Table 10.4. Manurial value of human-faeces as determined by their constituents is given in Table 10.5. Some pioneering studies on biomethanation of night-soil were carried out at NEERI (Nagpur) in past. Characteristics of night-soil used in these studies is given in Table 10.6. The COD : BOD ratio of 2.4 indicates that night-soil is readily biodegradable. For efficient biodegradation BOD : N : P ratio of 100 : 5 : 1 is considered to be optimum, and for a stable anaerobic fermentation process a ratio of 100 : 2 : 0.5 is regarded as optimum. Night-soil contains nitrogen and phosphorus much in excess of that needed for efficient aerobic or anaerobic treatment. Excessive nitrogen present in digester content causes ammonia toxicity which adversely affects fermentation. However, excessive phosphorus present has no harmful effect on biomthanation.

Table 10.4 Composition and Generation of Human-faeces on Dry Weight Basis

Characteristic	*Unit*	*Value of Characteristic*	
		Human Faeces	*Human Urine*
Generation Rate			
Moisture		135-270 g	1.00-1.31 kg
Dry weight		35-70 g	50-70 g
Composition			
Moisture	Per cent	66-80	93-66
Organic Matter	Per cent	88-97	65-85
Nitrogen	Per cent	5-7	15-19
Phosphorus	Per cent	1.3-2.4	1.08-2.16
Potassium	Per cent	0.83-2.1	2.6-3.6
Carbon	Per cent	40-55	11-17
Calcium	Per cent	2.9-3.6	3.3-4.3
C : N Ratio	Per cent	5-10	0.6-1.1

Source : Subrahmanyan, P.V.R., Digestion of Night-Soil and Aspects of Public Health, National Environmental Engineering Research Institute, Nagpur, 1977.

Table 10.5 Composition of Human Faeces Indicating by their Manurial Value

Constituents	*Per cent*	
	Faeces	*Urine*
Moisture	75.1	97.0
Organic Matter	22.1	2.0
Minerals	2.9	1.0
Nitrogen	1.5	0.6
Phosphorus (P_2O_5)	1.1	0.1
Potash (K_2O)	0.5	0.2
Lime (CaO)	1.0	0.3
Carbon	11.0	0.8

Sources : ICAR, Handbook of Manures and Fertilisers, Indian Council of Agricultural Research, New Delhi, 1971.

Table 10.6 Characteristics of Night-Soil used in Laboratory and Pilot-Scale Studies at the National Environmental Engineering Research Institute, Nagpur

Characteristic	*Unit*	*Average Values*
Moisture	Per cent	86.7
pH	Per cent	5.2-5.6
Total solids	Per cent	13.3
Volatile solids	Per cent	11.6
Total nitrogen (N)	Per cent of dry matter	4.0
Total phosphorus (P)	Per cent of dry matter	1.53
Potassium (K)	Per cent of dry matter	1.08
BOD (at 20°C)	g/g total solids	0.4467
	g/g volatile solids	0.5155
	g/g total solids	1.2344
COD/BOD ratio		2.4
BOD : N : P		100 : 9 : 3.4
Per capita night-soil contribution		
(a) on total solids basis	g	64.67 ± 2.13
(b) on volatile solids basis	g	56.2 ± 1.09

In past the NEERI (Nagpur) helped in setting up night-soil based biogas plant at the Nagpur Central Prison [29]. Feed rate was kept as 1.5 to 3.5 kg VS/m^3/day, digester temperature varied from 20 to 32°C and ambient temperature ranged between 10 to 45°C. The plant performance corresponding to the feed rate of 2.6 kg VS/m^3/day is given in Table 10.7. When the feed rate was increased beyond 3.5 in usual units, dewatering ability dropped substantially causing excessive delays in drying-up of sludge. Even at a feed rate of 2.6 kg VS/m^3/day filterability of the sludge was poor and it took 10 days to obtain a spadable cake. This also added to storage space problems as space requirement increases with increasing feed rate. Digested slurry from the plant is transferred to the sludge bed where it gets dewatered and dries up. The filterate from the drying bed is further treated in a stabilisation pond. A flow diagram of the integrated bioenergy system indicating all major components is shown in Fig. 10.1. The dry sludge is an effective biomanure and soil-conditioner on account of its being rich in nitrogen, phosphorus and potassium content

Table 10.7 Yield and Operating Parameters of a Night-Soil based Pilot-Scale Biogas Plant

Characteristic/Parameter	*Unit*	*Value*
Digester Volume	m^3	18
Gas storage	m^3	8
Detention time	Days	27
Digester temperature	°C	28
Feed rate of TS	kg/m^3/day	3
Feed rate of VS	kg/m^3/day	2.6
Solid concentration in feed	Per cent	8
VS destroyed	Per cent	50
Biogas yield	m^3/person/day	0.025
–do–	m^3/kg of TS added	0.394
–do–	m^3/kg of VS added	0.448
Sludge production rate (5 per cent solids)	1/person/day	0.71
Biogas composition		
(a) Methane	Per cent	60-65
(b) Carbon dioxide	Per cent	30-35
(c) Hydrogen sulphide	Per cent	0.05-0.1
Manurial value of sludge		
(a) Nitrogen (N)	Per cent	3.25
(b) Phosphorus (P)	Per cent	1.00
(c) Potassium (K)	Per cent	0.83
Characteristics of digester slurry		
(a) pH		6.5-7.8
(b) Alkalinity ($CaCO_3$)	mg/1	4500-16,000
(c) Volatile acids	mg/1	85-450
(d) Total ammoniacal nitrogen	mg/1	900-3300

Source : Subrahmanyan, R.V.R., Digestion of Night-Soil and Aspects of Public Health, NEERI, Nagpur, 1977.

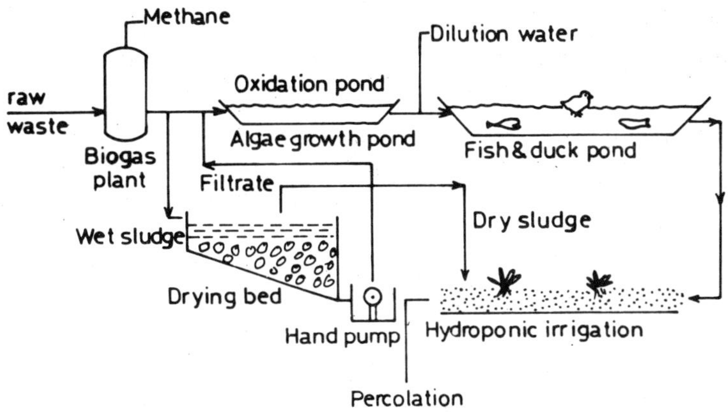

Fig. 10.1 Integrated bioenergy-cum-waste disposal system

S. Iwai, *et al.* [13] carried out studies on biogas production from night-soil. Based on experimental studies, it was reported that there occurred greater gas output during thermophillic digestion as compared to mesophillic digestion but this had no differential impact on gas composition which remained unaffected in the two cases. Heat requirement for thermophillic digestion was 2.5 times more than needed for mesophillic digestion and thus former can be economic only if inexpensive heat source such as solar energy is harnessed. Besides increased gas output, thermophillic digestion also ensures environmental improvement by destroying parasitic ova, pathogenic bacteria and other pollutants. J. Rammohan Rao [21] carried out experimental studies concerning thermophillic digestion of night-soil individually and in combination with cowdung in proportions of 2 : 1 at a temperature of 55°C. For a feed rate of 2.4 kg/VS/m^3/day, a volatile solid reduction of 45 per cent was reported for both the cases, and biogas yield of 0.45 m^3/kg.VS/day for night-soil and 0.35 m^3/kg. VS/day for night-soil and cowdung combination was observed. It was argued by the researchers that cowdung as a feedstock had a potential to suppress ammonia toxicity as the quantity of ammoniacal nitrogen released during mixed feed digestion was low compared to when night-soil was used alone [27]. S. Shanta [28] carried out experimental studies concerning survival of helmenthic ova during night-soil digestion. It was reported that on 20th day of detention period there occurred a reduction of 48 and 63 per cent in ascaris and hookworm population which rose to 70 and 93 per cent, respectively on 30th day of residence period.

C. Polprasert, *et al.* [19] described results of experiments conducted on four 3.5 m^3 ferro-cement digesters equipped with manually operated mixers. Digesters fed with a mixture of night-soil, water-hyacinth and rice-straw were operated on a semi-continuous basis at hydraulic retention times of 30, 50 and 70 days. The corresponding organic loading rates were 1.2, 0.75 and 0.53 kg m^{-3} day^{-1} of total volatile solids. Biogas production rate [m^3 (kg TVS)$^{-1}$ day^{-1}]

was 0.183, 0.210 and 0.164 at 30, 50 and 70 days of retention time, respectively. The respective volumetric production rates [m^3/m^{-3} day^{-1}] were 0.220, 0.158 and 0.086. During this study, common problems of excessive scum formation and accumulation were also faced. In another experiment 20-litre aspirator glass bottles were used for digestion in batch mode. Based on results of the experiment, at the end of 30 and 50 days biogas production rate was found to be 0.400 and 0.406 m^3 (kg TVS)$^{-1}$, respectively.

Night-soil-based biogas plants (NBPs) have become very popular and socially acceptable in Maharashtra in recent years. S. Kumar, *et al.* [32] carried out a survey of 40 NBSs installed under the aegis of the Maharashtra Energy Development Agency (MEDA). The survey of 8 to 85 m^3 capacity NBPs revealed that 75 per cent of them were working satisfactorily. In total about 715 m^3/day biogas was being produced which met cooking fuel needs of 1790 persons every day saving 1100 tonnes of fuelwood annually. Uninterrupted running of plant was found to largely depend on regular cleaning of slurry pit, maintenance of toilets and carrying out minor repairs locally from time to time. Failure of the plants were analysed to be due to underutilisation or poor upkeep of toilets leading to inadequate slurry and negligence of the owner towards occasional minor repairs. The institutional NBPs were found to be operating more efficiently following availability of improved infrastructural facilities with them as compared to those with *gram panchayats* and municipal councils.

Performance of a fixed dome type Krishna biogas plant of 6m^3 capacity using night-soil as feedstock was evaluated by *Shivasadan Sahakari Society*, Sangli, Maharashtra a type of which is owned and run by it (Fig. 4.11a). The plant is connected to three latrines and two urinals which are each provided with water-tap connected to an overhead water reservoir. The latrines were visited 6 days per week, 21 times per day and the urinals 32 times per day. This corresponded to hydraulic retention time of 7 days. The pH of the influent, BOD of the effluent and methane content of the gas were estimated intermittently. Gas production rate was measured as 0.427 m^3 per kg of faecal matter. Gas generation rate corresponding to a retention period of 7 days/week was found to be 0.727 m^3/day [33].

Dr Y.P. Anand, former Chairman, Railway Board, Govt. of India, has done pioneering work on low-cost sanitation systems which are very appropriate for India and developing countries at large. He has evolved designs for individual and community biogas plant-related sanitation systems based on extensive field work in railway complexes in Gorakhpur and Moradabad. According to him, system of biolatrines based on onsite biogas digesters is very economic and appropriate for a country like India [34].

10.6 Concluding Remarks

Sanitary landfills, municipal refuse, anaerobic lagoons and night-soil are valuable sources of biogas whose treatment in earlier times was only aimed at waste-disposal and environmental improvement. Several of these wastes contain parasitic ova and pathogenic organisms which cause roundworm, hookworm, tapeworm, typhoid, gastroentritis, hepatitis, poliomyelitis and several other diseases besides being the breeding ground for flies and mosquitoes. Unlike other organic wastes presence of these wastes is straight away harmful if these are not properly treated. Bio-treatment of these wastes can lead to the increased productivity of crops and minimise environmental degradation and fuel shortage in the countryside. All-out efforts need be made for treatment of these wastes through biomethanation and low-cost sanitation systems such as biolatrines.

REFERENCES

1. Allen, J.B. and Lowery, B., *Methane Gas Production from Dairy, Poultry and Swine Anaerobic Lagoons,* paper presented at the Southeast ASAE Meeting, Mobile, AL, 1976.
2. Anderson, D.R. and Cellinan, I.P., *Gas Generation and Movement in Landfills,* Proc. Natural Industrial Solid Waste Management Conference, Univ. of Houston, March 24-26, 1976, 311-316.
3. ASAE, *Design of Anaerobic Lagoons for Animal Wastes Management,* ASAE Engg. Practice EP 403, St. Joseph, MI, 394-396 1985.
4. Beluche, R., *Degradation of Solid Substrate in a Sanitary Landfill,* Ph.D. thesis, Univ. of West Virginia, Morgantown, West Virginia, 1968.
5. Bonhomme, M. and Pavia, A., A New System for Treatment of Urban Refuse: Valora Process, *Rev. Energie,* **37**, Aug.-Sep., 674-680 1986.
6. Bonhomme, M., Methane Fermentation Combined with Digester Residue Combustion, *Tech. Mod.,* **80(1)**, 49-54 1988 (in French).
7. Bowerman, F.R., *et al., A Case Study of the Los Angeles Sanitation District Palos Verdes Landfill Gas Development Project,* NERC Contract No 68-03-2143, U.S. Environmental Protection Agency, Los Angeles, California, 1976.
8. Bromley, J. and Parker, A., Methane from Landfill Sites, *Inter. Jour of Env. and Safety,* Aug. 9-11, 1979.
9. Chandler, J.A., *et al., A Low Cost 75-kw Covered Lagoon Biogas System,* paper presented at the Conf. on Energy from Biomass and Wastes VII, Lake Buena Vista, FL, 1983.
10. Chandra, J.P., *Delhi's Initiative: Electricity from Domestic Refuse,* A Report, New Delhi Municipal Committee, New Delhi, June, 1986.
11. Hill, D.T., Methane Productivity of the Major Animal Types, *Trans ASAE,* **27(2)**, 530-534 1984.
12. Humenik, F.J. and Overcash, M.R., *Design Criteria from Swine Treatment Systems,* EPA Report 600/2-76-233, 1976.
13. Iwai, S., *et al, Experimental Studies on High Rate Digestion of Night-Soil,* Proc. Inter. Conf. on Water Pollution Research, London, 1962.
14. Kumar, R.A., *Anaerobic Biogasification of Domestic Wastes and Direct Solar Energy Use to Produce Biogas, Biofertiliser and Distillate Water in a City: A Pilot Plant,* Proc. of Nat. Solar Energy Convention, IISc, Bangalore, Jan. 22-24, 1982.

15. Loll, U., *Microbial Energy Conversion*, Erich, Goltze, KG, Gottingen, Germany, 1976.
16. Mata-Alvarex, J. and Martinez-Viturtia, A., Laboratory Simulation of Municipal Solid Waste Fermentation with Leachate Recycle, *Jour. Chem. Tech. Biotech*, **36**, Dec., 547-556 (1986).
17. Merz, R.C. and Stone, R., Gas Production in a Sanitary Landfill, *Public Works*, **95**, 84-87 (1964).
18. Osada, Y., *et al.*, *Studies on Fuel Gas Production from Organic Wastes with Bioconversion in Japan*, Tokyo, 1979.
19. Polprasert, C., *et al.*, Integrated Biogas Technology in the Tropics-Performance of Small Scale Digester, *Solid Waste Management and Research*, **4(2)**, 197-213 (1986).
20. Ramaswamy, J.N., *Effects on Acid and Gas Production in Sanitary Landfills*, Ph.D thesis, Univ. of West Virginia, Virginia.
21. Rammohanrao, J., *Laboratory Studies on Thermophilic Digestion of Night-Soil*, M.Tech. thesis, Nagpur Univ., Nagpur, 1970.
22. Richards, K., *Landfill Gas Programmes in UK*, Proc. Inter. Conf. on Landfill Gas and Anaerobic Digestion of Wastes, Chester, UK, 1988.
23. Richards, K., Gas from Wastes Makes Growing Contribution to Fuel Supplies, *Energy Management (UK)*, Issue 1, April, 1988, p. 25.
24. Rovers, F.A. and Farquhar, G.J., Infiltration and Landfill Behaviour, *Jour. Env. Engg. Div.*, Amer. Soc. of Civil Engineers, 671-690 (1973).
25. Safeley, L.M. and Westerman, P.W., Biogas Production from Anaerobic Lagoons, *Biological Wastes*, **23(3)**, 181-193 (1988)
26. Safely, L.M. and Westerman, P.W., Anaerobic Lagoon Biogas Recovery Systems, *Biological Wastes*, **27**, 43-62 (1989).
27. Shanta, S., *Survival of Helminthic Ova During Anaerobic Digestion of Night-Soil*, A Report, NEERI, Nagpur, 1970.
28. Shanta, S. and Mohanrao, G.J., *Effect of Mixing Cowdung and Night-Soil Digestion*, A Report, NEERI, Nagpur, 1970.
29. Subrahamanya, P.V.R., *Digestion of Night-Soil and Aspects of Public Health*. Proc. Workshop on Biogas Systems, New Delhi, March 4, 1977.
30. West, C.E. and Ashare, E., *Feasibility Study for Anaerobic Digestion of Agricultural Crop Residues-1980 Update*, Report NO. TR 98175-1, SERI, Colorado, 1981, p. 2-7.
31. Wise, D.L., *et al.*, A Practical Village Biogas System for Wastes and Agricultural Residues, *Resource Conservation*, **13**, Dec. 23-36, 1986.
32. Kumar, S., *et al.*, *Implementation of Night-Soil based Large Size Biogas Programme in Maharashtra* in 'N.K. Bansal (ed.), Proc. Nat. Solar Energy Conv., Dec. 28-30, 1992, Tata Mc-Graw Hill Co. Ltd. New Delhi, 1993', p. 293-296.
33. Dandewade, L.N. and Joglekar, V.R., The Performance of a Night-Soil based Biogas Plants, *Biogas Forum (BORDA)*, **III** No. **46**, 4-6 (1991).
34. Anand, Y.P., *Study of the System of Management of Wastes in a Railway Complex*, Ph.D Thesis, Dept. of Chem. Engg., IIT, Delhi, 1989.

Chapter 11

APPLICATIONS AND USAGE OF BIOGAS

11.1 Introduction

Biogas being a clean gaseous fuel can be utilisied in several ways. It is extensively used in developing countries as a cooking and lighting fuel though in recent years its applications in several other areas have been attempted and commercial viability examined. Biogas provides a clean and efficient fuel for both cooking and lighting and in fact both biogas burner and lamp are structurally very much similar, both having a nozzle, an air inlet and a mixing chamber. India and China have taken lead in developing several designs of biogas burners and lamps some of which have become popular world over, and can be constructed by using locally available materials.

Biogas is increasingly emerging as an IC engine fuel for operating irrigation pump-sets. Attempts have also been made in India for using biogas for operating button-making, plastic-moulding and toy-making machines. Biogas can be used as a fuel for running both spark ignition and compression ignition engines. Whereas biogas in diesel engines is used in combination with diesel oil, petrol engines can be run exclusively on biogas. Dual-fuel engines provide easy changeover from dual-fuel to pure diesel operation and vice-versa. Biogas-fuelled engines can also be used for power generation.

Biogas can be used to produce electric power by coupling a dual-fuel engine to an asynchronous generator. Biogas-based power systems can provide decentralised power supply in several developing countries. In New Zealand biogas is also used as vehicle-fuel similar to CNG. Being gaseous, biogas ensures excellent cold starting for vehicles since unlike petrol it does not first have to be vapourised. Air-pollution from biogas emissions is much less as compared to petrol engines. As methane gas is only half the weight of air, gas dispersal and consequent detection is quicker in case of biogas. A major problem in biogas driven vehicles is the limited range per cylinder of the gas carried on the vehicle. In this chapter, it has been attempted to provide overview of biogas applications and principles of several biogas appliances.

11.2 Biogas as a Cooking Fuel and Some Common Indian Burner Designs

Biogas provides a clean and efficient fuel for cooking. Both biogas burner and lamp are structurally alike with identical nozzle, an air inlet and a mixing chamber in common. A burner has fire sieve element as an additional component. A nozzle is a hollow tube made of glass, metal, plastic or bamboo. Both in burner and lamp nozzle is of the size of a needle point with 0.5 to 0.7 mm. diameter whose other end is connected to the gas supply hose from the plant. As biogas passes through the nozzle air is allowed to be drawn into the mixing chamber. For obtaining desired flame temperature, nozzle adjustment is done by trial and error. Biogas stoves normally operate at gas pressure of 75 to 90 mm (3-3.5 inch) water column. Brightness and combustibility of gas can be controlled by regulating gas pressure and air-fuel ratio which is generally maintained as 10 : 1. Combustibility of gas is maximum when flame is slightly yellow, a bit bright and burnished. When the flame is turned blue and smokeless after air jet adjustment, it burns at a temperature of about 800°C. Cooking appliances are available in wide range both in single and double burner category with each burner consuming anywhere between 0.25 to 1.25 m^3 of gas per hour. From 0.28 to 0.42 m^3 (10 to 15 ft^3) of biogas is needed for meeting the cooking needs of one person per day. Biogas cannot be burnt on LPG/natural gas stoves as it tends to lift off due to slower flame speed factor. This problem can be somewhat overcome by reducing gas pressure but this result in higher gas consumption. Burning of biogas on country made crude stoves suffers from poor thermal efficiency of not more than 20 per cent and flame temperature seldom exceeding 500°C.

KVIC design. KVIC designed burners are made of cast iron with injectors made of gun metal (Figs. 11.1 and 11.2). These two types consume 6 ft^3 of gas per hour which are adequate for 1-2 member and 5-6 member family, respec-

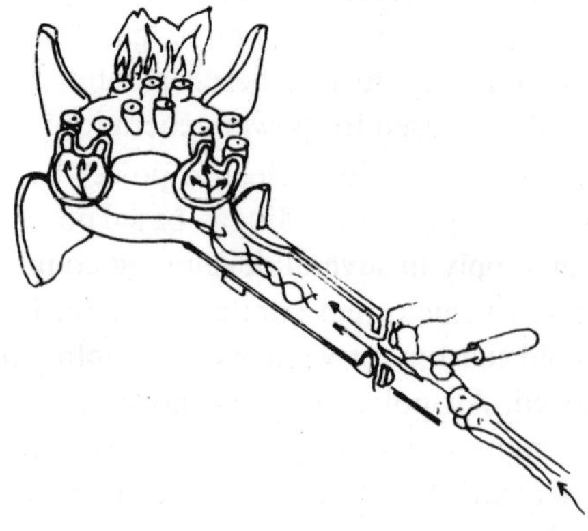

Fig. 11.1 Isometric view of utility type burners commonly used in India

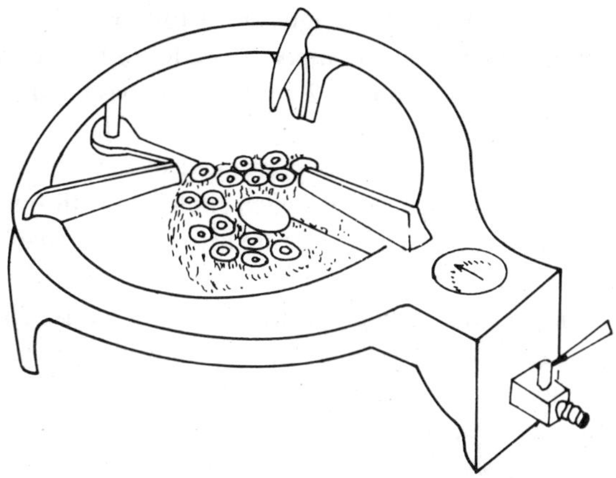

Fig. 11.2 A deluxe biogas burner used in India

tively. Initial pressure of gas in gasholder and burners vary from plant to plant. While designing these burners initial pressure of gas in gasholder is assumed as 8.8 cms (3.5 inch) water gauge and in burner between 7.6 cms (3 inch) and 8.3 cms (3.25 inch) water gauge. These burners attain flame temperature of 800°C and ensure complete burning of gas and consequently foul smell of unburnt gas rarely exists [3, 4].

Tin burners. These burners though not very efficient are made of tin to which gas is supplied from base through pipelines. Perforations are provided at upper portion of the tin to cnablc the gas to spurt out for burning. A number of small-sized stones are kept inside to facilitate improved air and gas mixing. Tin burner can be embedded inside an earthen '*Chulah*' to suit flat bottomed and round bottomed cooking vessels. The Indian Agricultural Research Institute (IARI) in past developed designs of tin burners which can be easily constructed using locally available materials (Fig. 11.3).

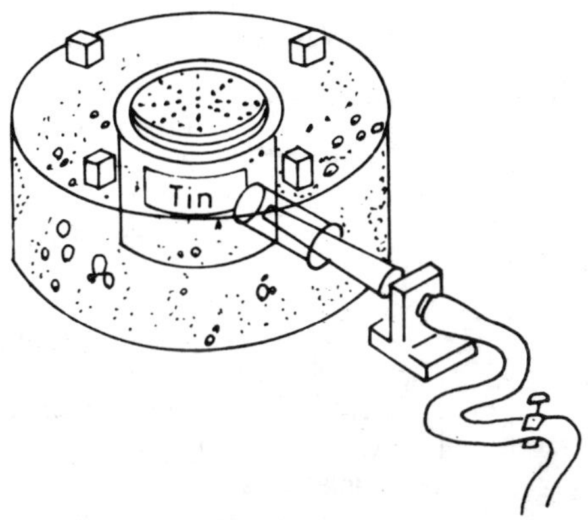

Fig. 11.3 A biogas burner made from food-tin

More details of two of such designs are as follows:

(i) This can be made from an empty cigarette tin by drilling 0.7 diameter hole from side at a height of 2 cm from the base and soldering 0.7 cm diameter and 7 cm long tube with 3 cm long portion inside and 4 cm projecting outside. The lid is perforated to have four holes of 2 mm diameter uniformly spaced on a pitch circle and with one additional hole in the centre. A number of stones are kept inside to enable proper air and gas mixing. Biogas is made to pass through the tube inside the tin which finally emerges out through perforations and burns. The burner is embedded in an earthern vessel to make it convenient to use it for cooking.

(ii) The second version commonly known as the '*Angithi*' burner can be made from boot polish tins or larger flat tins. A 0.7 cm dia metallic tube bent at right angles is soldered to the tin through a hole at its base. The tin cover is perforated with 2 mm dia. holes at a pitch of 2.5 cm on a uniform pitch circle diameter to ensure uniform heating from all around. This burner is fitted on a conventional iron '*angithi*' which makes it easy to use it for cooking.

11.3 Burner Designs Commonly Used in China

In China a number of designs of biogas burners/stoves have been developed which can be constructed by using locally available materials like clay. Three commonly used designs are shown in Figs. 11.4, 11.5 and 11.6. For improving

Fig. 11.4 Earthern Biogas Stove [IRDC, Compost, Fertiliser and Biogas Production from Human and Farm Wastes, Ottawa, Canada, 1980]

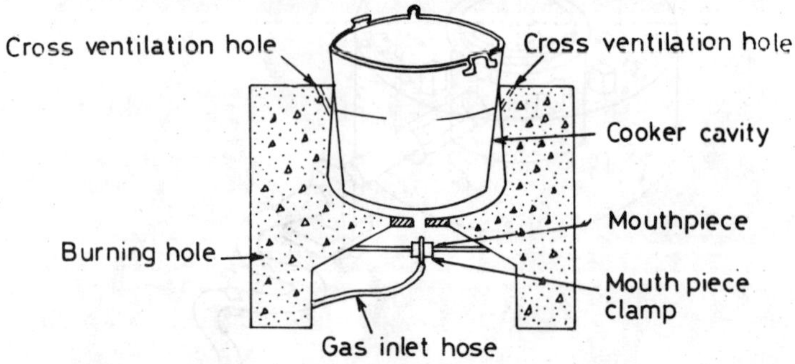

Fig. 11.5 A schematic view of biogas stove [Source: Buren, A.V., A Chinese Biogas Manual, Intermediate Technological Pub. Ltd., London, 1979, p. 106]

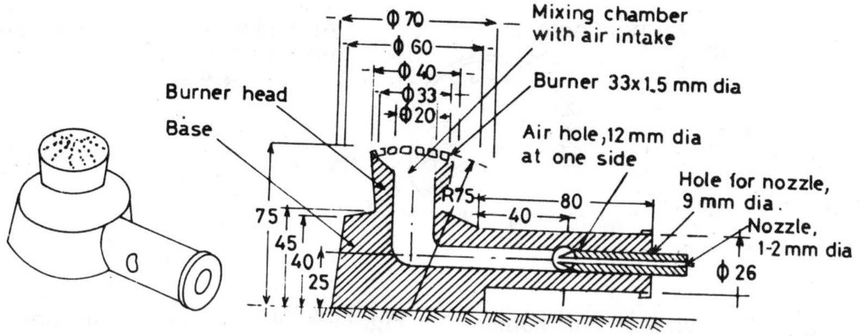

Fig. 11.6 Isometric and cross-sectional view of spreading-head clay burner [Source: BORDA, Biogas Plants: Building Instructions, GATE, P.O. Box 5180, 6236 Eschborn 1, Germany.]

heat resistance properties, clay is used after mixing with salt water, burnt rice-husk or fine coal powder. Design details of some commonly used Chinese clay burners are as follows [4, 5, 11].

Shower head burner. This type of burner is made of potters clay which provides a stable flame. Diameter of the shower head varies from 4 to 8 cm and its base diameter is almost twice the upper diameter. Fifty to sixty holes each of 1.5 to 2 mm in diameter are made in three to four rows. The air inlet is made 10 mm long and 5 mm wide. Air-fuel ratio is controlled by adjusting the position of gas nozzle relative to air inlet (Fig. 11.7).

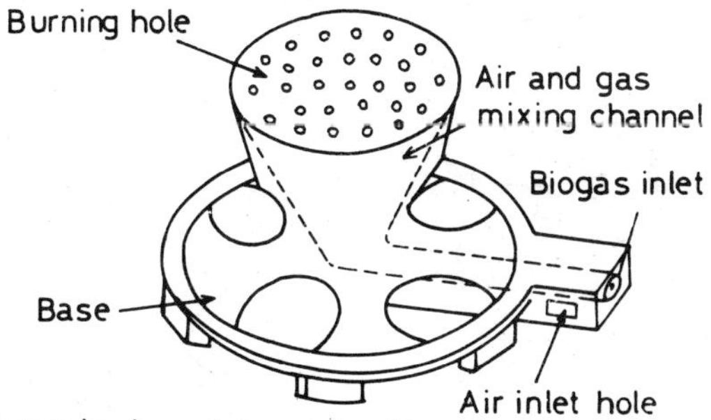

Fig. 11.7 An isometric view of shower head burner used in China [Source: Buren, A.V., A Chinese Biogas Manual, Intermediate Technological Pub. Ltd., London, 1979, p. 103]

Drum burner. This burner is considerably similar to the shower head burner with the added benefit that it can be made from any material easily available and not limited to clay only. It ensures efficient combustion as gas can be mixed properly in mixing chamber before burning. A metal gas pipe with nozzle is placed in gas and air inlet. A metal cover is fitted as an air control device (Fig. 11.8).

Long arm burner. This burner (Fig. 11.9) has an air and gas mixing tube at the top and is provided with a support. It is simple to construct and can be made from local materials. It provides strong and concentrated flame.

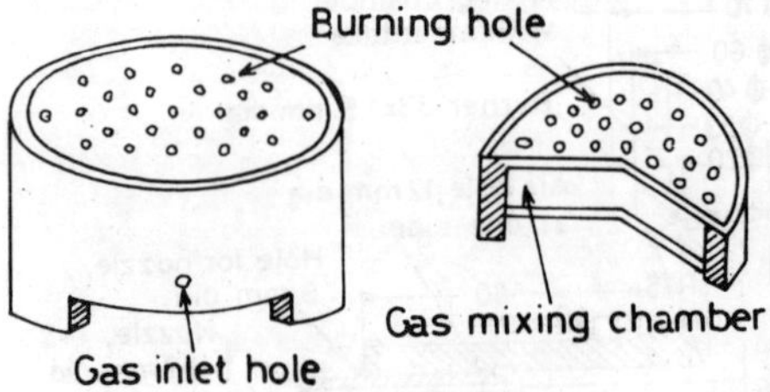

Fig. 11.8 A view of drum burner [Source: Buren A.V., A Chinese Biogas Manual, Intermediate, Technological Pub. Ltd., London, 1979, p. 105]

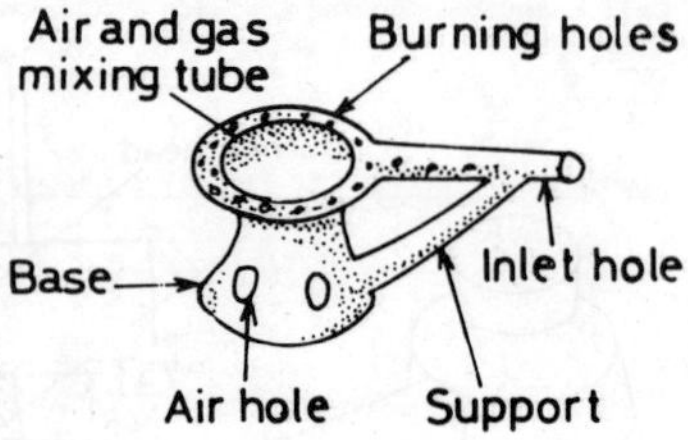

Fig. 11.9 A view of long arm burner used in China [Source: Buren, A.V., A Chinese Biogas Manual, Intermediate Technological Pub. Ltd., London, 1979, p. 103]

Double ringed burner. This burner which (Fig. 11.10) has gas and air inlet hole near upper surface. It has two annular air and gas mixing tubes at top having numerous holes through which air-fuel mixture emerges out and burns.

L-Shaped burner. This burner (Fig. 11.11) is a modified version of the shower head burner. It consists of a base shower head and a L-shaped gas inlet pipe. It also provides a strong and directed flame.

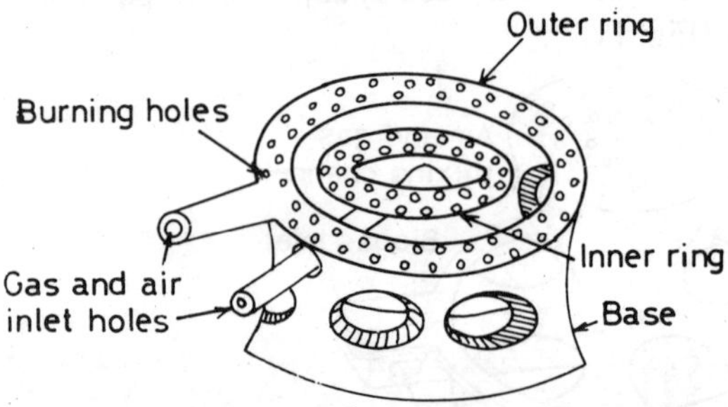

Fig. 11.10 A view of double ringed burner used in China [Source: Buren, A.V., A Chinese Biogas Manual, Intermediate Technological Pub. Ltd., London, 1979, p. 105]

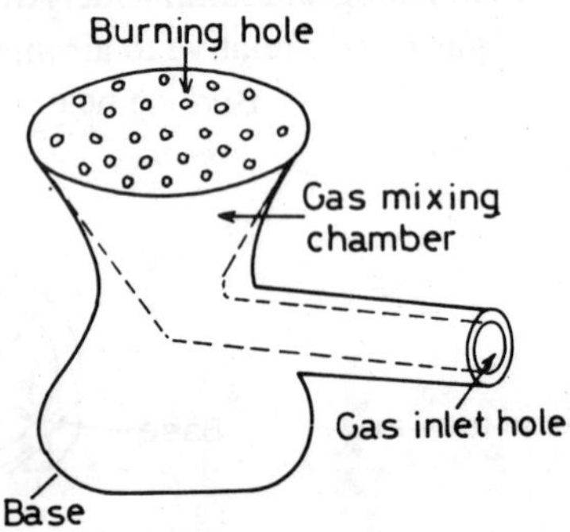

Fig. 11.11 A view of L-shaped burner used in China [Source: Buren, A.V., A Chinese Biogas Manual, Intermediate Technological Pub. Ltd., London, 1979, p. 104]

A stove of the following specifications is adequate for meeting the cooking needs of a family of 4 to 5 persons consuming 0.45 m^3 (16 ft^3) of gas per hour.

Jet size : 2.25 mm diameter

Area of jet : 3.98 mm^2

Flame-port size : 6 mm diameter

Number of ports : 20

Total area of ports: 565 mm^2

Ratio of jet area to flame port area : 1 : 1.42

Length of gas mixing pipe : 20 mm

Diameter of gas mixing pipe : 20 mm

11.4 Design of Biogas Burners

Several factor need be kept in view while designing biogas burners. These include desired efficiency, available construction facilities, ease of maintenance, ease of fitting and appearance. A hot plate consists of a burner with a facility to support a utensil. While designing hot plate burners, aspects like pot stability, ease of removal, mechanical strength, flame stability and thermal efficiency need be considered. A well-designed burner should have ease of replacement of its parts, sufficient approach space for cleaning and removal of spillage and lighting by a match-stick, quick and complete flame-travel the moment flame is applied to any burner port, supports to prevent any undue movement, pan supports to prevent tilting of vessels during heating, and corrosion resistant properties [23, 37].

Heat transfer efficiency is an important performance indicator of a burner. It represents the ratio of heat derived from gas for heating and heat actually available. For a reasonably efficient burner, its value is around 70 per cent. In mathematical symbols it can be expressed as:

$$\eta = \frac{(W + MS) \cdot (t_2 - t_1)}{VH} \times 100$$

where,

η = heat transfer efficiency of burner in per cent
W = Weight of vessel contents in kg
M = Weight of vessel along with lid
S = Specific heat of vessel and lid material
V = Biogas consumption in litres
H = Calorific value of biogas in kcal/litre
t_2 = Final temperature of vessel and its contents in °C
t_1 = Initial temperature of vessel and its contents in °C

If the biogas burner is to be used for '*chapati*' making, the corresponding heat transfer efficiency e can be estimated by the following expression.

$$e = \frac{m_p S_p (t_2 - t_1) + m_f S_f (t_2 - t_1) + m_m (t_2 - t_1) + m_{ml} (t_2 - t_1) + m_{ml}}{VH}$$

where,

S_p = specific heat of pan material
m_p = mass of the pan

m_f = mass of flour
S_f = specific heat of flour
m_m = mass of moisture in *'chapati'* after baking
m_{ml} = mass of moisture lost during baking
L = latent heat of steam
t_2 = final temperature
t_1 = initial temperature
V = gas consumption
H = calorific value

A schematic diagram of the biogas burner is shown in Fig. 11.12. It consists of an orifice for injecting biogas, an adjustable port for air entry, a chamber for thorough mixing of gas and air, and ports for exit and burning of gas-air mixture. Diameter of the orifice can be determined by the following equation [38].

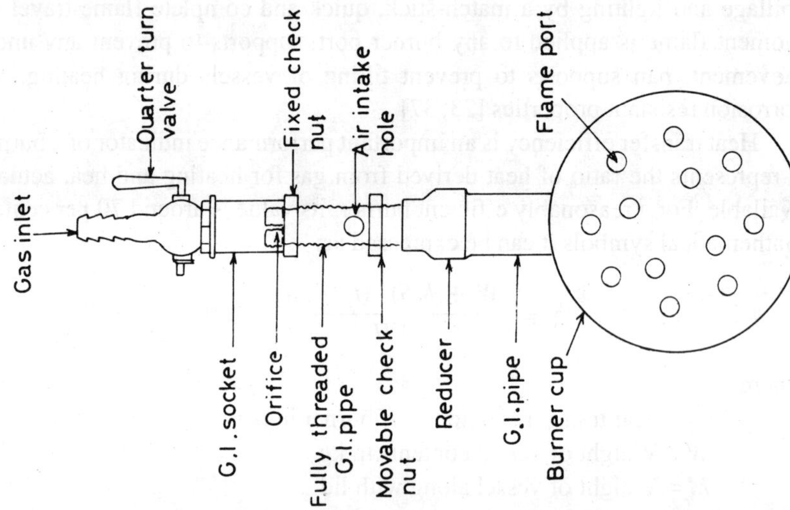

Fig. 11.12 Schematic view of biogas burner [Sources: TERI, Fixed Dome Biogas Plants, New Delhi, June, 1987, p. 41]

$$Q = 3.6\, C_d \cdot A_0 \cdot [2\, g\, (\rho_w/\rho_g)\, h_g]^{1/2}$$

where,

Q = gas flow rate in litres per hour
C_d = coefficient of discharge (about 0.6)
A_0 = area of orifice in cm^2
g = acceleration due to gravity (981 cm/sec^2)
ρ_w = density of water (1000 kg/m^3)
ρ_g = density of gas (1.128 kg/m^3)
h_g = pressure of the gas (cm in water gauge)

For gas pressures of 5-10 cm in w.g., the orifice area per unit flow rate (A_0/Q) is estimated as 0.01 mm^2 hr/litre.

Two 8 mm dia holes with an adjustable check-nut over them are provided near orifice for air entry. Mixing chamber is normally 15-20 cm long which can be suitably enlarged for high pressures. For proper flame propagation total area of flame ports should be about 300 times the orifice area. A total of 12 holes are generally sufficient. Thus, the diameter of each hole can be decided once the total number of holes is fixed. Distance of flame ports from the base of a utensil should range from 2 to 3 cm. A well-designed and fabricated stove burns with a thermal efficiency of over 55 per cent. Biogas burners approved by the KVIC and as per Bureau of Indian Standards specification, IS : 8749-1978, are commercially available.

Based on several studies carried out in past [7, 8, 30], norms of biogas consumption have been established for various cooking activities. Gas consumption for different types of burners is given in Table 11.1. Rate of biogas consumption for cooking according to family size is given in Table 11.2. Based on studies carried out by Ram Bux Singh [30], norms of biogas consumption for boiling water are given in Table 11.3, cooking rice in Table 11.4 and cooking '*arhar*' pulse in Table 11.5.

Table 11.1 Consumption Norms for Burners Under Different Combinations

S. No.	*Type of Burner*	*Gas flow rate (ft^3/hour)*
1.	One large standard burner	16
2.	One large and one small burner or one large burner and two lamps	24
3.	Two large burners or one large and one small burners and two lamps	32
4.	Three large burners	48
5.	Four large burners	64

Source : Dhussa, A., Designing Biogas Distribution System, Bioenergy Reviews, 2 (1&2), 70 (1983).

Table 11.2 Biogas Consumption Rate for Cooking

S. No.	*No of Family Members*	*Gas Pressure at Burners in mm of Water Column*	*Gas Consumption per Person per Day in ft^3 (m^3)*	*Efficiency (Per cent)*
1.	upto 4	63.5-87.5	16 (0.42)	60
2.	5 to 7	63.5-87.5	12 (0.34)	60
3.	8 and more	63.5-87.5	10 (0.28)	60

Source : Lapp, H.M., Fuel from Animal Waste, Paper No. 77-6531, Agricultural Engineering Department, University of Manitoba, Winnipeg, Canada, 1977, p. 7.

Table 11.3 Biogas Requirement for Boiling Water in 2.4 Litre Capacity Steel Utensil at 82.5°F Ambient Temperature

Gas Pressure in (Inches of Water Column)	*Boiling Time*		*Biogas Consumption in ft^3/hour*
	Minutes	*Seconds*	
0.75	6	10	1.81
1.00	6	2	1.92
1.75	5	8	2.07
2.75	5	8	2.11

Source : Singh, RB, Gobar Gas: Fuel of the Future, Gobar Gas Research Station, Ajitmal, Etawah, 1975, p. 97.

Table 11.4 Biogas Requirement for Cooking Rice at 84.5°F Ambient Temperature in a 671-g Steel Utensil of 2.4 Litre Capacity on a 6-inch Diameter Gas Stove

Quantity of Rice (g)	*Quantity of Water (g)*	*Gas Pressure (Inches of Water Column)*	*Cooking Time (minutes)*	*Biogas Consumption (ft^3/hour)*
100	500	2.5	13	3.15
230	870	2.5	15	5.85
400	1700	2.5	19	5.05

Source : Singh, R.B., Gobar Gas: Fuel of the Future, Gobar Gas Research Station, Ajitmal, Etawah, 1975, p. 69.

Table 11.5 Biogas Requirement for Cooking '*Arhar*' Pulse at 82.1°C Ambient Temperature in a 671-g Steel Utensil of 2.4 Litre Capacity on a 6-inch Diameter Gas Stove

Quantity of Pulse (g)	*Quantity of Water (litres)*	*Gas Pressure (Inches of Water Column)*	*Cooking Time (minutes)*	*Biogas Consumption (ft^3/hour)*
230	1.4	2.5	30	7.5
400	2.0	2.5	41	9.3
500	2.3	2.5	56	11.1

Source : Singh, R.B., Gobar Gas: Fuel of the Future, Gobar Gas Research Station, Ajitmal, Etawah, 1975, p. 96.

P.T. Palaniswamy and K.R. Swaminathan [21] converted a conventional kerosene stove into a biogas burner after introducing certain modifications. The converted stove is claimed to be fuel-efficient and inexpensive. The wicks were removed and holes kept as it is to facilitate upward movement of biogas. An inlet was made in the kerosene container on sides or top to enable biogas entry.

The perforated shroud around the wicks was replaced by circular light metal sheet to arrest biogas leakage from sides.

Design of a burner involves selection of requisite size of burner tube, injector orifice, venturi tube, ports, aeration facilities etc. Dimension of burner tube determines time needed to allow air and gas mixing and to bring it near flame port. If the diameter of the burner tube is too small, combustion mixture tends to travel faster destabilising the flame. A long tube on the other hand tends to limit aeration [23].

An injector is used to deliver gas for burning. The following mathematical relationship can be used for determining the diameter of injector orifice.

$$Q = 1300\, cd^2 \cdot \sqrt{\frac{p}{s}}$$

where,

Q = gas rate ft^3/hour
c = discharge coefficient
d = diameter of injector orifice in inches (to be determined)
p = pressure in water gauge
s = specific gravity of gas

Whence on rearrangement,

$$d^2 = \frac{Q}{1300\, c} \sqrt{\frac{s}{p}}$$

An example based on actual test observations carried out by the Gobar Gas Development Centre GGDC, KVIC, Borivli, Mumbai is given below for determining diameter and area of injector orifice [23]. Assuming

$$Q = 7 \text{ ft}^3\text{/hour}$$

$$c = 0.86$$

$$p = 3 \text{ wg}$$

$$s = 0.92$$

Following substitution, we get

$$d^2 = \frac{7}{1300 \times 0.86} \sqrt{\frac{0.92}{3.00}}$$

Thus, d = 0.05888 inch (1.475 mm)

$$\text{Area of injector orifice} = \frac{\pi d^2}{4} = \frac{3.14\,(0.059)^2}{4}$$

$$= 0.002735 \text{ in}^2 \ (1.707 \text{ mm}^2)$$

Venturi tube. This tube is necessary to maintain the laminar flow of combustion. Throat diameter can be determined from the throat area of venturi which is estimated by the following relationship.

Throat area of venturi = 0.4 × area of flame ports = $\frac{\pi \cdot d^2}{4}$

where, d is diameter of throat.

Distance between venturi throat and burner entry = 2 d

Distance between venturi throat and burner tube = 6 d

Diameter of burner tube d_1 can be determined by using the fact that burner tube area is equal to the area of flame ports, i.e.

$$\frac{\pi d_1^2}{4} = \text{area of flame ports}$$

whence it is easy to compute d_1.

An example based on actual test observations carried out by the Gobar Gas Development Centre, KVIC, Borivli, Mumbai is given below for determining diameter of throat, distance between venturi throat and burner entry, distance between venturi throat and burner tube, and diameter of burner tube.

Assumed area of flame ports = 0.5889 in^2

$$\text{Throat area} = 0.4 \times 0.5889 = 0.2355 \text{ in}^2 \ (1.520 \text{ cm}^2) = \frac{\pi d^2}{4}$$

whence, d = 0.5476 in (1.391 cm)

Distance between venturi throat and burner entry

$$= 2 \times 0.5476 \text{ in} = 1.0952 \text{ in (2.782 cm)}$$

Distance between venturi throat and burner tube

$$= 6 \times 0.5476 \text{ in} = 3.2856 \text{ in (8.346 cm)}$$

$$\text{Area of flame ports} = 0.5889 = \frac{\pi d_1^2}{4}$$

whence, d_1 = 0.866 in (2.2 cm)

Determination of port diameter. Gobar Gas Development Centre, Borivli, Mumbai has carried out studies in past for determining port diameter of biogas burners. In view of the fact that flame velocity of biogas is relatively less as compared to other gases it is preferable to keep biogas-air mixture velocity not less than 78 cm/sec with a view to minimise chances of flame lift-up and back- firing. As per the GGDC results for a burner receiving 225 litres of biogas every hour, port diameter should be such that its every port is able to burn 25 cc of biogas-air mixture (1 : 4.4) in order to get a well-defined high temperature stable flame free of any lift-up or back-fire.

Gas flow is governed by the relationship,

$$Q = a_0 \cdot \sqrt{2\,gh}$$

where

Q = gas flow in ft³/sec

g = acceleration due to gravity in ft/sec²

h = height in feet of gas column exerting pressure on gas flowing out from orifice.

a_0 = area of orifice in ft² $= \dfrac{\pi\, d_0^2}{4}$

d_0 = diameter of orifice

Velocity of gas flow can be determined by the following relationship

$$v = \sqrt{2\,gh}$$

Flame velocity of biogas-air mixture can be determined using the relationship,

$$V_T = V_{av} \cdot \sin \alpha$$

where,

V_T = Flame or transformation velocity

V_{av} = Average velocity of premixed gases leaving burner tube

α = Angle formed by inner flame contour with the verticle axis.

Ratio of area of flame port to area of injector orifice can be determined by using the following expression.

$$A_p = A_0 \cdot \frac{k}{c} \cdot (r + 1) \cdot \frac{(d + r)}{d}$$

where,

A_p = Area of flame ports

A_0 = Area of orifice

k = A constant representing the characteristic of burner used

c = Coefficient of discharge of burner ports

d = Density of gas

r = air-biogas ratio

An example based on actual test results obtained by the GGDC, KVIC, Borivli (Mumbai) is given below for computing h, v, v_T and A_p/A_0.

$$d_0 = 0.25 \text{ in}$$

$$a_0 = \frac{\pi\,(0.25)^2}{144 \times 4} = 0.1161 \times 10^{-6}\text{ft}^2$$

$$h = \frac{Q^2}{2\,a_0^2 \cdot g} = \frac{(25/28320)^2}{2 \times (0.1161 \times 10^{-6})^2 \times 32.2} = 0.1042 \text{ ft}$$

$$v^2 = 2\,gh = 2 \times 32.2 \times 0.1042 = 9.173$$

$$v = 2.590 \text{ ft/sec (78.94 cm/sec)}$$

Average velocity, v of gas flow is equivalently assumed as taking place when orifice diameter is 0.7 of the port diameter.

Port diameter = 0.25 in (6.350 mm)

Port radius = 0.125 in (3.175 mm)

∴ Orifice radius = 0.125 × 0.7 = 0.0875 in (2.2225 mm)

$$a'_0 \doteq \frac{3.14 \times 0.0875^2}{144} = 0.167 \times 10^{-3} \text{ ft}^2$$

$$h' = \frac{Q^2}{2\,a_0'^2 \cdot g} = \frac{(25/28320)^2}{2\,(0.167 \times 10^{-3})^2 \times 32.2}$$

whence,

$$h' = 0.4338 \text{ ft}$$

$$v = \sqrt{2\,gh'} = \sqrt{2 \times 32.2 \times 0.4338} = 5.284 \text{ ft/sec.}$$

$$v_T = v_{av} \sin \alpha = 5.284 \sin (12.5°) \quad [\alpha = 12.5°]$$

$$= 1.1444 \text{ ft/sec (34.86 cm/sec.)}$$

$$\frac{A_P}{A_0} = \frac{3.8\ (\text{cm}^2)}{0.01707} = 222.7$$

$$C = 0.6,\ r = 4.4,\ d = 0.92$$

$$\frac{A_P}{A_0} = \frac{k \cdot (r+1)}{c} \cdot \frac{(d+r)}{d}$$

$$227.7 = \frac{k\,(4.411)}{0.6} \cdot \frac{(0.92 + 4.4)}{0.92}$$

whence,

$$k = 4.281$$

Port diameter can thus be determined using the following relationship [23].

$$A_p = 4.281 \frac{A_0}{c} \cdot (r + 1) \cdot \frac{(d + r)}{d}$$

For efficient burner design, proper positioning of ports is an important factor. For optimum results, a distance of 0.75 in (1.9 cm) between two adjacent ports is recommended.

Apart from having optimum distance between adjacent ports it is also necessary to maintain proper distance between ports and utensils for minimising energy losses. For a biogas burner with port diameter of 0.635 cm, based on experimental findings GGDC, Mumbai, recommended a distance of 1 inch (2.54 cm) between port and cooking utensils [23].

Biogas domestic appliances are designed for a gas pressure of around 10 cm water column. In actual practice gas pressure may vary according to the type of biogas plant. V.K. Vijay, *et al.* [42] carried out a study to evaluate the performance of biogas burner and lamp for varying gas pressure. It was found that for fixed dome type of biogas plant under natural working, efficiency of burner increases with decrease in pressure. At around 10 cm pressure, maximum efficiency was recorded beyond which efficiency again declined. Biogas lamp on the other hand gave maximum intensity of illumination for pressure variation between 13-16 cm water column.

Pressure of the gas from biogas plant particularly fixed dome type plant is not uniform. Gas pressure from fixed dome type plant is much more during morning hours than during day time when gas usage is much more. Due to high gas pressure in the morning, more gas comes out than needed leading to reduced availability during day time. As per BIS standard for biogas stoves (IS : 87491), the inlet pressure of the gas should be 3 inches (7.62 cm) of water column while outlet pressure from fixed dome type plants is found to vary between 24 and 30 inches of water column during morning hours [43]. This causes more wastage of gas during morning hours within one hour and at noon not enough gas with requisite pressure is available for use though the quantity of the gas produced in the plant during night is sufficient enough to meet the household fuel needs [43]. For overcoming this problem, Patel Gas Crafters Pvt. Ltd, Mumbai, leading manufacture of biogas appliances, has designed and developed a pressure regulator which evens out pressure variations in biogas plants. The pressure regulator ensures operation with an even gas pressure for a longer time which meets the requirements for burners and lamps. According to manufacturer's instructions, pressure regulator needs to be mounted directly on the burner or lamp. If the gas is not utilised no gas should remain in the regulator to prevent shortening of its service life.

11.5 Use of Biogas as a Lighting Fuel

Biogas provides a clean fuel for lighting homes. Biogas lamp and burner are structurally alike both having a nozzle, an air inlet and a mixing chamber; but with additional mantle in case of lamp and fire-sieve plate in case of stove. In biogas lamp when gas is burnt, mantle of the lamp glows which causes lighting. Mantle in biogas lamp is similar to one used in a 'Coleman' or propane lamp. Mantle is normally made of Ramie fibre (which is also used in making glass cloth and liner) and is coated with thorium nitrate solution. While burning, the Ramie fibre reduces to ash forming a layer of thorium dioxide which emits dazzling white light at high temperatures. Nozzle is of the size of a needle point having 0.5 to 0.7 mm. diameter. The other end of the nozzle is connected to gas supply hose which is linked to a biogas plant. Biogas emerges out from the nozzle at very high temperature forming low pressure area around it. As biogas passes through the nozzle, air is allowed to be drawn into the mixing chamber to freely mix with it. Brightness of biogas lamp mainly depends on factors like gas pressure, and relative proportion of gas and air which is about 1 : 10, and thoroughness in mixing. For achieving bright intensity, nozzle need to be adjusted by trial and error. Low gas pressure causes poor light intensity, but higher pressure though it improves brightness tends to lower mantle-life. Biogas lamps can be either suspended (hanging type) or put on table (standing type). As biogas is almost half as light as air and as hot air flows upwards, brightness of a standing biogas lamp is somewhat greater than that of a hanging type. Biogas lamps are generally designed to produce 100 candle power and consume 0.11 to 0.15 m^3 (4 to 5 ft^3) biogas per hour.

Several designs of biogas lamps are popular but most common ones are what have come to be known as the Indian, Chinese and Pakistani designs. Indian lamps are designed with single or double mantles. Both single and double mantle types can be further categorised as internal or external types. An internal mantle lamp has a simple cover and is designed to produce 100 candle power which is equivalent to light intensity of a 60 watts electric bulb. An external single mantle lamp has an outside protective cover to safeguard it from wind and is also capable to produce same light intensity of 100 candle power. Designs of these two lamps are shown in Figs. 11.13 and 11.14 which work at gas pressure of 70 to 85 mm (2.75 to 3.25 in) water gauge. A double mantle lamp can also be both internal and external type. The internal type is fitted with a simple cover and can produce upto 100 candle power. The two mantle external type is equipped with a special cover to protect it from rain and wind and is also capable to produce 100 candle power. If despite higher gas pressure brightness of the lamp is found to be low, then this problem can be minimised by passing the gas through petrol or solvent oil kept in a container which leads to its enrichment. As per the available norms, 4 litres of petrol or solvent oil is sufficient to last for 1500 hours of lamp usage. This is equivalent to saying that four litres of enriching agent can last for one complete year, if the lamp is lit daily for four hours. A biogas-fuelled table lamp commonly used in Pakistan is shown in Fig. 11.15 the important features of which are self-explanatory.

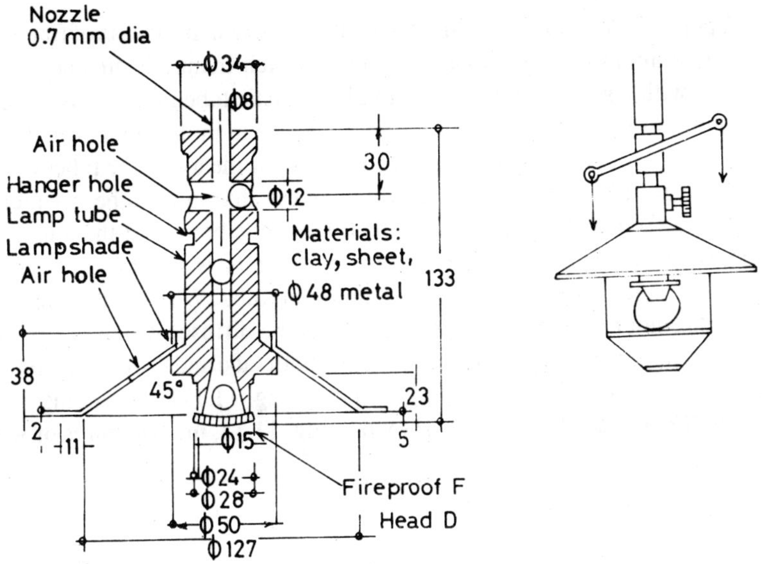

Fig. 11.13 Single-mantle hanging type biogas lamp commonly used in India [Source: BORDA, Biogas Plants: Building Instructions, GATE, P.O. Box 5180,6236 Eschborn 1, Germany]

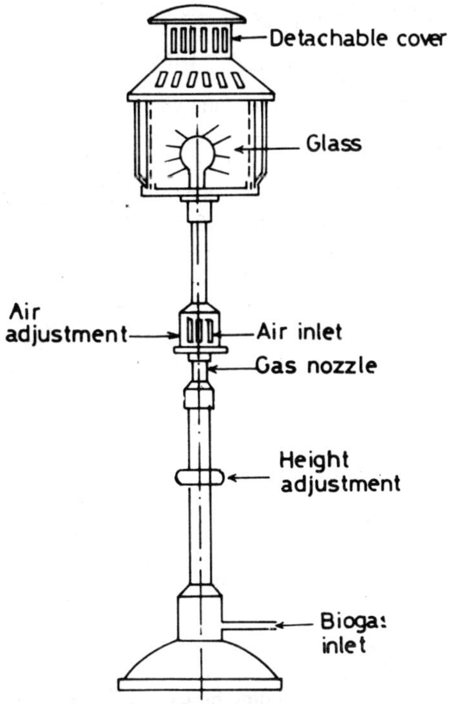

Fig. 11.14 Double mantle biogas-fuelled lamp [UN. Guide-book on Biogas Development, New York, 1980, p. 61]

Fig. 11.15 Biogas-fuelled table lamp commonly used in Pakistan [Source: UN, Guide-book on Biogas Development, New York, 1980, p. 61]

A popular Chinese lamp is shown in Fig. 11.16 which consists of four major parts: an aluminium tube, a gas diffuser to which mantle is attached, a disc reflector with a glass globe, and a thin glass/plastic tube that acts as a nozzle. The aluminium tube is normally 10 cm long 1.5 cm in diameter with four air holes of 0.3 to 0.4 cm diameter located little below the top of the tube which is open. The gas diffuser is held at the bottom of the aluminium tube to which the mantle is attached. The disc reflector is fitted at the bottom of the aluminium tube just above the diffuser. The glass or plastic tube is about 12 cm long and resembles a pipelle, and fits into the aluminium tube. Whereas the upper end of the aluminium tube is open, its base holds the diffuser which is similar to one in the Indian type. In lamp made of clay, both tube and diffuser are integral and made of this material including reflector (Fig. 11.17). A standing biogas lamp commonly used in China is shown in Fig. 11.18, main features of which are self-evident.

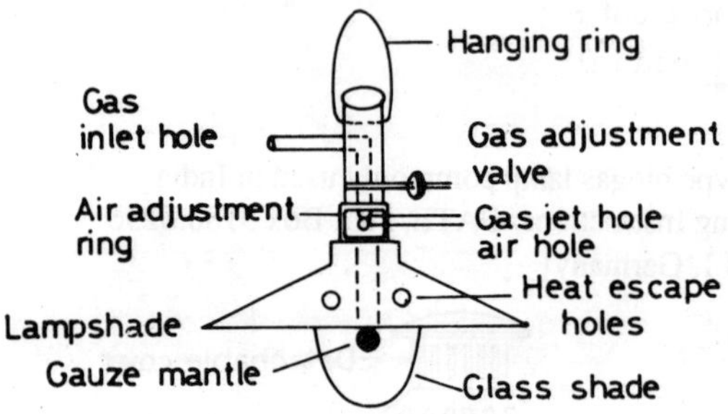

Fig. 11.16 A schematic view of 'Red Star' hanging lamp used in China [Source: Buren, A.V., A Chinese Biogas Manual, Intermediate Technological Pub. Ltd., London, 1979, p. 107]

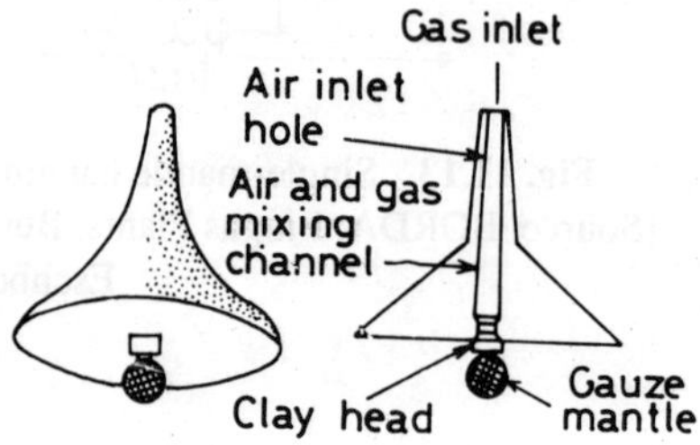

Fig. 11.17 A schematic view of clay hanging lamp used in China [Source: Buren, A.V., A Chinese Biogas Manual, Intermediate Technological Pub. Ltd., London, 1979, p. 107]

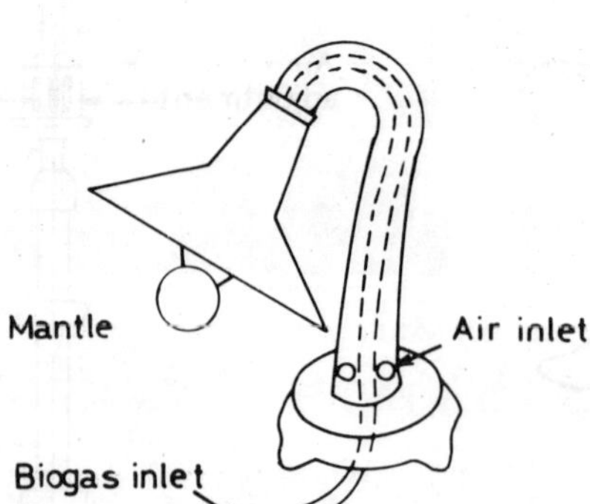

Fig. 11.18 A standing biogas lamp commonly used in China [F&AO, China: Azolla Propagation and Small-scale Biogas Technology, Rome, 1979, p. 57]

As for the starting sequence of the lamp, after opening it, clay nozzle (venturi) is fitted. This is then followed by opening of the mantle to form a hollow ball and subsequent joining to the venturi. After opening gas cock and regulator, mantle is lit to light the lamp. Lamp can be turned-off by operating the gas cock. Regulator is adjusted for achieving requisite brightness. For preventing the possibility of any hazard, the lamp should be lit by bringing the match-stick close to the mantle either via the hole in the base of glass globe or via the reflector after opening it.

11.6 Utilisation of Biogas for Pumping Water and Miscellaneous other Applications

Water-pumping by biogas-run IC engines is emerging as a very promising applications of biogas. R & D Division of Bharat Heavy Electricals Ltd. at Hyderabad has developed biogas-fuelled engines which on being coupled to a pumpset can be used for pumping water both for drinking water supply and irrigation purposes. One pumpset underwent field trials in past spanning over a period of five years and another pumpset gave trouble-free performance for a three year test period for supply of drinking water at Salojipally village in Medak district in Andhra Pradesh [28]. It was claimed that the cost of pumping 1000 litres of water upto a head of 10 metres with this type of pumpset (as per the prices prevailing during 1988) was 3.5 paisa as against 9.1 paise by diesel pumpset of equal capacity which clearly indicated the cost-effectiveness of a biogas operated pumpset. Estimates of biogas plant capacity for varying water head and duration of pump operation are given in Table 11.6. Size of agricul tural fields that can be irrigated with varying hours of pump operation are given in Table 11.7.

Table 11.6 Estimates of Biogas Plant Capacity for Varying Water Head and Duration of Pump Operation

Water head (m)	*Water-Delivery in Litres for Varying Operation Hours*				
	1 hr	*2 hr*	*4 hr*	*6 hr*	*8 hr*
5	64,800	1,29,600	2,59,200	2,88,800	5,18,400
8	40,500	81,000	1,62,000	2,43,000	3,24,000
10	32,400	64,800	1,29,600	1,94,400	2,59,200
15	21,600	43,200	86,400	1,29,600	1,72,800
20	16,200	32,400	64,800	97,200	1,29,600
Plant Capacity (m^3)	4	8	16	24	32
Cattle Needed	11	22	44	66	88

Source : Ramarao, A., Biogas Programme in India, Khadi Gramodyog, June, 1988, p. 410.

Table 11.7 Irrigation Potential of Biogas-fuelled Pumpset for Varying Operation Hours

Water head	*Cultivable Area of Rice-fields in Hectares*				*Cultivable Area of Non-Rice-fields in Hectares*			
	2 hr	*4 hr*	*6 hr*	*8 hr*	*2 hr*	*4 hr*	*6 hr*	*8 hr*
5	1.82	3.68	5.53	7.37	2.59	5.18	7.78	10.37
8	1.16	2.31	3.47	4.63	1.62	3.24	4.86	6.48
10	0.92	1.85	2.77	3.70	1.29	2.59	3.88	5.18
15	0.61	1.23	1.85	2.47	0.86	1.73	2.54	3.40
20	0.46	0.93	1.39	1.84	0.65	1.29	1.94	2.59

Source : Ramarao, A., Biogas Programme in India, Khadi Gramodyog, June, 1988, p. 410.

Following are some major technical specifications of BHEL designed biogas-fuelled pumpset.

Engine Type : Single cylinder, 4 stroke SI engine
Speed : 3000 rpm; shaft output : 3 hp
Pump : Centrifugal type; Suction : 8 metres
Water delivery : 35,000 litre/hour at 10 m total head
Gas consumption : 2.5 m^3 per hour
Cowdung requirement : 60 kg/hr during operation

V. Massey [28] of Division of Agricultural Engineering, IARI, New Delhi used two biogas-fuelled single cylinder two stroke petrol-cum-kerosene engines of 2.5 and 3 hp for chopping hay. First engine was run on petrol and then switched over on biogas by continuous injection of air-biogas mixture in suitable proportions. Mixing of biogas with air was regulated with the help of a modified carburettor. Each engine was rigidly installed on a wooden table which was coupled to a 450 rpm chaff-cutter with the help of a canvas belt. The two 2.5 and 3.5 hp engines were reported to run at 90 and 80 per cent efficiency and utilised 49 and 52 ft^3 of gas per hour at full load.

Biogas can also be directly utilised for operating some major button-making, plastic-moulding and toy-making machines. Ram Bux Singh [29] reported norms of biogas consumption in operating various plastic moulding machines (Table 11.8). Norms of biogas consumption for cooking, lighting, running IC engine, refrigerator etc. are given in Table 11.9 and 11.10.

Table 11.8 Norms of Biogas Consumption in Operating Different Plastic Moulding Machines

Machine	*Machine Capacity (Ounce)*	*Rounds per hour*	*Gas Consumption (ft^3 per hour)*	*Power Consumption (Watts per hours)*
1. Button Making Machine	2.25	120	4.8	500
2. Plastic Moulding Machine	0.50	100	5.0	600
3. Toy Making Machine	1.00	85	5.5	650
4. Toy Making Machine (large)	1.50	75	8.5	1000

Source : Singh, R.B., Biogas Plant: Generating Methane From Organic Wastes, Gobar Gas Research Station, Ajitmal, Etawah, 1974, p. 78.

Table 11.9 Rate of Biogas Consumption in Operating Various Appliances

Application	*Specifications*	*Rate of Biogas Consumption (ft^3/hour)*
1. Cooking	2" dia. burner	11.5
	4" dia. burner	16.5
	6" dia. burner	22.5
2. Lighting	1 mantle lamp	2.5
	2 mantle lamp	5.0
	3 mantle lamp	6.0
3. Refrigerator	10" × 18" × 12"	2.5
4. Incubator	18" × 18" × 18"	2.0
5. Boiling water	Per gallon	10.0
6. Running IC engine	Per BHP	16-18
7. Running table fan	Gas pressure 2" water column	3.5
8. Running heater	Gas pressure 2" water column	5.5

Table 11.10 Norms of Biogas Consumption for Selected Applications

Application	*Specification*	*Rate of Biogas Consumption (m^3/hour)*
1. Cooking[2]	5.1 cm burner	0.33
	10.2 cm burner	0.47
	15.0 cm burner	0.64
2. Gas Lighting[1]	single mantle lamp	0.07
	two-mantle lamp	0.14
	three-mantle lamp	0.17
3. Refrigerator	per m^3 capacity	1.00
4. Incubator	per m^3 capacity	0.5-0.7
5. Petrol or diesel engine converted to operate on biogas		0.60-0.68 per k.w.
6. Petrol	34.8 MJ/1	Equivalent Vol 1.43 l
7. Diesel	38.7 MJ/1	Equivalent Vol 1.59 l
8. Boiling water[2]	2.25 MJ/1	Equivalent Vol 1.59 l

Source : 1. Absolute volume of biogas (heat content = 24.2 MJ/m^3) required to provide energy equivalent of 1 litre of fuel.
2. Absolute volume of biogas (heat content = 24.2 MJ/m^3) required to change 1 litre of water into steam at 100°C.

11.7 Biogas as a Fuel for Running IC Engines

The idea of running internal combustion engines on gaseous fuels is not of recent origin. In fact the first gas engine was known to be tried as early as 1860. In 1868, Dr. Diesel, a legend in engineering, reportedly applied for a patent for a dual-fuel engine with one of the fuels being gaseous. These engines mostly utilised natural gas, producer gas, sewage gas and other gases generated by coal and chemical industries [27]. Some important properties of biogas, sewage gas and natural gas are sumarised in Table 11.11. IC engines which can work both on liquid and gaseous fuels can be broadly divided into two categories, namely, compression ignition (CI) and spark ignition (SI) engines. In CI engines, gaseous fuel is normally burnt at atmospheric temperature and pressure and fuel is ignited by injecting a small quantity of liquid fuel [40]. Injection for ignition of compressed air-fuel mixture is done while piston is at near top dead center. In SI engines, the gas is inducted through a gas carburettor or a mixing valve or by injection at low pressure depending upon the engine design [27]. A classification of IC engines of reciprocating types is given in Table 11.12.

Table 11.11 Some Combustion-Related Properties of Biogas, Sewage Gas and Natural Gas

Property	*Unit*	*Biogas*	*Sewage Gas*	*Natural Gas*
1. Composition:				
Methane	Per cent by volume	45-55	55-69	85-90
Ethane	–do–	–	–	3.40
Carbon diodixe	–do–	49-55	28-29	0.31
Air	–do–	0.2-0.4	3-4	9.10
2. Calorific value	kcal/m^3	4200	5640	7840
3. Specific gravity relative to air		1.04	0.88	0.63
4. Stochiometric requirement of air	m^3 air/4.76 m^3 of gas	4.76	6.40	9.55

Source : Pundir, B.P. and Singhal, S., Biogas Utilisation in Engines: An Appraisal, Proc. of the Workshop on 'Biogas Utilisation in Engines, Indian Institute of Petroleum, Dehradun, April 19-20, 1978,' p. 9-12.

Table 11.12 Broad Classification of Reciprocation Type Internal Combustion Engines

Engine type	*Fuel Used*	*Method of Injection*	*Compression Ratio*	*Method of Ignition*
1. Gas engines	Gas	Mixing valve or injection	Low	Spark ignition
2. Petrol engines	Petrol	Carburettor or injection	Low	Spark ignition
3. Diesel engines	Diesel or fuel oil	Injection	High	Compression ignition
4. Dual-fuel engines	dual-fuel	Injection	High	Compression ignition

Biogas provides a clean fuel for both SI (petrol) and CI (diesel) engines. Whereas biogas in diesel engines is used in combination with diesel petrol engines can be run on hundred per cent biogas. In diesel engines, temperature at the end of compression stroke is usually less than 700°C whereas the ignition temperature of methane and air mixture is 814°C. In view of this temperature difference, for igniting gas mixture, it is necessary to inject small quantity of diesel fuel before the end of compression stroke.

Use of biogas as an engine fuel offers several advantages. Biogas being a clean fuel causes clean combustion and reduced contamination of engine oil.

Use of biogas also results in reduced deposits on piston and combustion chamber. As gas needs are considerable, large capacity plants are needed to supply gas for running IC engines. It is to be noted that even a 5 hp engine for 8 hourly operation requires gas supply from a biogas plant of 500 ft^3/day capacity. Community plants with a gas output of 50 m^3 per day are generally used to drive biogas-fuelled electricity generating sets.

As per some rough estimates, nearly 575 million tonnes of cattle dung is generated per annum in the country. Assuming that about 75 per cent of it is available for biomethanation, an estimated 10,000 billion ft^3 of gas can be available for use. If two thirds of this gas is available for fuelling IC engines, with an assumed norm that 35 ft^3 of gas can replace 1 lb of diesel oil, an estimated 9 million tonnes of diesel can be saved annually which is much more than the total diesel needed for operating irrigation pump-sets and multifarious agricultural activities at large.

Normal consumption rate of biogas for running IC engines is 0.45 to 0.54 m^3/hour (16 to 19 ft^3/hour) per hp or 0.60 to 0.70 m^3/hour (21 to 25 ft^3/hour) per-kw if used for power generation. Biogas pressure is found to vary from 25 to 100 mm (1 to 4 inches) water gauge. If an engine consumes 0.50 m^3 of biogas per hp per hour, then quantity of gas needed for running a 10 hp engine for 10 hour operation per day becomes [39],

$$= (0.50\ m^3/hp/hour) \times 10\ hours \times 10\ hp$$

$$= 50\ m^3$$

Similarly, if an engine consumes gas at a rate of 0.65 m^3/kw/hour, then quantity of gas needed for running a 10 kw generator for 10 hour operation each day becomes

$$= (0.65\ m^3/kw/hour) \times 10\ hours \times 10\ kw$$

$$= 65\ m^3$$

Dual-fuel engines offer the flexibility of easy switch-over from dual-fuel to pure diesel operation and vice-versa. Schematic view of diesel engine utilising dual-fuel is shown in Fig. 1.4 (ch. 1). Important elements of a dual-fuel engine are air-filter, dual-fuel intake pipe, biogas choke and exhaust pipe. While working out the economics of biogas use in dual-fuel operation, it is necessary to consider relative costs of biogas and diesel, extent of diesel replacement with biogas, degree of additional wear and consequent higher maintenance and lubricant needs, and comparative thermal efficiency of the two systems. Thermal efficiency is governed by dual-fuel mix, engine output and speed. For a given engine output and speed, thermal efficiency first increases with increasing gas proportion and which beyond a limit again starts declining.

Dual-fuel ratio for maximum efficiency is different for different engine loads at a given speed. It is to be seen that minimum fuel cost does not necessarily correspond to a gas-fuel ratio which causes peak efficiency. Norms of biogas consumption for operating different capacity engines on dual-fuel operation are given in Table 11.13. These rates are based on 8 hourly operation with 75 per cent plant efficiency with biogas consumption rate being 9 ft^3/hp/hour having 50 per cent diesel replacement with biogas. Although replacement of diesel can be as high as 80 per cent, maximum thermal efficiency is achieved when gas proportion in mixed fuel ranges between 30 to 60 per cent. Advancing the injection timing by 5 to 7° BTDC over the standard advance helps to improve thermal efficiency of dual-fuel based CI engine.

Table 11.13 Rates of Biogas Consumption for Operating Different Capacity Engines on Dual-Fuel

Engine Rating H.P.	*Biogas Consumption (ft^3/day)*	*Capacity of Plant to Supply Biogas (ft^3/day)*
5	360	480
10	720	960
15	1080	1440
20	1440	1920

Conversion of diesel engine to run on biogas involves some technological changes. It can be achieved by installing an extra magneto and a spark plug to change the compression ignition system to spark ignition otto cycle. This modification enables diesel engine to work purely on biogas. Changeover is however difficult because in most CI engine designs no space is available for providing magneto and spark plug. Conversion is particularly difficult when engine size is small. With sufficient supply of biogas a large multi-cylinder diesel engine can be modified with greater ease. A CI engine can also work on dual-fuel with biogas forming the main fuel supply. When adequate quantity of biogas is not available, it can be quickly changed over to diesel fuel.

A dual-fuel engine is first started with diesel fuel only. After it has attained normal running for some time, biogas choke is opened to admit gas into the combustion chamber. Biogas admission can be controlled by adjusting the choke. For stopping the dual-fuel engine, biogas choke should be stopped first followed by throttle. It should be endeavoured to keep the engine speed uniform. If the engine does not pick up speed either due to engine overloading or due to higher intake of biogas, the engine should be allowed to run idle for some time. It can be again put on load after its running becomes normal. If during operation occasional sound or knocking is noticed, biogas choke should be adjusted, and if it is on account of small loads on engine it can be corrected by suitable increasing load on it. In a steady engine operation, dual-fuel achieves nearly 80 per cent saving in diesel consumption.

Dual-fuel operation is smoother when the compression ratio is kept low but it leads to low power output and gives rise to start-up problems. Diesel engines are generally set with an advance injection angle which makes them appropriate for running on dual-fuel as well as diesel alone. Furthermore, biogas possesses reasonably good anti-detonation properties. In view of these factors while running on dual-fuel, original compression ratio and advance injection angle set for diesel operation should not be disturbed. This also facilitates quick reversal to diesel fuel if some difficulties in running with dual-fuel are encountered [32].

Petrol engines running on biogas can run solely on gas without mixing with petrol. A specific design of a carburettor for 10 hp petrol engine running on biogas is shown in Fig. 11.19. A plastic bag is provided near the gas inlet to facilitate easy sucking of gas [39]. Petrol engines running on biogas however cause some maintenance problems. Valves, plugs etc. remain more clean and the sump oil needs less frequent replacement. However, engines working on biogas get heated up faster than petrol and accordingly latter need more effective cooling system.

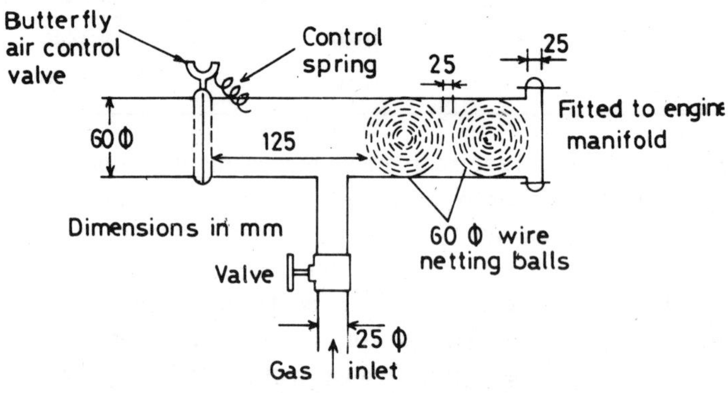

Fig. 11.19 Carburettor for 10 hp petrol engine running on biogas [Source: UN, Guide-book on Biogas Development, New York, 1980, p. 86]

For using biogas as a petrol substitute, small modifications in the fuel injection system are necessary. In a partially converted system, a gas tap is fitted to the inlet manifold of the engine. The engine is started initially on petrol and allowed to run until the running is smooth. This is followed by cutting off the petrol supply and slight opening of the gas tap. For minimising choking, air intake needs to be regulated. Opening of gas tap and adjustment of air supply is carried out simultaneously to achieve smooth operation. After its initial start with petrol and engine warming up, SI engines can solely run on biogas. Biogas run SI engines are capable to achieve 85 per cent of maximum BHP. Brake thermal efficiency of engine when it is run on biogas is slightly higher. Specific gas consumption is close to 0.67 m^3/bhp/hr and stochiometric ratio as 11.2. Advancing the injection timing by 20° to 25° BTDC over the standard advance helps to improve thermal efficiency of SI engines [2].

Attempts to use biogas for running IC engines in India date back to forties when gas from a sewage treatment plant in Mumbai was used for this purpose. Later some experimental work in this area was done by the KVIC in Mumbai in sixties and early seventies. Studies were also carried out concurrently in this field at the Gobar Gas Research Station of the PRAD at Ajitmal (Etawah) under the technical supervision of Ram Bux Singh. Investigations carried out established biogas potential as an engine fuel and identified major problems that need to be solved. Some experimental studies were also carried out at the Agricultural Engineering Division of the IARI, New Delhi. The Tamil Nadu Agricultural University conducted a trial for running a diesel engine with biogas. Biogas was injected through a nipple placed below the air cleaner. The trial showed that biogas could replace diesel oil by over 50 per cent. In 1975-76 a progressive farmer-cum-entrepreneur from Panipat (Haryana) conducted a field trial for the use of biogas through inlet manifold in diesel engine with the help of a carburettor designed by him. The Tractor Training and Testing Station, Budni in collaboration with the Tractor Training Centre, Hissar and the Indian Institute of Petroleum, Dehradun conducted similar trials on use of biogas as engine fuel. In 1976, Kirloskar Oil Engines designed and introduced biogas dual-fuel engines. The first 5 hp biogas engine capable to run at 1500 rpm was commercially released in June, 1977. From then onwards engines of higher capacity and rpm have been developed and used in the country [10].

IIP, Dehradun also carried out studies on use of sewage gas for dual-fuel operation. The studies were conducted at the experimental sewage gas plant of UP Jal Nigam at Dehradun. The plant has a capacity to produce 11.4 m^3 (400 ft^3) gas per day. Calorific value of sewage gas is marginally higher than biogas. Sewage gas consists of more methane as compared to biogas. Hence for a given diesel fuel flow rate and power output, less quantity of sewage gas need be admitted as compared to biogas. Higher compression ratio was found to give better engine performance. Maximum thermal efficiency of the engine was found to occur when nearly 40-50 per cent of the total energy input was derived from gas. Exhaust smoke was considerably less with gas induction while carbon monoxide slightly more. General engine cleanliness and lubricating oil performance were found to be more or less similar with and without sewage gas. Tractor Training and Testing Station at Budni (Madhya Pradesh) in association with the Tractor Training Centre, Hissar and the IIP, Dehradun carried out studies in seventies on dual-fuel operation of CI engines. Effect of varying dual-fuel proportions on engine speed and horse-power developed was studied. Biogas for these experiments was obtained from a 150 ft^3 capacity plant. No rough running of engine was reported for biogas share upto 76 per cent beyond which it occurred. Horse-power developed and engine speed declined with very high gas proportion in dual-fuel [10]. This indicates that dual-fuel operation is more ideal at medium than high speeds.

Dr. (Mrs) P.P. Parikh, *et al.* [22] of Dept. of Mechanical Engineering, IIT (Mumbai) carried out several studies in past on dual-fuel based IC engines. It was reported that light diesel oil (LDO) could substitute expensive high speed diesel (HSD). By keeping LDO at 20 per cent and raising biogas proportion gradually up, it was possible to raise the engine capacity to 120 per cent of its rated capacity. Smoke level in dual-fuel operation was less as compared to when LDO was used alone. Though at part loads the efficiency of LDO-biogas operation was slightly lower than with LDO alone, at full loads as well as at loads 20 per cent higher, the efficiency was higher with dual-fuel operation. This finding has a great practical utility as one can use relatively less expensive LDO in place of expensive HSD for dual-fuel operation. For all proportions of LDO and gas, dual-fuel operation is beneficial for the engine as well as the user. Dual-fuel systems do not warrant any specialised training for operation.

G.L. Patankar, *et al.* [23] at the GGRDC, KVIC, Mumbai carried out several experimental studies in seventies in association with the VJTI, Mumbai on use of biogas in running IC engines. Experiments were carried out on a single cylinder, horizontal, four stroke, 600 rpm engine which was run on fuel-oil in normal course. The engine was modified to work on dual-fuel. Rack in fuel injection pump was connected to the governor so that the engine could be started on fuel-oil, run on it by changing the lever position, and biogas by adjusting fuel-oil injection. Power developed and efficiency of the dual-fuel engine was comparable with that of solely fuel-oil. The engine was later on run at Tulshishyam village of Junagarh district where it was used to generate electricity and run a flour-mill. Dual-fuel engine was however found to corrode very soon, due to carbon dioxide, hydrogen sulphide and moisture for which some satisfactory solution was needed.

Kirloskar Oil Engines Ltd at Khadki (Pune) took lead in past to manufacture dual-fuel engines utilising a mixed fuel of 80 per cent biogas and 20 per cent diesel [15]. The company began with a 5 hp 1500 rpm engine which first rolled out in June, 1977. Subsequently, it developed engines of higher ratings upto 75 hp and beyond, with facilities for both water-cooled and air-cooled systems. The engine can quickly alternate between duel-fuel and diesel according to the need and convenience. When running on dual-fuel, nozzle and fuel pump of the engine is used to create pilot diesel spray in combustion chamber to initiate combustion. A special carburettor is provided to ensure proper mixing of biogas and air before admission to the combustion chamber. For operating a 5 hp engine, 2.25 m^3 of biogas is needed per hour. The KOEL dual-fuel engine consumes 15 ft^3 of biogas per bhp hour. This implies that 75 ft^3 of gas is needed to run a 5 bhp engine for one hour. The engine can be easily coupled to an irrigation pump set or used for operating farm machinery like crusher, thresher, flour-mill, power generating sets etc. R&D division of the company also analysed and compared the performance of diesel engines and dual-fuel engines. Factors analysed covered density of exhaust smoke, temperature of exhaust gas, engine-deposits and condition of lubricating oil. Density of exhaust smoke was found to be less with dual-fuel than when diesel was used alone. Temperature of the exhaust gas was however found the same with no significant difference. As for engine-deposits, it was found that engine run on dual-fuel was cleaner than diesel alone. Lubricating oil with dual-fuel was cleaner as compared to diesel.

A typical dual-fuel engine requires 15 per cent diesel and balance proportion of fuel is obtained from biogas. If a normal 5 hp engine is run on only diesel, it requires 185 g of diesel per bhp hour, then for 8 hour daily operation, monthly (25 days) diesel requirement becomes [15]:

$$\text{Monthly diesel consumption} = \frac{185 \times 5 \times 8 \times 25}{1000 \times 0.83} = 223 \text{ litres}$$

If the engine is run on dual-fuel only then 33 litres of diesel will be needed resulting in a saving of 190 (223-33) litres of diesel per month.

K. Kasturirangan, *et al.* [2] have carried out experimental studies concerning use of biogas as diesel engine fuel. The performance of a single cylinder dual-fuel engine with biogas and pilot diesel oil injection was studied under varying conditions. Tests were conducted at constant engine output with varying biogas diesel proportions. Tests were also conducted with variable output with fixed diesel oil flow rates. The peak output of the engine with dual-fuel was also analysed. These tests confirmed that it is possible to operate normal type of diesel engines used for irrigation purposes on biogas with minor modifications. The engine develops the same power as with diesel oil. Thermal efficiency was satisfactory at high loads but rather unsatisfactory at part loads. Throttling of air supply improved the efficiency to some extent. The engine could be operated over a wide range of gas to diesel ratios, depending on the availability of biogas and diesel oil, and offered the ease of switch-over to diesel oil when needed.

Several experimental studies were carried out in past by C.P. Kothandarman and S. Krishnamurthy [17] at the PSG College of Technology, Coimbatore on dual-fuel based CI engine and biogas based SI (petrol) engine. In a CI engine, biogas reduced brake thermal efficiency throughout the operating range with higher reduction at lower outputs. When gas proportion was increased to 50 per cent, peak pressure increased by 10 per cent with only marginal increase in exhaust gas temperature. At higher loads no vibration problems were encountered up to 50 per cent fuel replacement. Higher gas proportion beyond 50 per cent caused rough running (knocking) of the engine and high peak pressure. Biogas was found to reduce efficiency marginally at higher loads. At small loads dual-fuel caused less noise than diesel which was indicative of smoother combustion with biogas. As for SI engine operation, the engine was first run on petrol at varying loads at the rated 1800 rpm and overall efficiency analysed. The same engine was then run purely on biogas. It was reported that for the same power developed, SI engine run on biogas registered an improvement in overall efficiency [16].

Z. Cao [6] made a detailed veview of research on use of biogas in running single and multi-cylinder CI engines as dual-fuel involving study of compression ratio, selection of speed and thermal efficiency of engines. C.Q. Jiang, *et al.* [13] carried out studies on using compressed biogas as part of dual-fuel in running CI engines. They also studied problems of biogas compression, possibility as well as need of using compressed biogas, and economic viability of compressing biogas. According to B. Gehri [9], success of dual-fuel operation depends on using gas of fixed composition and in adequate supply at uniform pressure. For efficient running it is necessary to maintain dual-fuel engines by periodic replacement of filters, spark plugs and other vital parts.

According to D.J. Picken and H. Hassan [24] biogas operated engines perform poorly on account of harmful effect of hydrogen sulphide on copper-based engine parts. Whereas many engine parts can be substituted by other materials or protected from biogas, small end bearings pose particular problems [25]. On account of operational needs this bearing is more under high fatigue and stress, and cannot be easily replaced. Moreover, it is also not convenient in smaller engines to protect it by flooding it with oil. Increased wear of cylinder bore can be attributed to high hydrogen sulphide content in biogas, its leakage into crank case and harmful impact on end bearing [26]. For minimising this problem, copper small end bearings should be lubricated as far as possible.

11.8 Biogas as a Vehicle Fuel

Compressed natural gas (CNG) as also biogas are comercially used as a vehicle fuel in New Zealand and Italy. Conversion kit for biogas-fuelled vehicle is virtually the same as for CNG which is marketed as an automotive fuel in many parts of the world (Fig. 11.20). These kits permit changeover from gas to liquid fuel and vice-versa at a short notice. As methane is only half the weight of air, dispersal of the leaking gas is faster. This makes biogas a safer fuel than either LPG or petrol. A schematic view of entire biogas-driven vehicle system is shown in Fig. 11.21.

For use as a vehicle fuel, biogas is first compressed in a three or four stage compressor up to a pressure of 2500-3500 psi and stored in a high pressure gas storage cascade which helps to facilitate quick refuelling of storage cylinders. The number of cylinders in cascade depends on digester capacity and number of cylinders provided on each vehicle. For biogas from 45 m^3 capacity digester, at least three or four 9.2 m^3 capacity cylinders are required. A 9.2 m^3 of gas contains only 13.9 litres of petrol equivalent. Compressor size should be such so that supply rate of compressed gas matches with gas generation rate. For a vehicle of up to 25 litres of liquid fuel capacity and for a distance range from 100 to 150 km, a cylinder of length 1 m, diameter 300 mm and weight 50 kg is sufficient [34]. In order to provide sufficient range, two cylinders are generally needed in a motor car and up to four in a truck. Gas can be filled in about a minute's time with cylinder in place. Biogas needs to be used in compressed state for spontaneous flow otherwise space and volume requirements become

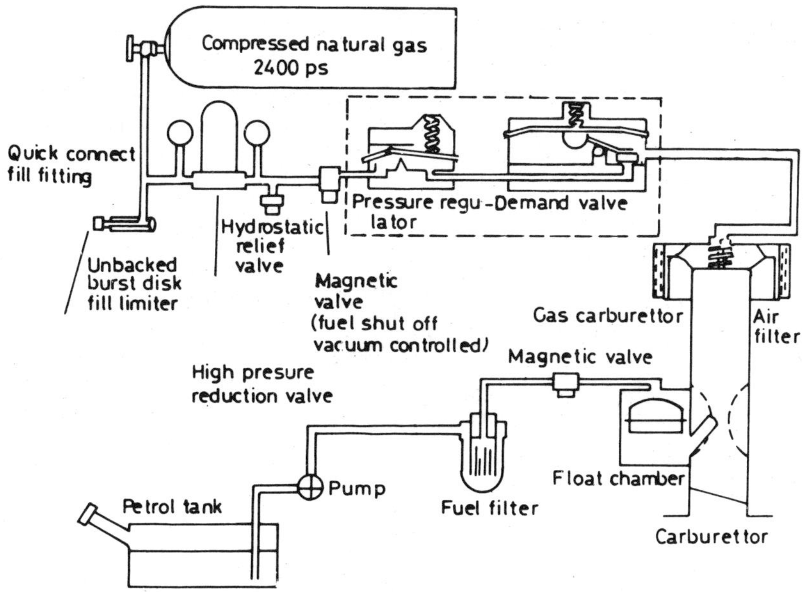

Fig. 11.20 Schematic diagram of a typical CNG conversion kit

excessive. Need for compression however adds to the costs which tends to offset the very purpose of using biogas as a vehicle fuel [41].

Being gaseous fuel, biogas provides excellent cold starting for vehicle as unlike petrol it need not first be vapourised. Once started the engine performs well instantly with no sluggish warm-up. This is because unlike petrol methane does not condense on cold intake manifolds cylinder heads and pistons. Exhaust emissions from biogas are much less as compared to petrol engines. Methane has a higher octane number as compared to petrol which causes slower combustion of the former. This higher octane rating makes biogas suitable for CI engines which can lead to greater power output. In a SI engine, higher power output and high compression can be obtained by machining the cylinder head or by fitting higher compression pistons. The permanent modification, however renders it unsuitable for any subsequent running on petrol. As biogas does not condense unlike petrol in cold areas it is free from problem of washing away expensive lubricating oil from top of piston ring. Consequently, there is no dilution of the lubricating oil either and accordingly inter-fuel replacement interval can be made larger by as much as two times. Spark plug servicing is also less critical as there is no lead pollution from exhaust gases and no lead-fouling of spark plugs. A major disadvantage of methane-based vehicles is the limited range per cylinder of gas carried on a vehicle.

Biogas-based vehicles in limited way were used in UK in 1942 from the gas produced by Mogden Sewage Works, London. One litre of petrol or diesel fuel could be replaced by 0.75 m^3 of methane. The fuel system of SI engine/petrol vehicle resembles a CNG vehicle. Each vehicle was fitted with one

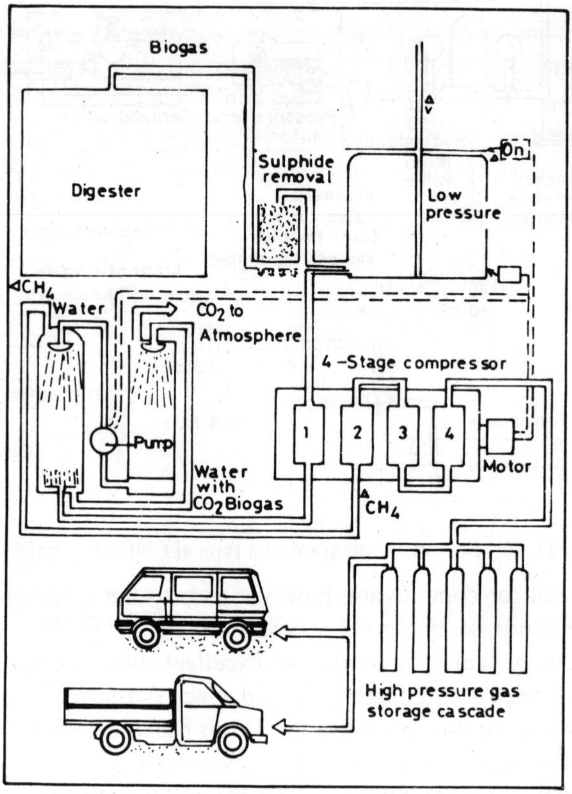

Fig. 11.21 A schematic view of entire system depicting biogas use in gas-driven vehicles [Source: Stewart, D. and Mc Leod, R., Producing Fuels on the Farm, NJ Jour. of Agriculture, Sep., 1980, p. 21]

or two 31.5 litre methane cylinders equivalent to 9 litres of petrol. Gas was supplied to carburettor via a pressure relief valve and other devices. The carburettor was in series between air filter and normal petrol carburettor. This enabled the throttle valve of the petrol carburettor to control the methane flow to the gas carburettor. As for the use of biogas in a diesel-fuelled vehicle as a dual-fuel, fuel pump maximum delivery was limited to 70 per cent of its normal maximum, and 30 per cent of the fuel was supplied by methane through a gas caburettor and main air inlet. A linkage was fitted to connect the gas carburettor throttle valve to the diesel governor input so that the ratio of methane to diesel remained constant at all speeds and loads. The vehicle was fitted with three 31.5 litre methane cylinders. In other respects the entire system was the same as that of a petrol-driven vehicle. Initiated in 1983, as part of a methane utilisation project in Sao Paulo (Brazil) biogas from three sewage treatment plants was used as automobile fuel. Sweetened gas following carbon dioxide removal with hot potassium carbonate solution was dried, compressed and bottled in 200 atmosphere cylinders, and used in running automobiles. A half tonne weight pick-up truck fitted with a petrol engine was run on compressed biogas for several years starting from mid seventies under the control of University of Manitoba, Winnipeg (Canada). Oxygen-type cylinders conforming to the Canadian Ministry of Transport Regulations were installed on the pick-up to carry the gas at a pressure of 2400 psi. While using in uncompressed state, biogas consumption was 0.19 m^3/km whereas the petrol consumption of the vehicle prior to changeover to biogas was 5.1 km/litre. For successful use of biogas as a vehicle-fuel, there should be adequate gas storage to cover a practical distance range. A comparative economic analysis of a vehicle capable to run on biogas and petrol indicated that frequent refuelling was preferable to carrying multiple high pressure tanks [35].

11.9 Applications of Biogas for Power Generation

Biogas can be used to produce electricity by coupling a dual-fuel engine to an asynchronous generator. Based on results of several studies carried out, 1 kwh of electricity can be generated from 0.75 m^3 of gas which can light 25 electric bulbs of 40 w rating whereas 0.75 m^3 of gas if directly burnt can light only 7 biogas lamps for one hour. Hence it is advantageous to first generate electricity and then light larger number of electric bulbs. In China asynchronous generators of 3, 3.5 and 7.5 kw rating coupled to biogas-based IC engines are commonly used. However, electricity generation from biogas is economical only when gas is supplied by large community plants but it requires high initial capital investment. Decentralised power systems can be cost-effective especially when they minimise transmission and distribution costs. For cost-effective system, unit cost of biogas-based electricity should be less than supplied by the electricity boards. Cost-effectiveness of biogas-based system improves with growing escalation in unit cost of conventional power.

Biogas is used to generate electric power as decentralised source of energy in several countries. Biogas-fuelled engines can be used for water pumping, crop processing and power generation. Economics of biogas-based decentralised energy systems is dependent on capital costs of the system and economic benefits derived from biogas as well as effluent produced. A study carried out by D.B. Machin [20] showed that cost-effectiveness of biogas-based power system was constrained by factors like inadequate sharing of information, lack of consensus on system design, low productivity of commonly used unheated digesters, and high cost of large digesters for providing sufficient gas at relatively low temperatures. T. Stahl, *et al.* [33] in the context of an integrated energy farm project at the University of Missouri in USA developed a mathematical model to assess the feasibility of locating a biogas-fuelled engine/generator on a farm to produce electricity and space heating. O.P. Singhal [31] emphasised that biogas-based power systems and biomass gasifiers are ideal considering that supply of conventional power to remote villages may still take many more years.

R&D Division (Hyderabad) of the state owned Bharat Heavy Electricals Ltd in past designed and developed biogas-based power systems. It successfully tested a 5 kw biogas power generator set. It estimated cost of biogas-based power as Rs 0.70 as against Rs 1.70 per unit with diesel generating set, which showed the cost-effectiveness of biogas-based power. Following are some major technical specifications of the BHEL designed biogas power system.

Type : Single cylinder 4 stroke, vertical, air-cooled engine directly coupled to 6 KVA generator.

Speed : 1500 rpm; speed control: by centrifugal governor.

Generator output : 5 kwh; 415 V; 3 phase.

Biogas consumption : 5 m^3 per hour.

Starting : Self starting DC motor and DC generator with cutouts for starting and charging battery.

C. La Torre [19] described a large-sized plant in Italy which produces biogas and electricity from animal wastes on a farm with 35,000 hens and 1,000 pigs. Before fermentation, wastes are purified by aerated lagooning. Biogas produced by the plant is used without further treatment as fuel for boilers and to produce both electric power and hot water using a cogeneration unit. While during winter energy needs of the farm matches the energy supply, during summer part of the gaseous energy becomes surplus which is compressed and used as fuel for tractors and farm vehicles. W. Tentcher [36] provided technical details of the construction of a full-scale plug flow pilot biogas plant and design of an associated engine-generator set for power generation. Red mud plastic (RMP) was used in constructing the gasholder. Wastes of about 800 pigs was used as feed material. The plant was operated at a maximum gas pressure of 10 cm water column. At ambient temperature and at 10 kw output, power generation was estimated as 1.4 kwh per m^3 of biogas. H.H. Jawurek, *et al.* [12] studied performance and combustion characteristics of a portable fixed ignition timing SI engine – alternator set fuelled with biogas. Maximum power output at optimally adjusted air-fuel ratio declined with increase in carbon dioxide content of the gas which also caused harsh running of the engine. Low electric power output, poor combustion and harsh running can be reduced with a marginally higher share of diesel in dual-fuel comprising a mixture of diesel and biogas.

Biomass Energy Institute Inc., Manitoba [18] in Canada evolved the following relationship for determining the number of animals needed for generating biogas-based electricity for a specified period in terms net fuel replaced with biogas.

$$\text{Number of animals needed} = \frac{1000\, l}{[f\, w\, d_f\, c_f + ewd_e\, c_e]}$$

where,

l = annual energy saving needed to recover investment, Rs. per year
w = average weight of animals, kg
f = net fuel replaced with biogas expressible as litre per day per 1000 kg liveweight
e = net electricity produced from biogas, kwh per day per 1000 kg live weight
C_f = cost of fuel in Rs. per litre
C_e = cost of electricity in Rs. per kwh
d = Number of days when biogas and electricity are produced, respectively.

11.10 Fuel Cell Linked Biogas Systems

A fuel cell is an electro-chemical device which directly converts chemical energy of a conventional or non-conventional fuel into low voltage electricity. Fuel cells provide low voltage D.C. power which can be converted to A.C. supply using high efficiency inverters which can be put to variety of household and industrial applications. Fuel cells react with the interacting fuel (biogas) in presence of oxygen from air to produce D.C. power. A single fuel-cell generates 100-200 watts, 1 volt D.C. electric power for each square foot of electrolytic cross-section. As voltage from individual fuel cell is low, a number of cells can be joined in parallel or series configurations to provide requisite voltage. In recent years, some attempts have been made to use biogas for generating electricity from fuel cells which can become a potential source of decentralised power supply in rural areas. Being modular in nature, a decentralised fuel-cell biogas system provides necessary flexibility in accommodating a wide range of load factors in any given situation. Starting from a very small plant to meet requirement for a few households, it can be easily expanded in size module by module to meet added needs of a large number of households. Furthermore, it provides a very high electrical conversion efficiency over a wide range of power levels. A mini plant providing 10 kw could operate at almost the same level of efficiency as a multi-mega watt plant. Typical efficiencies of fuel cell based biogas system range between 35-40 per cent which can be higher than from several decentralised power systems. The system provides higher efficiency over a wide range of load factors. This system is relatively inexpensive, more reliable with fewer breakdowns, and can have a life of more than 20 years. Moreover in a fuel-cell based biogas power system, resulting heat in the form of hot water or steam under pressure can be gainfully utilised for warming up digester contents for achieving higher gas yield.

Central Institute of Agricultural Engineering (CIAE) at Bhopal designed a biochemical fuel-cell which it named '*gobar* cell'. It consists of pair of electrodes and dung slurry to serve as electrolyte. As per CIAE design 60 g of dung is filled in a plastic cylindrical container of 42 mm diameter and 50 mm height. Two 6 mm dia and 50 mm long bars of carbon and zinc serve as pair of electrodes. A group of these biochemical cells when combined together in modules form '*gobar* cell panels' of varying voltage and current configuration. The ends of the electrodes are soldered to copper wire of 20 to 22 gauge specification. The panels are kept in wooden tray so as to provide operational ease without disturbing electrically soldered ends of the electrodes. A unit of gobar cell utilising fresh dung can generate 0.8 volt of electromotive force and 12-15 milliampere of current. A panel of 32 cells when grouped in series-parallel configuration can operate a transistor receiver set of 3V, 50 mw ratings for three days for daily five hourly operation. A 32-cells series-parallel configuration consist of eight parallel rows each containing four cells in series. Apart from operating transistors, these cells can also be used to operate relay controlled mechanisms, pocket calculators, call-bells, hearing-aids etc. which require very small currents for continuous or intermittent use. When not in use they can hold charge for over a month. However, they discharge rapidly, when in use. Mixing of 10 per cent sodium chloride (common salt) practically doubles their charge retention capacity.

11.11 Concluding Remarks

Commercial viability of use of biogas as a cooking and lighting fuel is established beyond any doubt. In several countries, biogas has also emerged as a fuel for space and water heating, crop-drying, refrigeration, irrigation and power generation. High proportion of methane in biogas and high heat value make it combustible and suitable fuel for IC engines. Biogas burns though with discontinuity even without a burner if an ignited match-stick is brought close the pipe end releasing gas. Biogas lamps have several features in common with biogas burners. A number of biogas burner and lamp designs have been developed and are in use in several countries. Potential for biogas applications is unlimited which can be exploited by having higher and higher ingenuity.

REFERENCES

1. Bendito, E.C., Biogas Clean-Up and Utilisation, *Water Science and Technology*, **18(12)**, 143-150 1986.
2. Bhattacharya, T.K., *et al.*, A Survey of New and Renewable Fuels for Running IC Engines, *Agricultural Engineering Today,* **10(5)**, 27-32 1986.
3. BORDA, *Biogas: Manual for the Realisation of Biogas Programmes*, GATE, P.O. Box 5180, 6236, Eschborn 1, Germany, 1980.
4. BORDA, *Biogas Plants: Building Instructions*, GATE, P.O. Box 5180, 6236 Eschborn 1, Germany.
5. Buren, A.V., *A Chinese Biogas Manual*, Intermediate Technological Pub. Ltd, London, 1979.

6. Cao, Z., *Research on Biogas and Methane Gas Fuel Engine,* Proc. Inter. Seminar on Rural Energy Technology, Beijing, China, Sep. 25-27, 1986, China Machine Press, 1986.
7. Dhussa, A., Designing Biogas Distribution System, *Bioenergy News,* **2 (1 & 2)**, 70 1983.
8. FAO, *China: Azolla Propagation and Small-Scale Biogas Technology*, Rome, Italy, 1979, p. 57.
9. Gehri, B., *Biogas-Fuelled IC Engines*, Proc. of the Information Meeting on Biogas, Moudou, Switzerland, Oct. 7, 1982.
10. Gupta, R.K. and Iyer, N.V., Use of Biogas in Diesel Engines, *Agricultural Engineering Today,* **1(9)**, 7-10 1977.
11. IRDC, *Compost, Fertiliser and Biogas Production from Human and Farm Wastes*, Ottawa, Canada, 1980.
12. Jawurek, H.H., *et al.*, Biogas-Petrol Dual Fuelling of SI Engines for Rural Third World Use, *Biomass,* **13(2)**, 87-103 1987.
13. Jiang, C.Q., *et al., Study on Compressed Biogas and its Application to the Compression Ignition Dual-Fuel Engine,* Proc. of the Inter. Seminar on Rural Energy Technology, Beijing, China, Sep. 25-77, 1986, Machine Press, China, 1986.
14. Khandelwal, K.C. and Mahdi, S.S., *Biogas Technology: A Practical Handbook,* Tata McGraw-Hill Pub. Co. Ltd., 12/4, Asaf Ali Road, New Delhi, 1986.
15. KOEL, *Kirloskar Gobar Gas Dual Fuel Engine*, Kirloskar Oil Engines Ltd, Khadki, Pune, 1980.
16. Kothandarman, C.P., *Some Investigations into the Use of Gobar Gas in IC Engines*, A Research Report, Dept. of Mech. Engg., PSG College of Technology, Coimbatore, 1980.
17. Kothandarman, C.P. and Krishnamurthy, S., *Performance of a Diesel Engine with Increasing Proportions of Gobar Gas Induction*, A Research Report, Dept. of Mech. Engg., PSG College of Technology, Coimbatore, 1980.
18. Lapp, H.M., *Fuel from Animal Waste*, Paper No 77-6531, Agricultural Engineering Dept., Univ. of Manitoba, Winnipeg, Canada, 1977.
19. La Torre, C., Biogas and Electricity from Animal Wastes, *Condiz. dell Aria,* **31(5)**, 603-608 1987 [In Italian].
20. Machin, D.B., *Biogas Power System in Remote Locations in Developing Countries*, Proc. of the Sixth SOLERAS Workshop, Las Cruces, New Mexico, USA, May, 243-249 1985.
21. Palaniswamy, P.T. and Swaminathan, K.R., Tamil Nadu Agricultural University Biogas Stove, *Biogas Newsletter, Nepal,* 27, April, 2-3 1988.
22. Parikh, P.P. and Jain, C.K., Biogas as Supplementary Fuel for Diesel Engines, *Urja,* **1(4)**, 27-29 1977.
23. Patankar, G.L., *Recent Developments in Biogas Technology*, A Report, Gobar Gas Development Centre, Borivli, Mumbai, 1977.
24. Picken, D.J. and Fox, M.F., *Use of Biogas for Thermal, Electrical and Mechanical Power Generation,* Proc. Ist E.C. Conf., Brighton, Nov. 1980, Applied Science Pub., London, 910-914 1981.
25. Picken, D.J., *Biogas in Engines,* Proc. 2nd E.C. Conf. on Energy from Biomass, Berlin, Sep. 20-23, 1982, Applied Science Pub., London, 372-375 1982.
26. Picken, D.J. and Hassan, H., *Factors Affecting the Life of Engines Operating on Biogas,* Proc. 2nd E.C. Conf. on Energy from Biomass, Berlin, Sep. 20-23, 1982, Applied Science Pub., London, 1982.

27. Pundir, B.P. and Singhal, S., *Biogas Utilisation in Engines,* Indian Institute of Petroleum, Dehradun, April 19-20, 1978, 9-15.
28. Ramarao, A., Biogas Programme in India, *Khadi Gramodyog,* June, 1988, p. 410.
29. Singh, R.B., *Biogas Plant: Generating Methane from Organic Wastes,* Gobar Gas Research Station, Ajitmal, Etawah, 1974.
30. Singh, R.B., *Gobar Gas: Fuel of the Future,* Gobar Gas Research Station, Ajitmal, Etawah, 1975.
31. Singhal, O.P., Energy Alternatives for Irrigation in India, *Urja,* **21(2)**, 125-128 1987.
32. SPOBD, *Biogas Technology and Utilisation*, Proc. of the Chengdu Seminar, Sichuan Provincial Office of Biogas Development, Sichuan, P.R. China.
33. Stahl, T., *et al.*, Thermal Energy Component of Biogas Cogeneration, *Trans. Amer. Soc. of Agri. Engineers,* **29**, 1237-1242 1986.
34. Stewart, D.J., *Biogas Utilisation,* A Report, No Ag. Link FPP 425, Research Division, Invermay Agricultural Research Centre, Mosgiel, New Zealand, 1980.
35. Stewart, D.J., *et al.*, Producing Fuels on the Farm, *New Zealand Journal of Agriculture,* Sep., 1980.
36. Tentcher, W., Electricity from Pig Manure, *RERIC News,* **10(2)**, 1-3 1987.
37. TERI, Novel Biogas Plant, *Biogas News,* Inaugural issue, Feb., 1982.
38. TERI, *Fixed Dome Biogas Plants*, TERI, Delhi, June, 1987.
39. UN, *Guidebook on Biogas Development,* United Nations New York, 1980.
40. Viswanathan, G., *et al.*, *Combustion Characteristics of Gobar Gas Diesel Oil,* Conf. on IC Engines and Combustion, Madras, E2-51 1977.
41. Mital, K.M., Scope of Industrial Engineering in Promoting CNG as Vehicle Fuel, *Udyog Pragati* (NITIE Journal), **15 (3 and 4)**, 3-9 (1991).
42. Vijay, V.K., Pandey, S. and Jain S., Performance Evaluation of Biogas Domestic Appliances on Different Pressure Conditions in 'N.K. Bansal, Proc. Nat. Solar Energy Conv., Dec. 28-30, 1992, Tata McGraw-Hill Pub. Co Ltd., New Delhi, 1993, p. 407-416.
43. Patel, U., Biogas Pressure Regulator, *Biogas Forum (BORDA),* **I** No. **56**, 25-26 1994.

Chapter 12

POTENTIAL OF BIOGAS PLANT EFFLUENT AS ENRICHED FERTILISER

12.1 Introduction

Effluent from a biogas plant provides rich fertiliser. Utilisation of animal wastes in a biogas plant whereas on one hand it generates a clean fuel, on the other it provides a high quality fertiliser. Not only it provides improved fertiliser but also it generates more in quantity. Higher quantity results from addition of certain crop-residues and night-soil in feedstock which otherwise remain unutilised. Digested slurry has a higher fertiliser potential due to its having larger nitrogen content which is an important ingredient for plant growth. In case of air-drying of animal wastes, 30 to 35 per cent of nitrogen is lost in air, whereas through anaerobic fermentation route it is limited to 10 to 15 per cent. If animal urine is also used in slurry preparation, fertiliser properties of plant sludge improve further following increased availability of nitrogen and potassium [3]. Like compost and farmyard manure, plant effluent improves soil-fertility, soil-porosity and water-holding capacity. It supports bacterial growth in soil that facilitates release of vital nutrients which contribute to increased crop productivity. Digested slurry can be used effectively for any type of soil and crop. Effluent can be applied at a norm of 10 tonnes per hectare in irrigated areas and 5 tonnes per hectare in dry land areas. Effluent can be either directly used in liquid form or after composting in a pit. Composting of effluent with other organic wastes is a common practice. However, it requires adequate land available near the biogas plant. Effluent can also be used in semi-dry form which can be done by spreading effluent in shallow pits and allowing partial sun-drying. It is then scraped and stored in piles and covered with plastic sheets or mats. Effluent can also be used as supplementary fish meal in fish ponds [9].

Plant effluent also helps to improve village hygiene. Despite the fact that certain micro-organisms continue to exist in effluent which are injurious to health, effluent does not cause any harmful effect. In the absence of oxygen and presence of ammonia, pathogenic organisms and parasitic ova cannot survive, develop and multiply for long which contributes to improved hygiene though resulting health benefits are not ultimate hygienic standards. Organic fraction of effluent may contain 30 to 40 per cent of lignin, undigested cellulose and lipid materials on dry-weight basis depending upon type of raw material used. Remaining portion consists of substances originally present in raw material but protected from bacterial decomposition and small amounts of volatile fatty acids. There is less risk of creating odour and insect-breeding problems when anaerobically digested sludges are stored or spread on land than when untreated or partially treated organic waste materials are disposed off or stored [7]. One of the direct benefits of anaerobic fermentation process is that nutrient elements in plant residues and animal wastes such as nitrogen all of which are conserved. The only materials that are removed from the system are gases comprising methane, carbon dioxide and hydrogen sulphide. Biogas plant digested slurry retains about 99 per cent of nitrogen present in input slurry.

12.2 State-of-art of Effluent Utilisation as Enriched Fertiliser

Biogas plant is generally regarded as a source for generating gaseous fuel. In the context of a developing country like India, biogas plant can also be viewed as a source of rich fertiliser. In China biogas plants are also sometimes referred to as 'mini-manure factories'. Plan effluent can be basically used in three ways. It can be applied to the field directly without any treatment, it can also be applied after enriching with liquid ammonia and phosphorus, or it can be applied after composting in a pit with straw and silt. Biogas plant effluent regardless of the method of application leads to higher crop yield than obtainable from direct use of animal manure. Different crops respond differently to application of animal manure or biogas plant effluent. Crops which are more responsive cover vegetable crops like potatoes, tomatoes, sweet potatoes, water-melons, radishes, carrots, cauliflowers, onions, garlic etc; fruit trees and vines such as orange, grape, apple, guava, mango etc.; and other crops like sugarcane, rice, jute etc [7]. Crops which are less responsive include cereal crops such as *jowar, bajra*, wheat, *ragi*, barley and oats; oilseed crops such as groundnuts, linseed, sesame, castor beans, coconut; and other crops such as cotton [10]. Unlike chemical fertiliser, the effect of effluent is carried over a longer period of three years or more. Effluent is source of plant nutrients and provides an excellent soil conditioner as it has requisite quantities of different elements all essential for healthy soils and plants. Effluent has more nutrients than farmyard manure as the latter loses much nitrogen during air-drying. Plant effluent contains more ammoniacal nitrogen than farmyard manure which is readily absorbed by crops. However, much of its ammoniacal nitrogen can be lost if digested slurry is allowed to dry for long in air. Like other compost materials effluent provides humus and supports microbiological activity in the soil [10].

Nitrogen in digested slurry is present in two forms-ammoniacal nitrogen and organic nitrogen the exact proportion of which depends on feed composition. M.A. Idnani and S. Varadarajan [8] carried out experiments to study balance nitrogen left following anaerobic fermentation process (Table 2.1).

According to their findings, more than 17 per cent of total nitrogen is in ammoniacal form and a small amount of 0.8 per cent is the form of volatile compounds. In another experiment, anaerobic digestion of rice straw with its lower nitrogen content of only 0.8 per cent produced 8 to 10 per cent of ammoniacal nitrogen. Thus, as much as 18 per cent of the original nitrogen can be lost if it is allowed to volatilize. As only a very small amount of nitrogen in crop-residues and animal wastes is utilized by anaerobic bacteria to synthesize proteins, the greater is the nitrogen content of feed material, the greater will be the concentration of ammoniacal nitrogen in sludge. It is possible to keep nitrogen losses through volatilization minimum by proper storage of sludge and by following proper application technique. To minimize losses, digested sludge should be stored in deep lagoons or tanks that provide a minimum of surface area for volatilization. Nitrogen can be conserved maximum if the sludge is injected below the soil surface a few days before crop planting or just prior to cultivation. If the sludge is spread on the soil surface and allowed to dry without interruption by rainfall, large proportion of ammoniacal nitrogen is likely to be lost by volatilization.

Table 12.1 Nitrogen Balance in an Anaerobic System

Measure	*Total Kjeldahl Nitrogen (TKN)*	*Ammoniacal Nitrogen*	*Organic Nitrogen*
Nitrogen Input:			
In 1.5 kg dung (wet)	3.805	0.130	3.675
In 500 ml inoculum	0.798	0.126	0.672
Total input	4.603	0.256	4.347
Per cent	0.31	0.017	0.29
Nitrogen Recovered:			
In gases evolved	0.005	0.005	
In digested slurry	4.536	0.797	3.739
Total recovered	4.541	0.802	3.739
Per cent of total TKN input	98.7	17.4	81.2
In digested slurry after evaporation to dryness	3.731	0.028	3.703
(Per cent of total TKN input)	81.1	0.006	80.4

Source : Idnani, M.A. and Varadarajan, S., Fuel Gas and Manure by Anaerobic Fermentation of Organic Materials, Technical Bulletin No 46, Indian Council of Agricultural Research, New Delhi, 1974.

In many situations crops are specially cultivated for being used as feed in biogas plants. One of the benefits of harvesting crops for producing biogas is saving of valuable plant nutrients that are otherwise lost by leaching to depths below root zone and beyond reach for absorption by crops used as food crops.

Many nutrient elements of plants are transferred to soils to form sparingly soluble compounds and thus their availability to succeeding crop becomes less and less. Difficulties arising out of this kind can be alleviated if leaching is carried out within the root zone. For instance, some elements such as potassium can be easily leached from crop residues if these are left to decompose on soil surface. On permeable soils, these elements may be leached below the root zone of the next or the succeeding crop. Other nutrients, such as phosphorus and some trace-metal elements, mineralized from agricultural plant residues during composting or at soil surface, often become less available for plant absorption with time as they react with inorganic soil constituents to form compounds with relatively low solubilities. Therefore, if agricultural crop residues are harvested and properly stored as feedstock for a biogas plant, many important plant nutrients are used more efficiently as they are transferred back to the soil to be used by next crop achieved by using plant effluent as fertiliser. Several trace elements that are otherwise destroyed during microbial decomposition of crop residues in digesters, can be maintained in this state until they can be used by the succeeding crop. If exposure period of these elements to the soil environment is shortened by reducing storage period of sludge of which these elements are part, availability of plant nutrients to succeeding crop becomes much higher [2, 7].

According to some entomologists and pathologists, recourse to underploughing of agricultural crop-residues is beneficial for decreasing the incidence of insect and plant pathogen infestation in succeeding crop. On the other hand, where soils are subject to severe erosion by water or wind, soil conservationists recommend the ploughing should be delayed as long as possible before planting, or it be eliminated altogether. Anaerobic fermentation of agricultural crop residues can be viewed as a way of controlling some insects and diseases more effectively than what is possible if these crop residues are allowed decomposition aerobically. For example, according to some scientists, fungi that tend to cause plant disease can be more effectively controlled by anaerobically digesting crop residues than by ploughing them under the soil. As the anaerobic surfaces of cultivated soils contain abundant supplies of various kinds of micro-organisms, readily degradable organic constituents of digested sludge are rapidly decomposed to carbon dioxide and water or are transformed to humus compounds [10].

There occurs some partial degradation of lignin and lipid components of sludge when it is applied to soil as fertiliser. According to the results of a study when anaerobically digested municipal sludge was annually applied for 6-7 years at various rates and on different types of soil, the soil organic matter was found to reach a more or less dynamic equilibrium state. About 20 to 40 per cent of the organic matter in sludge decomposed each year depending on difference in soil type. Within 5 years after sludge application began soil organic content reached a more-or-less stable level, maximum soil organic matter content at equilibrium depended on amount of sludge applied annually.

In other words, the greater the amount applied each year was, the greater soil organic matter content was found on reaching equilibrium [7, 10].

M.A. Idnani, *et al.* [7] proposed enrichment of biogas plant effluent for improving its fertiliser properties. It can be enriched in nitrogen and phosphorus content by taking 11 kg of urea and 31 kg of superphosphate and dissolving them in requisite quantity of water. This solution is then absorbed in 48 kg of dry low grade manure thoroughly mixed and spread out in the sun to dry. The enriched manure would then contain 5 per cent nitrogen and 5 per cent phosphoric acid in addition to its original strength of these plant nutrients. These impregnated manures which have been found to be highly effective in field experiments can provide impetus for effective utilisation of biogas plant effluent as improved quality fertiliser [5].

Results of Studies carried out at the Guangdong Province Agri-Cultural Institute at Canton (China). Effluent from a biogas plant is an efficient fertiliser as following anaerobic fermentation, nitrogen content in animal and agricultural wastes is transformed into ammonia which crops can more easily absorb. According to result of studies carried out at the Agricultural Institute of Sichuan Province in China [4] ammonia content of organic fertiliser fermented for 30 days in a pit increases by 19.3 per cent and the useful phosphatic content by 31.8 per cent. Sealing this organic fertiliser in pits prevents evaporation and loss of ammonia. As per the studies carried out at the Guangdong Province Agricultural Institute at Canton [4], ammonical content of common fertiliser stored in these pits was found to increase by 144.7 per cent in 30 days, whereas the same manure if stored in traditional heaps and unsealed pits would undergo a loss of 84.1 per cent in ammonical content. Manure that was fermented in pits was found to substantially increase agricultural yield. According to the results of experiments carried out by the fourth work-team of the third brigade of Yungxing commune in Mianyang County of Sichuan Province, the maize yield was found to increase by 28 per cent. Similarly according to the results of experiments carried out by the seventh brigade or Qingdao commune in Mianzhyu county, rice yield increased by 10 per cent. In Pengsi county, the Agricultural Science Station of Dongfen commune reported that the wheat yield increased by 12.5 per cent and according to Jinghua brigade of Chunguan Commune in Gunnan County, Hubei Province, the cotton yield increased by 24.7 per cent. Experimental results comparing yield of four crops when fertilised with unfermented excreta and digested slurry are summarised in Table 12.2. The fertiliser value of cow wastes as determined by nitrogen effectiveness index is basically the function of its freshness and interval after which it is applied to fields (Table 12.3).

Results of Studies Carried Out at the IARI, New Delhi for Analysis of Manurial Efficiency of Effluent. A.B. Ghosh and M.N. Dutt [6] carried out studies at the Division of Soil Science and Agricultural Chemistry, IARI,

Table 12.2 Yield of Four Crops when Fertilised with Unfermented Excreta and Digested Slurry

Fertiliser	*Maize*	*Rice*	*Cotton*	*Wheat*
Unfermented excreta	100	100	100.0	100.0
Biogas plant digested slurry	128	110	124.7	112.5

Table 12.3 Nitrogen Effectiveness Index of Cow Wastes under Different Field Conditions

Field Practice	*Nitrogen Effectiveness Index*
Dung spread and ploughed in immediately	100
Dung piled for 2 days before spreading and ploughing.	80
Dung piled for 14 days before spreading and ploughing.	55
Dung piled for 30 days before spreading and ploughing.	50
Effluent from gas plant introduced immediately into irrigation water.	100
Dried biogas plant effluent spread and ploughed.	85

New Delhi for analysing manurial efficiency of biogas plant effluent vis-a-vis other fertilisers. In their experiments, 15 kg of identical soil, slightly alkaline (pH 8.2), medium in phosphorus and potassium contents and low in nitrogen and organic matter was used for observing plant growth under varied conditions. Various fertiliser samples used in their experiments included biogas plant effluent containing 0.12 to 0.14 per cent nitrogen, another effluent sample with 2.03 to 2.08 per cent nitrogen and sun-dried farm yard manure containing 0.56 to 0.61 per cent nitrogen, and ammonium sulphate containing 20.8 per cent nitrogen. These samples were applied separately under identical conditions to IR-8 variety rice, and berseem crops and their growth observed first during a period of three years and later these were extended for a period of further two years [21, 22]. Comparative nutritive value from fresh and fermented manure for varying combinations of fertiliser application rates is given in Table 12.5. Relative manurial efficiency of fresh and dried biogas plant effluent, farm yard manure etc. measured in terms of productivity of rice and berseem crops grown with their applications is given in Table 12.6. From the results summarised, it is to be seen that both wet and dried effluent compared very well with farmyard manure, caused 15 to 37 per cent yield of rice grain, and 15 to 44 per cent higher yield of berseem during the first three years vis-a-vis when no manure was applied. Oil cake and ammonium sulphate was found to produce much higher yield of rice but their effect on berseem yield was found to be considerably less as compared to that of dried effluent and farmyard manure. As for the relative performance of fresh and sun-dried effluent, the latter was found to achieve 20 to 28 per cent higher productivity of the two crops. All the fertiliser samples used were found to increase crop uptake of nitrogen. The soil tested after three years of manuring with the effluent showed a significant improvement in total nitrogen and available phosphorus content. In case of dried effluent, nitrogen and phosphorus uptake changed from 0.055 to 0.066 per cent and 18 to 20 kg per hectare, respectively; and in case of fresh effluent from 0.055 to 0.07 per cent and 18 to 22 kg per hectare, respectively; and in case of FYM from 0.055 to 0.069 per cent and 20 kg per hectare, respectively. No such improvement was observed in case of groundnut-cake whereas in case of ammonium sulphate-phosphate it was observed in decline substantially. These experimental results thus confirmed that direct and residual manurial value of biogas plant effluent is generally superior to farmyard manure, a fact which is widely underscored during campaigns for biogas programme [6].

Table 12.4 Comparative Nutritive Values of Fresh and Fermented Manure from Different Animals

Source of Manure	*Fat*	*Ash*	*Calcium*	*Phosphorus*	*Dry Matter*	*C : N*
Fresh manure						
Swine	8.9	10.7	1.35	0.92	28.2	32
Layer	1.2	12.8	2.02	1.82	31.8	42
Dairy cow	0.3	2.7	0.77	0.54	16.2	102
Buffalo	2.2	3.9	0.16	0.46	17.4	77
Sludge						
Swine	9.2	10.8	1.62	1.40	25.8	40
Layer	2.0	18.9	2.41	2.57	29.2	35
Dairy cow	0.4	2.8	0.92	0.92	10.8	24
Buffalo	2.8	4.2	0.23	0.81	10.5	50

Source : Tumwasron, S., Biogas Technology and its Applications at Different Farm Levels in Central Thai Village, A Report, Animal Science Department, Kasetsart University, Bangkok, Thailand, March, 1981, p. 5.

Table 12.5 Comparative Nutritive Values of Fresh and Fermented Manure for Varying Combinations of Application Rates

Treatments	*Volatile solids*	*Fixed solid*	*Nitrogen*	*P_2O_5*	*K_2O*	*C*	*C/N*	*DM*
Fresh manure	12.1	2.4	1.62	1.82	7.45	46.2	28	14.5
Fermented manure:								
a_1b_1	9.6	3.3	2.42	1.01	6.08	37.6	16	12.9
a_1b_2	5.7	2.1	2.55	1.02	7.65	35.0	14	7.8
a_1b_3	2.6	3.4	3.15	1.98	6.13	36.5	12	6.0
a_2b_1	6.5	2.3	3.16	1.01	5.52	39.8	13	8.8
a_2b_2	6.1	1.5	1.19	0.39	0.79	37.1	31	7.6
a_2b_3	2.3	0.9	2.80	0.93	4.97	38.1	14	3.2
a_3b_1	6.8	2.2	2.02	0.67	4.04	38.1	19	9.0
a_3b_2	6.1	1.8	1.71	0.49	2.68	40.0	23	7.9
a_3b_3	1.6	1.1	1.84	1.10	1.47	42.2	23	2.8

Note : $a_1 = 10$ kg/cm^2, $a_2 = 50$ kg/cm^2, $a_3 = 100$ kg/cm^2
b_1 = once, b_2 = every third day, b_3 = every day

Source : Tumwasorn, S., Biogas Technology and its Application to Different Farm Levels in Central Thai Village, Animal Science Department, Kasetsart University, Bangkok, Thailand, March, 1981.

Table 12.6 Relative Manurial Efficiency of Fresh and Dried Biogas Plant Effluent, Farmyard Manure etc. Measured in Terms of Productivity of Rice and Berseem

Fertiliser	*Crop Yield$^{(1)}$ (gm/pot)*		*Nitrogen Uptake$^{(1)}$ (mg/pot)*		*Crop Yield$^{(2)}$ (gm/pot).*	
	Rice	*Berseem*	*Rice*	*Berseem*	*Rice*	*Berseem*
Fresh effluent	20.4	18.4	383	590	11.8	14.4
Dried effluent	24.4	23.1	410	650	12.9	15.0
Farmyard manure (FYM)	21.9	22.8	383	610	11.7	14.2
Groundnut cake	50.8	18.3	877	557	7.6	7.3
Ammonium sulphate	43.2	13.4	813	406	7.5	8.3
No manure	17.8	16.0	330	507	6.6	10.1
O.D. at 5 per cent level	2.5	2.2	52	72	4.1	4.0

Note : 1. Values are based on average observed during first three year period.
2. Values are based on average observed during the next three year period.

Results of studies carried out at the Institute of Soil and Fertiliser in Sichuan (China). Scientists at the Institute of Soil and Fertiliser (ISF), Academy of Agricultural Science Research (AASR), Sichuan, during late seventies carried out extensive experiments to analyse the fertiliser value of biogas plant effluent [20]. Studies were inspired by the fact that plant effluent as fertiliser improves soil fertility more than farmyard manure. Advent of biogas programme in China has greatly revolutionised the system of collecting and producing manure in the countryside. As is known, sludge consists of all main nutrients of influent and organic matter. There is no fixedness with regard to proportion of nutrients present in effluent which largely depends upon feed composition.

Proportion of nutrient elements present in digested slurry (effluent) and sludge, as analysed by the ISF scientists are given in Table 12.7. ISF scientists also studied manurial efficiency of anaerobically treated digester effluents and aerobically treated farmyard manure. Both these fertilisers were applied to the rice, maize, wheat, cotton and rape crops (Table 12.8) under identical conditions. Experiments were also carried out for determining the relative manurial efficiency of applying digester sludge after air-drying, and on continuous basis directly, keeping the amount of sludge applied same in both the cases. Crops whose growth were studied under two variants were sweet potato, rice, maize and cotton (Table 12.9). Effect of adding a chemical fertiliser such as ammonium bicarbonate to plant effluent were also studied (Table 12.10) which had a beneficial effect to improve crop yield and minimise soil erosion.

Table 12.7 Proportion of Nutrient Elements Present in Effluent and Sludge

Item	*Organic Matter (Per cent)*	*Humic Acid (Per cent)*	*Total Nitrogen (Per cent)*	*Total Phosphorus (Per cent)*	*Total Potassium (Per cent)*
Effluent	–	–	0.03-0.08	0.02-0.06	0.5-0.1
Sludge	30-50	10-20	0.8-1.5	0.4-4.6	0.6-1.2

Source : SPOBD, Biogas Technology and Utilisation, Chengdu Seminar, Sichuan Provincial Office of Biogas Development, Sichuan, China, 1979.

Table 12.8 Relative Manurial Efficiency of Anaerobically Treated Digester Effluents and Aerobically Treated Farmyard Manure

Crop	*Crop Yield in jin/mu with Fertiliser as*		*Increase in Crop Yield*		*Total Tests for each Crop*
	Digester Effluent	*Aerobically Treated Manure*	*Actual (jin/mu)*	*Per cent*	
Rice	636.4	597.5	38.9	6.5	18
Maize	555.9	510.4	45.5	8.9	9
Wheat	450.0	390.5	59.5	15.2	29
Cotton	154.5	133.5	21.5	15.7	2
Rape	258.4	233.6	24.8	10.6	15

Source : SPOBD, Biogas Technology and Utilisation, Chengdu Seminar, Sichuan Provincial Office of Biogas Development, Sichuan, China, 1979.

Note : 1 jin (Chinese) = 0.550 kg.
1 mu(Chinese) = 660 m^2 = 0.66 hectare.

Table 12.9 Relative Manurial Efficiency of Applying Digester Sludge After Air Drying and of Applying Directly without Air Drying on Continuous Basis

Crop	*Amount of Digester Sludge Applied (jin/mu)*	*Crop Yield in jin/mu with Fertiliser as*		*Increase in Crop Yield with Fertiliser as*	
		With Air-drying	*Without Air-drying*	*Actual (jin/mu)*	*Per cent*
Sweet Potato	2250	3236.0	2863.0	373.0	12.1
Rice	2000	871.9	798.9	73.0	9.1
Maize	3000	667.5	617.7	49.8	8.3
Cotton	3000	166.6	154.3	12.3	7.9

Source : SPOBD, Biogas Technology and Utilisation, Chengdu Seminar, Sichuan Provincial Office of Biogas Development, Sichuan, China, 1979.

Table 12.10 Relative Manurial Efficiency of Digester Effluent Applied Individually and in Combination with Chemical Fertiliser Such as Ammonium Bicarbonate

Treatment	*Crop Yield (Rice)*			*Crop Yield (Maize)*		
	Yield (jin/mu)	*Increase*		*Yield (jin/mu)*	*Increase*	
		Actual (jin/mu)	*Per cent*		*Actual (jin/mu)*	*Per cent*
Digester Effluent (4500 jin/mu)	736.6	39.9	5.7	677.3	106.0	18.8
Ammonium Bicarbonate (20 jin/mu)	736.0	39.9	5.7	620.0	53.3	9.4
Digester Effluent (4500 jin/mu) + Ammonium Bicarbonate (20 jin/mu)	781.3	84.5	12.1	780.0	213.3	37.6

Source : SPOBD, Biogas Technology and Utilisation, Chengdu Seminar, Sichuan Provincial Office of Biogas Development, Sichuan, China, 1979.

In another experiment to alleviate problem of low sulphur content in certain soils which reduces crop productivity, a new kind of fertiliser called biogas phosphohumate was developed which when applied to soils improved their sulphur content. This chemical was obtained by mixing air-dried sludge with phosphorite powder in proportion of 10 to 20 : 1 and composting it for 1 to 3 months (Table 12.11). Effect of applying this fertiliser on purple soil and yellow clay which are common in China are summarised in Table 12.12. Application of plant effluent as fertiliser helps to improve soil properties [11]. With successive cultivation soil nutrients are increasingly diminished which lower soil fertility. Plant effluent as fertiliser which consists of most of the lost nutrients of soil, helps to make up the deficiency of these elements in soils (Table 12.13). Chemical and physical properties showed significant improvement after 1 to 2 years application of digester sludge. As a result, organic matter content, total nitrogen, total phosphorus, available phosphorus and porosity increased whereas soil density showed some decline. Following improvement in soil properties, total yield of some crops went up by 20 per cent [20].

Table 12.11 Relative Manurial Efficiency of Applying Digester Sludge Individually and in Combination with Phosphorite Powder

Treatment	*Crop Yield (Rice)*			*Crop Yield (Maize)*		
	Yield (jin/mu)	*Increase*		*Yield (jin/mu)*	*Increase*	
		Actual (jin/mu)	*Per cent*		*Actual (jin/mu)*	*Per cent*
Digester Sludge (400-1000 jin/mu)	634.3	53.8	9.1	581.4	72.8	13.8
Phosphorite Powder (40-50 jin/mu)	620.0	38.5	6.6	558.6	60.0	11.4
Biogas Phosphohumate (440-1050)	653.3	71.8	12.3	611.7	83.1	15.7

Source : SPOBD, Biogas Technology and Utilisation, Chengdu Seminar, Sichuan Provincial Office of Biogas Development, Sichuan, China, 1979.

Table 12.12 Relative Manurial Efficiency of Applying Digester Sludge Individually and in Combination with Phosphorite Powder on Wheat Crop in Different Types of Soils

Treatment	*Purple Soil*			*Yellow Clay*		
	Crop Yield (jin/mu)	*Increase in Yield*		*Crop Yield (jin/mu)*	*Increase in Yield*	
		Actual (jin/mu)	*Per cent*		*Actual (jin/mu)*	*Per cent*
Digester Sludge	542.3	66.6	14.0	706.0	80.0	12.8
Phosphorite Powder	515.5	39.8	8.4	666.0	40.0	6.4
Biogas Phosphohumate	572.9	96.2	20.2	758.0	132.0	21.1

Source : SPOBD, Biogas Technology and Utilisation, Chengdu Seminar, Sichuan Provincial Office of Biogas Development, Sichuan, China, 1979.

Table 12.13 Improvement in Physical and Chemical Properties of Soil following use of Biogas Plant Sludge as Fertiliser

Property	*Unit*	*At Chyu County After Two Years*			*At Dayi County After One Year*		
		Without Sludge	*With Sludge*	*Per cent Increase*	*Without Sludge*	*With Sludge*	*Per cent Increase*
1. Organic matter	Per cent	1.040	1.210	0.17	1.035	1.286	0.25
2. Total nitrogen	Per cent	0.064	0.068	0.004	0.071	0.101	0.03
3. Total phosphorus	Per cent	0.096	0.110	0.014	0.109	0.110	0.001
4. Available phosphorus	ppm	13.2	14.4	1.2	16.3	0.04	4.1
5. Density	g/cm^3	1.44	1.41	– 0.03	1.27	1.16	– 0.11
6. Porosity	Per cent	45.66	46.59	0.93	52.59	57.35	4.76
7. pH		6.85	6.80		8.30	8.35	

Source: SPOBD, Biogas Technology and Utilisation, Chengdu Seminar, Sichuan Provincial Office of Biogas Development, Sichuan, China, 1979.

12.3 Techniques of Effluent Usage as Fertiliser

Considering enormity of agricultural operations in a country like India chemical fertilisers alone cannot meet the total demand. Demand for fertilisers is required to be met by alternative sources of supply [11]. Wastes obtained from animals, agricultural plants and human beings can provide six to eight times more nutrients than available from chemical fertilisers. Among organic fertilisers, effluent from biogas plant provides enriched fertiliser. Very little of NPK in plant slurry is lost during anaerobic fermentation and whatever little occurs is in the form of gaseous ammonia. It is possible to keep this loss minimum by using proper storage methods such as tanks or lagoons. As far as application methods are concerned, when sludge is injected from below rather than spread on soil surface, loss of nitrogen is found to be much less. Generally 50 to 70 per cent material of plant slurry which are biodegradable are decomposed during anaerobic fermentation and only a small proportion of 10 to 20 per cent of carbon is converted to cellular matter. Thus problems arising from using sludge on soil are much smaller than those from using aerobically treated wastes [10].

Effluent from a biogas plant can be utilised in a number of ways. It can be utilised in liquid form which can be removed daily from outlet chamber or in the form of semi-dried sludge which settles at bottom and can be taken out when digester is periodically emptied. Liquid form of effluent can be taken in buckets and used for added productive growth of vegetable and fruit trees. Liquid digested slurry can also be applied in sufficiently diluted forms which can be carried to fields after mixing with water. Biogas plant effluent is, however, normally used after composting. It is quite easy to compost digested slurry by filling it in usually 2 ft 8 in (80 cm) deep pits. Plant effluent is very effective composting material having necessary micro-organisms which cause its easy break-down into stable matter. Composting is carried out in a particular way. Inside the pit, first a layer of crop residues comprising straw, leaves etc. is spread on which the first layer of effluent is poured. This is followed by another layer of crop residues and then by one more layer of effluent. Proceeding in this way alternate layers of crop-residues and effluent are built one over another till the composting pit is completely filled in. Pits need to be in sufficient number of requisite size so that they can hold entire compost before being put to use. When effluent is obtained from dome type biogas plant (Janata plant), it is directly used without composting. However, even in this case the sludge obtained twice a year when digester is emptied, is preferred to be composted in the above manner.

12.4 Application of Plant Effluent as a Protein Source for Farm Animals

Promotion of non-conventional energy sources has led to the installation of large number of biogas plants in India and many other developing countries. These plants not only produce biogas but also generate rich fertiliser which apart from use as fertiliser can also be used to supplement animal feed. This application is particularly well-suited for countries which are chronically short of animal feeds and fodders. K.K. Saxena [14, 15, 16, 17, 18, 19] has carried out a number of studies at the Indian Veterinary Research Institute, Izatnagar concerning use of slurry as feed to ruminants in limited amounts. Plant slurry used in his experiments was produced from dung of concentrate-fed animals which contained 8-9 per cent crude protein (CP). It was shown that fermentative processes in the biogas plant enhance the protein content of dung from 8 per cent to about 15 per cent in the digested slurry. Animals subsisting on poor quality fodders without any concentrate yield a dung of only about 5-6 per cent CP. K.K. Saxena [19] showed that even low CP dung when digested in a biogas plant can be upgraded to a digested slurry of high protein content. For effective results this however required mixing of digested slurry with molasses, wheat straw, mineral mixture and salt. Ration of this composition could be made easily available during scarcity conditions when concentrate feeds and high quality fodders are not available. Ability of animals under study during the short period of 21 days confirmed the feasibility of using such rations during periods of scarcity.

Biogas plant effluent can be used as valuable feed for fish, poultry and cattle. Balasubramanian and R. Kasturibai [1] working at the School of Energy Sciences, Madurai Kamaraj University, Madurai analysed the nutrient status of cattle dung before and after biomethanation. According to their findings, ammonia content in effluent increased by 70 per cent following anaerobic fermentation. On account of these changes, effluent can prove valuable fertiliser for aquatic plants and feed for fish culture which tends to improve yield.

Studies at the Madurai Kamaraj University covered several varieties of fish culture namely Tilapia fish (*Oreochromismossambicus*), Rogu (*Labeorohita*), Mrigal (*Cirrhina mrigala*) and the Common Carp (*Cyprinus Caris*). These fish were grown in biogas plant effluent without any supplementary feed thereby avoiding the cost of additional feed and fertiliser. Fish weight was found to increase five to seven times in this effluent than through conventional production with recommended doses of chemical nutrients.

12.5 Selection of Pit Size for Composting

Volume of pit is equal to product of volume of animal wastes added per day and average period for which it remains in pit before being removed. Area of the pit can be determined by dividing the pit volume by average height of the pit.

Assuming depth of pit to be 80 cm, and if compost pit is emptied every 7 month and biogas plant is fed with 120 kg (120 litres appx.) of animal wastes per day, area of composting pit can be calculated in the following manner [7, 10].

Volume of compost collected in 7 months

$$= \frac{120 \text{ litres} \times 7 \text{ months} \times 31 \text{ days per month}}{1{,}000 \text{ litres/m}^3}$$

$$= 26 \text{ m}^3$$

If pit depth is 80 cm, then area required

$$= \frac{26}{0.8} = 33 \text{ m}^2 \ (350 \text{ ft}^2)$$

Therefore, three composting pits of size 2.5 m × 4.5 m should be large enough to hold effluent for composting.

12.6 Water Separation from Effluent

Effluent coming out from a biogas plant has high water content. For its more effective utilisation, it is generally desirable that it be dried first. Drying improves its properties and makes the transportation easier. Separated water can be re-used for slurry-preparation which leads to higher gas yield than fresh water. Due to reduction in volume following water-separation, storage needs are also somewhat reduced. Although there are distinct advantages of water separation from effluent, nevertheless it adds to the costs and poses some practical difficulties. For instance, bringing water to the upper surface of the sludge due to its being widely scattered in all layers poses practical difficulties. Some of the possible methods of water separation are [13].

Water absorption by using agricultural crop residues. In this method, agricultural crop residues are spread over digested slurry (sludge) which help to absorb its water content. This resulting waste is somewhat acidic in nature which helps to conserve ammonical nitrogen which during sun drying is substantially lost. This method however has the limitation that separated water is permanetly lost which cannot be reused in slurry preparation.

Water separation by chemical sedimentation. This is another practical method but it suffers from added costs following use of several chemicals. For making the process inexpensive, it is necessary that as far as possible locally available chemicals should be used in the sedimentation process.

Water separation by centrifugation. Water separation by centrifugation is another effective method but it also suffers from high operation costs. Based on laboratory studies it is possible to separate out water by rotating the slurry at a speed of 3000 rpm. In centrifuge-separation a basket type of bucket is normally employed. Operating difficulties are faced at times when centrifuge gets clogged with the thick rotating effluent.

G.L. Patankar [13] carried out several experiments at the Gobar Gas Development Centre, Borivli, Mumbai in centrifuge-separation. For this purpose 100 g of effluent was made to rotate centrifugally for 10 minutes which separated 52 g of water leaving 48 g of dry sludge. Separated water consisted of 0.8973 per cent total solids, 0.0367 per cent total nitrogen and 0.0005 per cent ammonical nitrogen. Dried sludge contained 13.92 per cent total solids, 0.357 per cent total nitrogen and 0.011 per cent ammonical nitrogen. As for the bacterial count, nutrient and thioglycollic agar were counted as 1.9×20^8 and 10^9 in separated water and dry sludge, respectively.

Water separation by filtration. In sewage purification plants, filter beds are mostly used for separating water and solids in effluents. The process originally developed for sewage plants can also be effectively used for separating water and total solids from biogas plant sludge. However, use of filter beds makes the process expensive which is one of its limitations. The process also requires large space. In place of filter beds, it is possible to use GI mesh for filtration. While using wire mesh it is necessary that it be kept cleared of accumulated solids possibly with the help of a stirrer or rubber wiper.

G.L. Patankar [13] carried out several experiments for separating out water from effluent by filtration process. The filtration was carried out both without and with agitation of effluent. In the first case, the effluent was allowed to filter through 1 sq. ft. nylon mesh. Two kg of effluent was made to filter for a period of 24 hours which separated 0.75 kg of water leaving 1.24 kg of dry sludge. The separated water contained 1.2 per cent total solids, 0.031 per cent total nitrogen, and 0.0078 per cent ammonical nitrogen. Separated sludge contained 11.25 per cent total solids, 0.32 per cent total nitrogen and 0.013 per cent ammonical nitrogen.

In second case, i.e. with agitation 2 kg of effluent was made to filter through the same mesh and for the same period which separated 1.047 kg of water leaving 0.953 kg as dry sludge. Separated water contained 2.36 per cent total solids, 0.097 per cent nitrogen and 0.011 per cent ammonical nitrogen. Separated sludge consisted of 11.85 per cent total solids, 0.30 per cent total nitrogen and 0.0083 per cent ammonical nitrogen. From these experiments several conclusions can be drawn. Due to presence of anaerobic bacteria in separated water, its re-use in slurry preparation can lead to higher gas yield. However further experiments need be carried out for determining yield in precise terms under varying purity levels of separated water. Efficiencies of water separation by filtration with agitation and by centrifugation were found to be sufficiently close. In case of filtration with agitation, amount of total solids in separated water was found to be 2.5 times more than when it was done by centrifugation.

In India many simple methods have been tried to remove water such as those involving use of grass beds, pebble beds and sand beds. However they get choked after a few days resulting in complete loss of slurry water rich in nitrogen and other nutrients. R.K. Malik [12] described the use of a simple basket of about 5 ft dia made of mulberry plant sticks to separate water from the plant effluent. These baskets require no special techniques for their production and are already made by local craftsmen in villages which are used for handling dung manure and fodder.

Biogas plant slurry is a rich manure which can be effectively used as a partial substitute for chemical fertiliser for increasing soil productivity by following proper techniques for its application. From a small biogas plant of capacity range 5-8 m^3, following high slurry discharge rate of 80-120 litres/day and with high water content of the order 93-94 per cent, filtration of plant slurry for dewatering poses practical difficulties particularly if the plant is located in low temperature zones or hilly areas. It is neither possible to store the slurry in liquid form in large '*kacha*' pits for months till it is applied in agricultural fields nor it can be easily dried due to low ambient temperatures. Solution to this problem lies in using a properly designed dewatering pit. Dewatering pit not only helps in speedily preparing biogas plant manure but also ensures availability of separated water for re-use in slurry preparation which is rich in anaerobic flora and thus conducive for added biogas yield. A.K. Kalia and S.S. Kanwar [23] developed the design of biogas plant slurry dewatering pit which can be easily constructed in rural areas. They carried out series of experiments to analyse the impact of different filtration materials on dewatering efficiency. They tested various filters made of twigs and plant matter. The volatile solid content of the slurry could be increased after 5th day to 11.5 per cent in the drying bed using a filter made of willow twigs. Depending upon the capacity of biogas plant, two to three dewatering filtration pits can be constructed one after another to enable continuous dewatering of plant effluent.

12.7 Concluding Remarks

The end result of applying anaerobically digested sludge to soils has if not more the same beneficial effect as that by applying any organic matter. The humus materials formed in sludge improve physical properties of soil. For example, aeration, moisture-holding capacity, water-infiltration capacity, and cation-exchange capacity are improved. Furthermore, sludge serves as a source of energy and nutrients for development of microbial populations that directly or indirectly improve the solubility of essential chemical nutrients contained in soil minerals. For making effective use of plant effluent, it is first necessary to remove water and dry it. Some commonly used methods involve use of sloping screen separator, pressure roller separator and screen centrifugation separator [12]. Apart from using rich fertilisers, it is also side-by-side necessary to undertake soil-productivity improvement programmes including recourse to multiple-cropping. If trees are grown along with crops, the former provides a mild micro-climate with increased atmospheric humidity and reduced thermal fluctuations. Due to their deep roots, trees absorb soil nutrients and transfer them to agricultural field in form of fallen leaves. Layer of organic matter such as straw grass and leaves on earth surface tend to lower soil-erosion and water loss. It also helps to build rich insect life in soil which in turn helps to maintain nitrogen balance following linkage with atmospheric nitrogen. Roots and leaves secretions of some weeds are also known to improve soil-productivity following certain bacterial activity [3].

REFERENCES

1. Balasubramanian, P.R. and Kasturibai, R., Biogas Plant Effluent in Aquaculture, *The Hindu*, Sep. 4, 1991, p. 19.
2. Barnett, A., Pyle, L. and Subramanian, S.K., *Biogas Technology and the Third World: A Multidisciplinary Review*, International Development Research Centre, Ottawa, Canada, 1978, p. 22.
3. BORDA, *Biogas: Manual for the Realisation of Biogas Programmes*, Bremen Overseas Research and Development Association, Bremen, Germany, 1979.
4. Buren, A.V. (ed.), *A Chinese Biogas Manual*, Intermediate Technology Publications Ltd, London, 1979, p. 17.
5. Chawla, O.P., *Advances in Biogas Technology*, Publications and Information Division, Indian Council of Agricultural Research, New Delhi, 1986.
6. Ghosh, A.B. and Dutta, M.N., *Manurial Efficiency of Digested Slurry from Cow Dung Gas Plant*, paper presented at the National Symposium on Biogas Tech nology and Usage, IARI, New Delhi, Nov. 29-30, 1977.
7. Idnani, M.A., Laura, R.D., and Chawla, O.P, Utilizing the Spent Slurry from Cow Dung Gas Plant, *Indian Farming*, **19(1)**, 35 1969.
8. Idnani, M.A. and Varadarajan, S., *Preparation of Fuel Gas and Manure by Anaerobic Fermentation of Organic Materials*, ICAR Bull. No. 46, New Delhi, 1974.
9. Khandelwal, K.C. and Mahdi, S.S. *Biogas Technology: A Practical Handbook*, Tata McGraw-Hill Publishing Company Ltd., New Delhi, 1986.

10. NAS, *Methane Generation from Human, Animal and Agricultural Wastes*, Report of an Ad. Hoc. Panel of the Advisory Committee on Technology Innovation, National Academy of Sciences, Washington, DC. 1977.
11. NCST, *Draft Status Report on Utilisation and Recycling of Wastes,* National Committee on Science and Technology, New Delhi, 145-146 1971.
12. Malik, R.K., Simple Device: Mulberry Flat Basket for Separating Water from Biogas Plant Slurry, *Urja*, **21(6)**, 373 1987.
13. Patankar, G.L., *Recent Developments in Biogas Technology,* A Report, Gobar Gas Development Centre, Borivli, Mumbai, 1977.
14. Saxena, K.K., *Effect of Feeding Sun-dried Digested Slurry from Biogas Plant on the Growth Rate and Nutrient Utilisation in Crossbred Calves*, Annual Scientific Report of Animal Nutrition Division, IVRI, Izatnagar, 1980.
15. Saxena, K.K., *et al.*, Note on the Utilisation of Digested Slurry from Biogas Plant as a Source of Protein in the Maintenance Ration for Sheep, *Indian Jour. of Animal Science*, **50**, 896-898 1980.
16. Saxena, K.K., *Nutritive Value of Sun-dried Digested Slurry from a Biogas Plant in Buffaloes* , Annual Scientific Report of Animal Nutrition Division, IVRI, Izatnagar, 1982.
17. Saxena, K.K., *Nutritive Value of Fresh Digested Slurry from a Biogas Plant in Buffaloes*, Annual Scientific Report of Animal Nutrition Division, IVRI, Izatnagar, 1983.
18. Saxena, K.K., and Ranjhan, Nutritive Value of Digested Cattle Slurry from a Biogas Plant in Sheep, *Agricultural Wastes*, **8**, 119-124 1983.
19. Saxena, K.K., *et al.*, The Effect of Using Dung from Cattle Fed High, Low or No Concentrate Rations on the Quality and Nutritive Value of Slurry from a Biogas Plant, *Biological Wastes*, **28**, 73-79 1989.
20. SPOBD, *Biogas Technology and Utilisation*, Proc. of the Chengdu Seminar, Sichuan Provincial Office of Biogas Development, Sichuan, P.R. China, 1979.
21. Taiganides, E.P. and Haza, T.P., Properties of Farm Animal Excreta, *Trans. American Society of Agricultural Engineers,* 374-376 1966.
22. Tumwasorn, S., *Biogas Technology and its Applications to Different Farm Levels in Central Thai Village*, Department of Animal Sciences, Kasetsart University, Bangkok, Thailand, March, 1981.
23. Kalia, A.K. and Kanwar, S.S., *Filtration Pit for Dewatering Biogas Plant Slurry in Hills, Biogas Forum (BORDA),* **I** No. **56**, 1994.

Chapter 13

APPROACHES FOR IMPLEMENTING BIOGAS PROGRAMME, AREAS FOR FURTHER RESEARCH AND CONCLUDING OBSERVATIONS

13.1 Overview of Biogas Technology

Advancement and promotion of biomethanation should receive high priority as part of strategy to promote new and renewable sources of energy. Promotion of biogas essentially supports the doctrine of a decentralised and self-reliant rural economy which can meet multifarious needs of a village such as cooking, lighting, running of flour-mills, irrigation and power generation.

Renewable energy resources occur in nature in the form of energy flows of indefinite duration as compared to depletable fossil fuels of finite magnitude. Recent emphasis on exploitation of renewable energy resources in developing countries has been prompted by two main factors: increased international price of oil and gas, and growing concern for environment protection. The impact of the higher prices on the balance of payments, alarmingly rapid depletion of fuelwood supplies, need for rapid economic growth, and growing demand for energy have underscored the need to develop renewable energy alternatives [23]. According to some estimates, renewable energy can substitute 5 to 15 per cent of the oil consumed in developing countries by the turn of the century. This implies that renewable energy, while no panacea, can substitute for a small but significant share of oil consumption in the developing countries in long run. Traditional renewable energy resources such as fuelwood, charcoal, crop residues and animal dung account for nearly the entire fuel used in many rural areas, and may account for about 20 to 25 per cent of the total energy consumption in the developing world. Africa is most dependent upon traditional renewable resources, followed by Asia, and Latin America. About 75 per cent of the population of the developing countries depends on conventional fuels for meeting household cooking and lighting needs [40].

It is to be noted that developing countries have been consuming their fuelwood supplies far more rapidly than renewing them thus causing grave environmental degradation.. With growing dwindling of fuelwood supplies, households are increasingly burning dung and agricultural residues as a fuel-wood surrogate. This option however deprives soil of valuable nutrients and organic conditioning material. It is in this backdrop that biomethanation offers attractive option to provide a clean gaseous fuel, improve sanitation, and increase the fertiliser and soil conditioning value of animal and agricultural wastes [3].

Biomethanation helps to conserve continually depleting forest resources which if not controlled may lead to soil-erosion, flash-floods, landslides and river-siltation in plains. It is vital for any new energy policy to recognise that growing dependence on renewable energy sources as depletable reserves of coal, oil and gas may not last for ever. Task-forces and committees set up by the government from time to time should act as 'think tanks' to keep the content and output of existing policies under constant review. There should be a national level cabinet committee on energy policy serviced by energy-related ministries to catalyse action on new possibilities, lay down priorities for achieving coordination and improvement in supply and demand of energy from all available sources [9].

Biomethanation has made rapid strides in India considering the fact that 21.65 lakh plants were already installed by 1994-95. Furthermore, as per MNES estimates, there exists a potential to instal upto 40 million biogas plants which is nearly 20 times the plants upto 1993-94. A total of 23.05 lakh biogas plants installed upto Feb 29, 1996 can generate gas equivalent of 72 lakh tonnes of fuelwood valued at Rs. 360 crores per annum. In addition these plants can also produce 245 lakh tonnes of enriched organic manure per annum. Economic gains accrued to the country from biomethanation are substantive in view of the fact that nearly 85 per cent of them are reported to be functioning satisfactorily [7, 10].

Biogas-based power systems can provide viable alternative to electric power for a substantial number of India's 5,67,000 villages especially those which are difficult to be connected through transmission lines [11, 29]. Amount of fertiliser obtained from a 2 m^3 plant in an year is equivalent to 109 kg of urea, 275 kg of superphosphate and 50 kg of potash. This can meet the fertiliser needs of one acre of land for two crops in a year. It is to be realised that cost-effectiveness in plant operation can be achieved with efficient burners which as per Indian standard specifications IS: 8749-1978 should have a minimum thermal efficiency of 55 per cent. It is to be seen that gas intake for running a 5 hp pump set for irrigation purposes is 18 m^3/day which is difficult to be fulled by family units. Hence for commerical success of biogas-fuelled pump-sets, it is important that adequate gas is available from higher capacity plants or plants which utilise high gas yield feed materials [16, 28].

Mr P.V. Narasimha Rao, the former Prime Minister of India, while addressing a conference of state ministers incharge of non-conventional energy sources (NCES) in New Delhi on Oct 28, 1992 observed that the country's ambitious programmes to generate more electricity could face serious problems due to their adverse impact on the environment. In this context alternative sources of energy such as biogas, solar energy and wind-power assume significance. The shortage of power is a dampener to the government's plans to woo more foreign investors into the country, and development of renewable energy sources can alleviate these problems substantially if these are adequately harnessed. In a message to a special supplement of Indian Express (Oct. 28, 1992) on theme 'Non-Conventional Sources of Energy: A Decade of Achievement', Mr Rao noted:

> 'Energy from the sun, wind and biomass can alleviate the pressure on traditional sources that have been considerably depleted. Practical and economical means to harness such new energy sources need to be developed. India's initiative in this regard must succeed.'

Mr. S. Krishna Kumar, the then Minister of State for NCES in a Financial Express Symposium (May 18, 1994) observed that one should look the cost effectiveness of non-conventional energy technologies, not only in terms of the initial cost but also in terms of other factors such as short gestation, favourable environmental impact, no cost excalation etc. [45]. Recent experience has shown that certain non-conventional energy sources such as biofuels, windpower and small hydro projects are economical and competitive even today [45]. Other technologies are fast becoming cost-effective as technology advances and volume production take place.

The former Union Minister of State for NCES in the inter-state ministerial meeting held on Oct 28, 1992 requested the state governments to popularise NCES of which biogas is an important part. He emphasised the need for setting realistic and ambitious targets. He underscored the need to motivate entrepreneurs to promote NCES particularly in coastal and hilly regions where there is considerable untapped potential. He suggested that the NPBD should be reoriented by involving the women in rural areas and taking help of *Mahila Mandals* (women's groups) and voluntary organisations. In his message to a special supplement on NCES brought out by the MNES in Indian Express (Oct 28, 1992), he observed:

> 'There is a considerable potential for the development of biogas plants which utilise agricultural and cattle wastes. R & D efforts have to be undertaken to achieve cost-reduction both through economies of scale and technological innovation and refinement.'

Mr. K.C. Pant, Chairman, Tenth Finance Commission and former Chairman of the erstwhile Advisory Board on Energy while addressing a seminar on Energy Management organised by the HRD Foundation and the Indian Renewable Energy Development Agency (IREDA) in New Delhi on Dec 21-22, 1992 emphasised the need for NCES. He stressed the need to tap unlimited energy from the sun and to step up research and development in NCES-related fields. He suggested that the government should make all-out efforts to tap energy from the sun, wind, sea and biological wastes, along with the conventional sources of energy.

Mr. B.R. Prabhakara, Secretary, MNES, in his key-note address at the National Workshop on the 'Policy Issues in the Biogas Programme' held in New Delhi (Aug. 21-22, 1995) observed that scope for biomethanation in India is enormous considering the number of rural families who have one or two cattle and unlimited need for household energy and enriched fertiliser. Biogas offers multifarious benefits such as cleaner environment, rural employment generation, improvement in quality-of-life for women, and conservation of scarce fuelwood. Social benefits from biogas are also substantial as every biogas plant directly saves trees and raises the status of women in society. Funds from external agencies can also be harnessed for promoting biogas programme which can be implemented as a collaborative programme by the Ministry of Rural Development and Ministry of Environment and Forests which stand to gain by its success. There is a need to intensify awareness compaigns in north-eastern states which have tremendous biogas potential.

According to Mr. B.R. Prabhakara, MNES has adopted a multi-pronged approach for development of renewable energy in India by creating awareness, promoting research and development, extension work and commercialisation through provision of soft-loans. While the goals to be reached are still distant, it is gratifying to note that awareness at various levels has increased considerably in the recent past and grid-connected renewable energy already constitutes nearly 1 per cent of total generation capacity in the country. The number of biogas plants is over two million which in terms of conventional electricity is equivalent to about 1100 mw of generation capacity with a difference that these capacities have been created in a decentralised manner in the rural areas throughout the country with hardly any transmission and distribution losses [50].

Biogas plants, if not an ultimate solution to the country's rural energy needs, can at least serve as transitional source of energy till conventional supplies are easily and adequately available. Small villages in remote areas are ideal choice for using biogas as an effective alternative to conventional energy. Decentralised energy systems based on local renewable resources can be a viable alternative to many conventional energy systems particularly when the hidden subsidies for the latter and the environmental degradation they cause are also taken into account. For instance, for a village 40 km from the grid with a minimum load of 10 kw, photovoltaic systems are more economical than laying tie-line from the grid [24]. Dr. R.K. Pachauri, Director, TERI, while addressing the national convention of Solar Energy Society of India at Jadavpur on Dec 26, 1990 observed that historical changes in the consumption pattern of fuels in the world have made the transition from fossil fuels to renewables absolutely necessary. For certain categories and situations, economic viability of NCES such as wind-farms in coastal areas, photovoltaic power in several centres, and community biogas plants at all locations is no longer a matter of doubt [41].

MNES launched a 'National Pilot Programme on Energy Recovery from Urban, Municipal and Industrial Wastes' in 1995 covering all states and union territories. The programme aims at promotion, development, demonstration, dissemination and adoption of appropriate conversion technologies for both solid and liquid wastes through improved waste management and introduction of financial incentives. On the occasion of launching this programme, Mr B.R Prabhakara, Secretary, MNES observed [49]:

> 'It is indeed an exuberant moment for me to announce to all concerned that the MNES have launched a nation-wide campaign for the promotion and commericalisation of electricity generated from the wastes – both liquid and solid. This ministry has introduced a National Pilot Scheme for giving impetus to the waste-to-energy projects in the country. Environment-friendly, as this scheme is' it will go long way in improving waste management practices and thereby reduce greenhouse gas emissions in the environment, besides widening the scope of NRSE and bringing into this sector additionality of investment.'

Prof. T.K. Moulik, an eminent agricultural economist, while addressing the National Workshop on 'Policy Issues in the Biogas Programme' in New Delhi (Aug. 21-22, 1995) highlighted the need for authentic data on biogas programme to be presented to policy-makers. In defence of continuity of subsidy vis-a-vis that available for LPG, it is necessary to perform cost-benefit analysis using hard data which is, however, not readily available. Biogas programme must be packaged in a manner so that it is easily marketable. There is a need to develop effective financial arrangements and evolve alternative institutional mechanism to promote biogas technology. Biogas should not merely be seen as beneficial in rural areas but it has also considerable potential in urban areas where it can be linked to municipal waste treatment programmes.

Administrative problems involved in biomethanation are those typical of any new programme. Any initial administrative or technical failure may result in loss of villagers faith and confidence in the programme and finally they may turn unresponsive towards biomethanation. A technically sound plant design is key to a successful biogas programme and research efforts need be directed to develop a design which can function under wide range of operating conditions [14].

The Government of India under the Chairmanship of the Secretary, MNES, set up National Bioenergy Board (NBB) on April 1, 1994 as the national apex agency for providing policy guidlines and directions for harnessing bioenergy from urban, municipal and industrial wastes. NBB is represented by members from concerned ministries/departments/agencies such as Planning Commission, UNDP, Ministry of Urban Affairs and Employment, Ministry of Environment and Forests, Dept. of Science and Technology, Dept. of Bio-Technology, Dept. of Economic Affairs, Council of Scientific and Industrial Research,

IREDA (Indian Renewable Energy Development Agency), HUDCO (Housing and Urban Development Corporation), CII (Confederation of Indian Industry), FICCI (Federation of Indian Chambers of Commerce and Industry), ASSOCHAM (Association of Chambers of Commerce and Industry) and individual experts [49].

Methane yield is the most important economic parameter for assessing the performance of a biogas plant. Two basic approaches for increasing gas yield are (i) controlling feedstock composition and (ii) improving digestion systems. In anaerobic digestion of biomass for gas generation, high conversion efficiency is needed for economic viability. Biologically sensitive steps which can be regulated in waste treatment are difficult to control in biomethanation. This underscores the need to use appropriate feed material, efficient and economic plant design and adopt novel approaches like multi-phase digestion. In multiphase digestion number of phases may be quite large to control temperature, pH, microbial life-cycles, oxidation-reduction potentials, and production and removal of inhibitory levels of organic acids. Rate control affects several aspects of anaerobic digestion process. It may be used to keep various components of the fermentation in balance, to maximise overall rate of the fermentation and to alter end products to help produce biogas with maximum methane content [42]. Biogas is often hailed as best example of appropriate technology as it meets basic need as a cooking fuel, makes optimal use of local resources such as cowdung and other organic wastes, helps to promote indigenous growth using local skills and technologies, and leads to environmental improvement [41]. Biogas programme should be integrated into an overall energy plan for a particular region instead of deciding targets in isolation [20].

MNES promotes renewable energy through R&D, development of standards, operation of test facilities, simplification of manufacturing, development of markets, commercialisation of technologies, operational programmes, awareness creation, information dissemination and international cooperation. During 1993-94, it evolved a new 'Strategy and Action Plan' by giving market orientation to its programmes and involving private entrepreneurs and NGOs in its efforts. The Eighth Plan (1992-97) accords high priority to renewable energy programmes through universalisation of the ongoing programmes of biogas and improved wood stoves; expansion of the Integrated Rural Energy Programme for meeting the diverse energy needs for subsistence and production in the diverse energy needs for subsistence and production in the most effective manner at the micro level; promotion of solar energy for meeting cooking, heating and lighting needs; and, recovery of energy from agricultural, industrial, urban and human wastes. MNES on its part has assumed a new role of 'mission leader' and 'facilitator' for mobilising a mass movement for renewable energy the country. The implementation of these programmes is carried out with involvement of state governments, electricity boards, industry and NGOs, and a new community of energy entrepreneurs who provide dynamism to the renewable energy movement in India [48].

The Government of India has placed growing emphasis on using NCES as is evident by increasing financial outlays earmarked for sixth (Rs. 50 crores), seventh (Rs. 415 crores) and eighth plan (Rs. 860 crores). During eighth plan out of an outlay of Rs. 860 crores, Rs. 557 crores were allocated for central sector and Rs. 303 crores for the state sector. During eighth plan period Rs. 320 crores were allocated for promotion of biogas which represented 37 per cent of the total budget earmarked for MNES. This implied that Rs. 64 crores would be available annually to the MNES for spending on biogas programme which was not a very large sum in wake of vast untapped potential in the country. As per the suggestion made in a recent TERI document [44], one of the measures to overcome this constraint is to promote biomethanation on commercial lines. For adopting this approach, it is necessary to identify fuel-scarce regions of the country where it can work effectively to be followed by requisite steps for implementation. In keeping with this strategy, MNES has already started laying emphasis on turnkey approach in setting up biogas plants. For speedier implementation, biogas programme can also be linked with other development activities. For example, rural housing and rural industry are the two sectors where biogas promotion can be simultaneously pursued [44]. Biogas programme also needs to be linked with dairy development programme with a view to maximise utilisation of animal.wastes for biomethnation [33].

MNES has initiated a project to be completed by Sep. 1, 1999 for developing high rate biomethanation process for reducing greenhouse gas emission from select substrates involving upscaling, validating and nearly full-scale demonstration of mature technology [49]. Aided by several institutions including UNDP/GEF (Global Environment Fund), this scheme covers sixteen demonstration projects in different areas: sewage (3), pulp and paper effluents (2), vegetable market wastes (3), leather effluents/abattoir wastes (5) and biogas utilisation for power (3).

Applications of computers in biogas technology are rapidly growing. Computer usage in biomethanation is generally in two areas, (i) planning and implementation of biogas programme by central and state government agencies and (ii) expanding frontiers of biogas technology. In India planning and control for NCES development is carried out by the MNES and various state agencies. Computer applications in biogas technology relate to data collection and data management for optimisation of design parameters, selection of design parameters for maximising biogas yield under a given set of operational/environmental conditions, project management for setting up large capacity plants, modelling of microbiology and process biochemistry, and evaluation of plant economics. In developed countries, computer-aided design and manufacturing (CAD/CAM) are being used for over a decade for monitoring the operation of community biogas plants and sanitary landfill gas projects. Computer softwares have been developed for simulating fermentation process for achieving process control, evaluating economics of plant operation, and analysing the impact of environmental factors on biogas plant performance [43]. For instance in Denmark, farm digesters with a total volume of 2500 m^3 are monitored and managed with CAD/CAM. In 1992, a computer aided plant with three large sized biogas digesters for anaerobic treatment of municipal refuse from Chengdu city was set up at Shanwayao (P.R. China). In the Second International Exhibition on New and Renewable Energy held in May, 1992, Prof. A. Hamzech and Prof. M. Horanl presented their newly developed design software for large and medium sized biogas digesters for power generation which contains field tested standards for 50-500 m^3 capacity digesters [43]. The main technical parameters estimated include daily electricity load, performance data of digester, climatic conditions, gas yield, feed requirement, digester volume, and heating needs for digester. A Chand, *et al.* [4] developed a computer-based management information system for monitoring the implementation of the NPBD at the district level in Maharashtra on trial basis for a cluster of villages in the Marathwada region. The system was developed with a view to identify operational problems and constraints in the effective functioning of biogas plants.

13.2 Recommendations of the UN Conference on New and Renewable Sources of Energy

The late Prime Minister, Mrs Indira Gandhi, in her key-note address at the UN conference on new and renewable sources of energy in Nairobi in 1981 observed:

> 'Measures must be taken to increase the use of mature technologies in renewable energy such as the fuller harnessing of hydro-power potential, coal technology, large-scale programmes of reforestation and energy-plantation, **biogas** production, use of solar energy for low and medium temperature requirements like water pumping and of wind energy for pumping... Let a powerful call go out from this Conference to make energy a focal point of contemporary international cooperation and collaboration.'

The Nairobi Conference achieved a global consensus concerning framework for action at national, regional and international levels which are fully relevant even today in the context of biogas programme. The programme of action and the resolutions passed include following proposals and initiatives[7].

- Importance of recognising the need for a smooth energy transition to an era of less dependence on hydrocarbons and greater reliance on renewable sources of energy.
- Emphasis on global cooperation and maximum consensus on energy issues.
- Identification of a concrete programme of action to provide a framework for national, regional and international action and issuance of guidelines for activities in priority areas.
- Intensification of R&D efforts in renewable energy.

- Wider application and utilisation of mature NCES technologies.
- Larger efforts for more efficient utilisation and conservation of energy.
- Reinforcement of infrastructural network for large scale utilisation of renewable energy.
- Strengthening regional cooperation as well as economic and technical cooperation among developing countries.
- Formation of an inter-governmental body to supervise and monitor the implementation of the Nairobi programme of action.
- Establishment of a flexible consortium and consultative mechanism for the mobilisation of financial resources to support renewable energy programmes in developing countries.
- According priorities to afforestation programmes along with intensive development and utilisation of renewable energy such as **biogas**, wind and solar energy in rural areas to minimise fuelwood consumption and improve standard of living.

India presented a national paper at the Nairobi Conference which reviewed the energy scene and country's efforts in developing renewable energy. It made several recommendations which are very appropriate for the success of biogas programme. It sought cooperation of developed nations to [7].

- Support the efforts of developing countries to strengthen their R&D base.
- Provide financial, material and technical support to developing countries for setting up pilot projects, production facilities etc.
- Participate in bilateral and multilateral arrangements to ensure technology transfer to developing countries on equitable terms.
- Exchange scientific and technological information with developing nations.
- Undertake R&D activities which are more relevant for the needs of developing countries.
- Remove tariff and non-tariff barriers against renewable energy-related products exported by developing countries.

A workshop on 'Integrated Rural Energy/Environmental Protection for Sustainable Rural Development' was organised in 1995 by the Chinese Ministry of Agriculture and the Department of Environmental Protection and Energy in association with ESCAP/UNDP at the Chinese Centre of Rural Energy Research and Training (Beijing). It was attended by representatives from 12 countries of the Asian region six participants from ESCAP (Economic and Social Commission for Asia and the Pacific), UNDP (United Nations Development Programme), F&AO (Food and Agricultural Organisation), UNESCO (United Nations Educational, Scientific and Cultural Organisation), BORDA (Bremen Overseas Research and Development Association), and APCTT (Asia Pacific Centre for Technology Transfer); and numerous representatives from different Chinese ministries and institutions. Asian countries which sent their representatives included Bangladesh, China, India, Iran, Laos People's Democratic Republic, Malaysia, Mangolia, Myanmar, Nepal, Philippines,

Thailand and Vietnam. All countries presented rural energy outlook and policies, and environment planning and development endeavours of respective countries which was interspersed by talks from international experts, presentation of two Chines case studies and site visits. The Workshop, among others, reviewed sustainable development and rural energy planning, drew work plan for 1995-96, and identified regional training needs for energy and sustainable development [47].

Federal Association of Renewable Energy (Germany) has formed a permanent 'Biogas Group' in response to the growing worldwide interest in biogas and the concern for environment protection particularly with reference to large farms in eastern Germany. Dr M. Winkler (Address: Steghauser 39a, 84359 Simbach/Inn) was appointed as the Chairman of the four-member 'Biogas Group' [47].

13.3 Common Approaches for Implementing Biogas Programme

Biogas programme in a country should be pursued by a nodal agency which can be supported by agencies identified for promoting biomethanation. In India, this programme is also recommended to be pursued after linking with integrated rural development programme and dairy development programmes [24]. Some important aspects of this programme are: short term and long term target-setting, programme monitoring, information dissemination, training and development, arranging financial support for meeting plan targets, conducting workshops and seminars, and pursuing R&D activities. The programme also needs to be evaluated periodically on the basis of following activities [25, 26].

- Socio-economic evaluation of the impact of biomethanation on living standards and quality of work-life of the affected population [34].
- Assessment of biogas programme as a public health measure for waste treatment.
- Economic evaluation of impact of biogas programme on improvement in agricultural productivity following application of enriched fertiliser.
- Programme evaluation for meeting integrated energy needs of an area.
- Evaluation of programme impact on deforestation following reduced dependence on fuelwood as cooking fuel.

In this section programme implementation through a centralised agency, turn-key approach, and biogas cooperatives are discussed.

Programme Implementation through Centralised Coordinating and Planning Agencies: For popularising and implementing the biogas programme during the sixth plan period, Government of India set up the NPBD and assigned overall responsibility to MNES with following goals [32]:

- To advance in consultation with central and state governments, grants, subsidies and loans directly to the public corporate bodies like state renewable agencies, KVIC, KVIB, state agro-industries corporations, dairy development corporations, tribal development corporations etc.; and to secure voluntary agencies support in programme implementation [21].

- To promote biogas programme by providing training and extension facilities [21].
- To liaise with national and international research organisations and act as focal point for information dissemination.
- To liaise with banking and other financial institutions for arranging bank loans and institutional finance for setting up biogas plants.
- To liaise with the concerned government departments for release of special quotas of cement, steel etc. earmarked for the programme.
- To interact with concerned government agencies in solving implementation problems.
- To sponsor research projects on biomethanation at agricultural universities and other institutions.
- To develop, test and popularise low cost models of biogas plants developed by various agencies from time to time.
- To undertake R&D activities on various aspects of biomethanation such as improvements in design of plants, use of locally available materials in plant construction and operation, supply and portability of gas, and the like.

Promoting Programme on Turn-key Basis. Installation of biogas plants on turn-key basis is one of the important approaches in implementing biogas programme. A limited experiment of plant installation on turn-key basis was initially tried out in Gujarat during early eighties. This approach was later extensively followed during late eighties and nineties. Adoption of turnkey approach is largely responsible for phenomenal success of biogas programme in Inida in recent years resulting in installation of nearly 2 lakhs plants a year. Necessary guidelines and framework for this approach are as follows [27].

- Implementing agency is required to work out feasibility report giving estimates of plant capacity, finances needed, resource inputs, and identification and availability of locally available materials both in plant construction and plant operation.
- Following approval of the feasibility report by the beneficiary, implementing agency intiates action for arranging loan from banks and subsidy from government departments as applicable.
- Implementing agency keeps a track of skilled artisans locally available for undertaking plant construction on daily or piece rate basis.
- Implementing agency should have requisite facilities at its disposal such as a small workshop to manufacture steel components like gasholder, classroom facilities to impart necessary training to the concerned staff, infrastructure support for plant maintenance, adequate space for storing cement and steel, and personnel and accounting division for keeping record of administrative and financial activities [15, 17].

Promoting Programme by Forming Cooperatives. Biogas plants in a region can also be set up by forming a cooperative society. In this context it should be remembered that the cooperative law aims at incorporating the principles of cooperation that distinguish them from industrial enterprises. Distinctive features of a cooperative are: open and voluntary membership, democratic control, savings belong to members, limited interest on capital, cooperative education, and cooperation among cooperatives. A. Mazumdar and N.K. Gopalakrishnan [18] proposed *modus operandi* of a biogas cooperative. Cooperatives can be formed at village, block and district levels and there can be a national level organisation to coordinate and monitor their activities. For individuals to become members of a cooperative, it is necessary that they live within the area of operation of society and they should be competent to contract. The responsibility of audit of cooperative institutions vests with the registrar of cooperative societies while in companies members at the annual general meeting appoint auditors on the recommendations of the Board of Directors. A cooperative functions like any other organisation except that its by-laws differ. A biogas cooperative is expected to possess expertise in preparing feasibility reports, arrange loan from banks and subsidy from the concerned government departments, own workshop facilities and trained workforce for setting-up plants and work relating to their day-to-day maintenance and operation. It should also have facilities for training and development of individuals in construction, operation and maintenance of biogas plants. A cooperative can receive financial assistance from several cooperative financial institutions and technical and managerial input from the State Cooperative Union (SCU), District Cooperative Union (DCU), and National Cooperative Union of India (NCUI).

Apart from forming separate cooperatives, biogas programme can also be pursued by linking with dairy and milk cooperatives already in existence. The National Dairy Development Board (NDDB) which helps to establish village dairy cooperatives can also work towards biomethanation. Dairy cooperatives can be particularly helpful in popularising biogas plants among milk producers. The integrated dairy-cum-biogas cooperative can ensure that each milkshed has at least one community plant and one demonstration family unit. Visit of farmers to demonstration units can help to promote biomethanation [5, 6].

A National Workshop on the 'Policy Issues in the Biogas Programme' which focused on the role of biogas technology for the development of rural women and protection on natural resources was organised by the Aga Khan Rural Support Programme at New Delhi during August 21-22, 1995. The Workshop was co-sponsored by the Aga Khan Foundation and the European Commission. It recommended that the NPBD should launch awareness campaign among women about benefits of biomethanation. It also sought involvement of women's rights and women's health activists in obtaining women support for biogas programme. The Workshop stressed collaboration of biogas programme with forestry, health care and rural employment programmes. The Workshop also stressed the need to continue government support for the biogas programme by continuing to provide subsidies. The Workshop felt that in the government's drive to promote market-driven economy, withdrawing subsidies will hurt marginal and landless farmers most and hard hit already overburdended women and forest resources. Instead a more careful and cautious approach with gradual introduction of market forces need be tried. The Workshop also proposed that for at least five years the subsidy should not fall below its current level. During that time, agencies can experiment with alternative marketing and production strategies. The promotional incentive of Rs. 50 per plant for motivators should continue and be available with implementing agencies. Government subsidy to support service charges and turnkey fee should also continue [46].

Biogas programme suffers from lack of hard data needed to perform cost benefit analysis for use in policy-making. The Workshop also felt that it was necessary to evaluate biomethanation vis-a-vis other energy sources, estimate cost burden on economy and quantify total benefits. Promoters of biomethanation should also devise different ways to finance the programme for harnessing its full potential and penetration into rural areas. More focus need be placed on gender aspects of biomethanation which should be seen as a social programme based on appropriate technology rather than a mere technical programme. Biogas plant design should be standardised into a design which can be moulded, fabricated and delivered to the beneficiary. Government should have a minimum 20 year time-frame for achieving a long range target of 10 million plants [46].

13.4 Areas for R&D Endeavours in Biomethanation

There is considerable scope to pursue R&D to improve plant design, construction methodology and operation, and make them more cost-effective. Some important areas for research are as follows:

1. Site specific studies for identifying optimum mix of locally available feed materials containing efficient strains of micro-organisms for obtaining higher gas yield at low cost need be carried out prior to setting up community or cluster of biogas plants in a particular region. This may include evaluation of locally available materials for anaerobic fermentation, pre-treatment of influent slurry, and impact of environmental conditions on fermentation efficiency.
2. Microbiology of bacteria in anaerobic (able to live without oxygen), aerobic (able to live under oxygen) and facultative (able to live under different conditions) environment need be studied for analysing their behaviour, growth rates and nutrient needs; and impact of environmental factors on their metabolism activities [36].
3. Addition of hydrogen producing bacteria during anaerobic fermentation improves biogas yield. Further studies need be carried out to establish their full potential in improving the gas yield [36].

4. For plants utilising municipal refuse as feed, it is necessary to develop low cost techniques for separating inert and non-digestible constituents before feeding to a digester which clogg and create scumming problems during digestion [22].
5. Even after achieving considerable success, more efforts need be directed for developing low cost plant designs which simplify construction and use locally available materials. Optimisation techniques can be used for maximising gas yield for varying loading rate, retention time, digester temperature, and pH of digester contents. Specific studies for reducing gasholder size by varying system parameters to suit applicable rates of gas utilisation can be conducted explored on plant-to-plant basis [8, 22].
6. In view of increasing automation of biogas plants, there is a need to develop low-cost technologies for manufacturing accurate and sensitive instruments for measuring parameters like feed rate, effluent outflow rate, digester temperature, pH of digester contents and gas flow rate [16].
7. For any large-scale commercial application of biogas whose output varies with seasons, it is important that adequate storage facilities be created. Storage of biogas unlike LPG is particularly troublesome following difficulties in liquefaction and higher proportions of carbon dioxide compared to other gaseous fuels [22, 35].
8. Success of both small-scale and large-scale biogas plants, among others, depends on their trouble-free operation and therefore efforts need be directed to ensure that plant operation be made as much trouble-free as possible.
9. Apart from making increasing use of locally available materials as feedstock, handling techniques also need to be improved for using effluent as enriched fertiliser under varying conditions of soil and environment. Technologies for producing unconventional products like vitamin B_{12} from plant effluent also need to be developed further [12].
10. Biogas apart from being used as cooking, lighting and engine fuel, can also be used for making products like ammonia and methanol through multi-stage decomposition and successive re-cycling. There is thus scope in developing technologies to produce these products on commercial basis.
11. As proportion of methane in biogas is less compared to other gaseous fuels, its usage as dual-fuel needs to be studied on engine life, selection of engine parameters for efficient performance, and redesign of engine carburettor and ignition system. Furthermore, in view of large proportion of carbon dioxide, hydrogen sulphide and other impurities in biogas, it is necessary to develop low-cost anti-corrosive measures to protect vulnerable parts from their harmful effect [13, 19].
12. Most agricultural farms need electric power for operating farm machinery, energy requirement of which, can largely be met by power supply from biogas-based decentralised energy systems. However, energy

supply from decentralised sytems in many cases is not cost-effective. Efforts can thus be made to improve the economics of power generation by such systems. Gas-based decentralised energy systems can even supply power to the state power-grids in the event of surplus power generation [26, 39].

13. Concept of integrated farming systems involving cultivation of food and energy crops, algal cultivation, pisciculture and biogas generation wherever feasible needs to be popularised and promoted [37, 38].
14. There is a need to retain and possibly enchance the subsidy amount offered by the government for promotion of biogas programme. This should be backed up by launching of effective nationwide demonstration programmes to promote intensive utilisation of biogas plant fertiliser [32].
15. For reviewing biogas programmes in different countries and making international comparisons, detailed surveys need be conducted periodically by international organisations. It may involve study of : number and type of plants set-up, number of plants in operation and under break-down; plant category whether individual, institutional or community type; pattern of plant ownership and beneficiary's economic background, construction costs incurred, pattern of annual operating costs, annual yield of biogas and effluent produced; pattern of biogas usage in cooking, lighting, and operating farm machinery etc.; and socio-economic and behavioural aspects of programme implementation [30, 31].
16. Significance of analysing and controlling feed rate, gas flow rate, temperature, pH, and pressure needs to be promoted among beneficiaries [16].

13.5 Concluding Remarks

Biogas is ideal for introducing small scale decentralised energy systems which very well fit in with country's pattern of rural and agricultural economy. Considering that organic wastes are anyway generated plentifully in any nation, it is always advantageous to harness them to produce biogas which as a renewable energy source can go a long way in supplementing energy supplies [1]. It should, however, be realised that production of biogas from organic wastes will not completely solve the energy problems of a nation even if the wastes are available adequately [2]. With a high rate of deforestation which must be prevented for ecological reasons, scope for biomethanation in rural areas is high. Success in biomethanation is also linked with success in afforestation. Some experts believe that biogas programme cannot make real headway unless the biomass availability is also simultaneously increased so that it can substitute for cattle dung which is preferred as direct fuel [37]. An integrated energy system based on biomethanation, recycling of waste and biomass agro-forestry while meeting rural energy needs can also help to prevent soil-erosion, and improve hygienic conditions and rural health [30, 31]. Reliance on biogas as a fuel involves neither dependence on foreign exchange reserves nor dependence on diminishing fossil fuel supplies. Whereas fuelwood, charcoal, kerosene, petroleum and coal may not last forever, biogas shall be available as long as cattle play a major role in a country's agriculture and milk production.

REFERENCES

1. ABE, *Towards a Perspective on Energy Demand and Supply in India in 2004/05*, Advisory Board on Energy, New Delhi, May, 1985.
2. Brown, N., Biogas Systems in Development, *Appropriate Technology*, **14(3)**, 5-7 (1987).
3. Buren, A.V., *Transferring Biogas Technology to the Rural Areas of the Third World*, International Institute of Environment and Development, London, 1980.
4. Chand. A., *et al.*, District Level Management System for Biogas Programme: An Analysis, *Economic and Political Weekly*, 23, May, M 80-84 (1988).
5. Chottani, A.A., *Linking Biogas Programme with Dairy Development Programmes*, paper presented at the National Workshop on Biogas Development, New Delhi, Feb., 1982.
6. Datta, K., Energy Use in Agriculture, *Economic Times*, New Delhi, July 21, 1986, p. 5.
7. Dayal, M., Development of New and Renewable Sources of Energy, *Financial Express*, Dec 20, 1982.
8. Dholakia, J.R., Biogas to the Fore, Special Supplement on Energy, *Economic Times*, New Delhi, Jan 17, 1981, p. 11.
9. *Economic Times*, Need to Harness Non-Conventional Energy Sources, A Report, July 4, 1988.
10. *Economic Times*, Energy: Long Term Outlook, A Special Supplement on Energy, New Delhi, June 9, 1988, p. 2.
11. *Economic Times*, Renewable Sources of Energy (Infrastructure), Economic Survey 1990-91, New Delhi, July 21, 1991, p. 14.
12. FAO, *China: Azolla Propagation and Small Scale Biogas Technology*, FAO Soils Bulletin No 41, Food and Agriculture Organisation, Rome, 1978, p. 69.
13. Gupta, R.K. and Iyer, N.V., Use of Biogas in Diesel Engines, *Agricultural Engineering Today*, **1(9)**, 7-10 (1977).
14. *The Hindu*, Role of Renewable Sources of Energy, Visakhapatnam, Dec 27, 1990.
15. *The Hindu*, Renewable Energy Sources: Need for Greater Priority, *'The Hindu'* Survey of Indian Industry, 1990; Feb, 1991, p. 177-179.
16. Kishore, V.V.N., Rangarao, V.V. and Raman, P., *Some Problems of Implementation of Biogas Technology in Rural Areas*, A Report, TERI, New Delhi, Oct., 1987.
17. Leach, G., Household Energy in South Asia, *Biomass*, **12(3)**, 155-184 (1987).
18. Mazumdar, A. and Gopalakrishnan, N.K., *Role of Cooperatives in the Diffusion of Biogas Technology*, A Brochure, TERI, Mumbai, 1982.

19. Mehta, M.M., *et al.*, Biogas Utilisation in Tractor Diesel Engines for Stationary Applications, *Agricultural Engineering Today*, **3**, 13-16 (1979).
20. Murthy, R.K., The Energy Crunch: An Appraisal, *The Hindu*, June 18, 1991, p. 17.
21. Myles, R.M., *A Practical Guide to Janata Biogas Plant Technology*, A Report, Action for Food Production, New Delhi, 1985.
22. NAS, *Methane Generation from Human, Animal and Agricultural Wastes*, Report of an Ad. Hoc. Panel of the Advisory Committee on Technology Innovation, National Academy of Sciences, Washington, DC, 1977, p. 85.
23. ONGC, *Energy Profile*, A Published Brochure, Economic Environment Scanning Group, ONGC, New Delhi, 1987.
24. Pachauri, R.K. and Srivastava, L. (eds.), *Energy Policy Issues*, Wiley Eastern Ltd., New Delhi, 1986.
25. Panditrao, Y.A., Energy Crisis and Biogas, Special Supplement on Energy, *Economic Times*, New Delhi, Jan 17, 1981, p. II.
26. Pani, N., Energy Planning: Fresh Approach Needed, *Economic Times*, New Delhi, May 3, 1988, p. 5.
27. Parikh, M.P., *Installation of Plants on Turn-key Basis*, paper presented at the National Workshop on Biogas Development, New Delhi, Feb., 1982.
28. Prakash, G., Rastogi, S.K., Gadgil, A., and Kishore, V.V.N., *Comparison of Various Renewable Energy Options to Meet Energy Demand of an Urban Community*, Proc. National Solar Energy Convention, Solar Energy Society of India, Madurai, Sep. 13-15, 1986.
29. Raghavan, B.S., Non-Conventional Energy: Need for Policy Shifts, *The Hindu*, Jan 1, 1988.
30. Ramana, G.V., Energy Policy Issues-I, *Economic Times*, New Delhi, May 2, 1988, p. 5.
31. Ramana, G.V., Energy Policy Issues-II, *Economic Times*, New Delhi, May 2, 1988, p. 5.
32. Rao, G., *National Project for Biogas Development: Role of ARDC and Bank*, A Report, Agricultural Refinance Development Corporation, Mumbai, 1982.
33. Reddy, A.K.N., Energy Planning: Time for Tough Decisions, *The Hindu*, Visakhapatnam, Nov. 23, 1990.
34. Shah, A.C., Impact of Drought on Biogas Programme, *Economic Times*, Mumbai, June 13, 1988, p. 11.
35. Sobel, A.T. and Muck, R.E., Energy in Animal Manures, *Energy in Agriculture*, **2**, 161-176 (1983).
36. SPOPD, *Proc. of the Chengdu Seminar on Biogas Technology and Utilisation*, Sichuan Provincial Office of Biogas Development, Sichuan, P.R. China, 1979.
37. Thukral, K., Kishore, V.V.N., Gadgil, A. and Pachauri, R.K., *State of Renewable Energy Technologies in India*, A Report, World Bank, Washington, DC, Sep., 1986.
38. Thukral, K., Kishore, V.V.N., Hossain, J., Gadgil, A. and Pachauri, R.K., *Problems and Prospects for Expanded Commercialisation of Mature Non-Conventional Energy Technologies*, A Report, World Bank, Washington, D.C., Dec., 1986.
39. Vinodkumar, T.M. and Ahuja, D.R. (eds.), *Rural Energy Planning*, Wiley Eastern Ltd., New Delhi, 1986.
40. WB, *Renwable Energy Sources in the Developing Countries*, World Bank, Washington, DC, 1980.

41. Ramana, P.V., Biogas Programme in India, *TIDE*, **1(3)**, 1-18 (1991).
42. Smith, W.H., *et al.*, Methane from Biomass and Waste: A Program Review, *TIDE*, **2(1)**, 1-18 (1992).
43. Lisbon, W. and Maegaard, P., Computerisation in Anaerobic Digestion Technology, *Biogas Forum*, **IV No (55)**, 13-17 (1993).
44. TERI, *Biogas Technology: A Promise for the Future*, A monograph sponsored by the MNES, New Delhi, 1994.
45. *Financial Express*, Non-Conventional Energy Sources Must Become Conventional Sources, F.E. Symp., New Delhi, May 18, 1994, p. 5.
46. AKRSP, Proc. of the Indian National Workshop on '*Policy Issues in the Biogas Programme*' held during Aug. 21-22, 1995, Aga Khan Rural Support Programme (India), Ahmedabad, April, 1996.
47. Reeps, U., UNDP/ESCAP Regional Workshop on 'Intergrated Rural Energy/Environmental Planning for Sustainable Rural Development', *Biogas Forum*, **I** No. **60**, 23-24 (1995).
48. CII and IREDA, *Indo-US Partnership in Renewable Energy and Sustainable Development*, India Habitat Place, New Delhi, Dec., 1994.
49. MNES, *Energy from Waste: National Programme on Energy Recovery from Urban, Municipal and Industrial Wastes*, 14, CGO Complex, Lodhi Road, New Delhi, 1995.
50. Prabhakara, B.R., Foreword to the Author's book, 'Mital, K.M., *Non-Conventional Energy Systems: Principles, Progress and Prospects* Wheeler Publishing, New Delhi, 1996', June 5, 1996.

Appendix I

BIBLIOGRAPHY

Abbasi, S.A. and Nipaney, P.C., Generation of Biogas from *Salvinia Molesta* (Mitchell) on a Commercial Digester, *Env. Tech. Letters,* **5**, 75-80 (1984).

Abbasi, S.A., *et al.*, Bioenergy Potential of Eight Common Aquatic Weeds, *Biological Wastes,* **34**, 359-366 (1990).

Abbasi, S.A., *et al.*, Use of Aquatic Weed Salvinia as Full Partial Fuel in Commercial Biogas Digesters, *Indian Jour. of Tech.*, **30(9)**, 451-457 (1992).

Abdeldayem, A.M. and Hamad, M.A., *Design and Operation of Small Biogas Plants,* A Report, Nat. Res. Centre, Cairo, 1980.

Abeles, T.P., *Energy Reclamation from Agricultural Wastes,* Conf. Proc. on Energy, Agri. and Waste Management, Cornell Univ., Ithaca, NY, 247-484 (1975).

Abeles, T.P., *Energy and Economic Analysis of Anaerobic Digesters,* Research Report, College of Env. Sci., Univ. of Wisconsin, Green Bay, Wisconsin, 1975.

Abeles, T.P. and Atkinson, P., Economic and Energy Considerations for Anaerobic Digestion of Farm Wastes, in 'W.J. Jewell (ed.), *Energy, Agri. and Waste Management,* Ann Arbor Sci. Pub., Ann Arbor, MI, 1975,' p. 353-360.

Abeles, T.P., Design and Engineering Considerations in Plug Flow Farm Digesters, in '*Clean Fuels from Biomass and Wastes,* Inst. of Gas Technology, Chicago, 1977.'

Abeles, T.P. and Dale, E., *Safety and Design Considerations for Farm Digesters,* Proc. First Inter. Conf. on Energy Use Management, C. Smith (ed.), Pergamon Press, Vol 1, 781 (1977).

Abeles, T.P., *et al., Bibliography of Anaerobic Digestion,* US Env. Protection Agency, Washington, 1978.

Abeles, T.P., *et al., Energy and Economic Assessment of Anaerobic Digesters for Rural Waste Management,* EPA 600/7-78-174, US Env. Protection Agency, Washington, 1978.

Abeles, T.P., *System Design Consideration in Farm Digesters,* Proc. Conf. on Humanistic Choices in Alternative Energy, 1978.

Abeles, T.P., *et al., Anaerobic Digestion of Agricultural Residues: A Technology Assessment,* Proc. Fourth Annual Conf. on Solid and Hazardous Waste, US Env. Protection Agency, Washington, 1979.

Abeles, T.P., *et al., A Technology Assessment of Anaerobic Digesters Using Agricultural Residues,* Proc. Second Inter. Conf. on Energy Use Management, 1979.

Abeles, T.P., *Farm Integrated Utility Systems : Methods for Farm Scale Recovery of Methane,* Proc. of a Seminar on Biomass Energy for City, Farm and Industry, Emmaus, J.G. Press, 234-243 (1980).

Abeles, T.P., *et al., A Guide Book for Farm Scale Methane Digesters,* Ozarks Regional Commission, 1981.

Abelson, P.H., Methane : A Motor Fuel, *Science,* **218**, 641 (1982).

Aburas, R., Hammad, M., and Abu-Reesh, I., Construction and Operation of a Demonstration Biogas Plant: Problems and Prospects, *Bioresource Tech.*, **53(2)**, 101-104 (1995).

Acharya, C.N., Studies on the Anaerobic Decomposition of Plant Materials, *Biochemical Jour.*, **29(3)**, 528-541 (1935).

Acharya, C.N., Some Factors Influencing the Anaerobic Decomposition of Rice-Straw, *Biochemical Jour.*, **29(4)**, 953-960 (1935).

Acharya, C.N., Decomposition of Rice-Straw Under Anaerobic, Aerobic and Practically Aerobic Conditions, *Biochemical Jour.*, **29(5)**, 1116-1120 (1935).

Acharya, C.N., Decomposition of Plant Substances of Varying Composition, *Biochemical Jour.*, **29**, 1459-1467 (1935).

Acharya, C.N. and Subramanyan, V., Hot Fermentation Process for Composting Town Refuse and Other Wastes Material, *Indian Jour. of Agri. Sci.*, **9**, 741 (1939).

Acharya, C.N., *Organic Manures*, Res. Revision Series, Bull. 2, ICAR, New Delhi, 1952.

Acharya, C.N., Cow Dung Gas Plants, *Indian Farming*, **3(9)**, 16-18, 28 (1953).

Acharya, C.N., *Prospect of Compost Manure from Town Wastes*, ICAR, New Delhi, 1954.

Acharya, C.N. and Juneja, P.C., *Studies on the Production of Combustible Gas and Manure from Cattle Dung*, Proc. Indian Sci. Congress Assoc., **41**, 246 (1954).

Acharya, C.N., With This Change the Cow Dung Gas Plant Works Better, *Indian Farming*, **6(9)**, 10-11, 13 (1956).

Agarwal, K.C., *et al.*, *Ecological Effects of Production of Biogas from Human Excreta on the Enteriopathogens*, paper presented at the 18th Annual Conf. of Assoc. of Microbiologists of India, Madurai, Dec 21-23, 1977.

Agnihotri, S.B., *Biogas from Water hyacinths*, paper presented at the Panel Meeting on Biogas Tech., Bardoli, Gujarat, Nov 5-6, 1981.

Aili, Y., *et al.*, Study and Design of a High Performance Biogas Project Treating Chicken Manure, *Biogas Forum (BORDA)*, 2 Special Issues on China, 14-19 (1991).

Ainea, E.K., Cylinder Capped Biogas Plants, *Biogas Forum (BORDA)*, **48**, 22-24 (1992).

Ainea, E.K. and Nasser, R., Research Result in Gas-tight Plastering, *Biogas Forum (BORDA)*, **52**, 21-26 (1993).

Alaa El-Din, M.A., *Recycling of Organic Wastes in Agriculture*, Proc. Inter. Conf. on Biopotentials for Dev., Houston, Texas, Nov 5-11, 1978, Pergamon Press, NY, 1980.

Alaa El-Din, M.A., Biogas Technology with Respect to Chinese Experience, *FAO. Soils Bull.*, **45**, 175-186 (1982).

Alexander, D.W., *Compression, Absorption and Storage of Methane Gas*, A Report, Dept. of Mech. Eng., Univ. of Manitoba, Winnipeg, Manitoba, Canada, 1972.

Alicbusan, R.V., *et al.*, Preparation and Utilisation of Methane Bacteria as Starter in Biogas Production, *World Science News*, **12(9)**, 415-416 (1975).

Alicbusan, R.V., *Recent Developments in Philippines in Production and Utilisation of Biogas,* Proc. NRCP-NIST Symp. on World Dev. of Biogas Tech., Manila, 1976.

Alone, B.Z., *et al.*, Anaerobic Digestion of City Refuse and Dung. *Indian Jour. of Env. Health*, **20**, 134-140 (1978).
Amaratunga, M., Integrated Biogas Systems, *Regional Jour. of Energy, Heat and Mass Transfer*, **2(1)**, 45-49 (1980).
Amartunga, M., Role of Integrated Biogas Systems in Sri Lanka, *Khadi Gramodyog*, **27(3)**, 190-192 (1980).
Anand, R.C. and Singh, E., A Simple Technique, Charcoal Coating Around the Digester Improves Biogas Production in Winter, *Bioresource Technology*, **45(2)**, 151-152 (1993).
Anand, Y.P., *Study of the System of Management of Wastes in a Railway Complex*, Ph.D thesis, Dept. of Chem. Engg., IIT, Delhi, 1989.
Andersen, C.G., Overhauling and Repairs to Methane Gasholder, *Sewage Works Jour.*, **15(3)**, 535-541 (1943).
Andrews, J.F., Cole, R.D. and Pearson, E.A., Kinetics and Characteristics of Multistage Methane Fermentation, SERL Report No. 64-11, Sanitary Eng. Res. Lab., Univ. of California, Berkeley, 1964.
Andrews, J.F., Dynamic Model of an Anaerobic Digestion Process, *Jour. of Sanitary Eng.*, Div. of ASCE, Feb., 95-116 (1969).
Anglo, P.G., *Recent Developments in Biogas Technology*, Proc. of the Seventh Miami Inter. Conf. on Alternative Energy Sources, Miami Beach, Dec 9, 1985, Coral Cables, Florida, 1986, p. 313-314.
Anonymous, Microbiology and Chemistry of Biogas Production, Summer Institute of Waste Mgt. by Recycling, Nat. Dairy Res. Inst., Mumbai, *Jour. Indian Chem. Manufacturers*, **12(12)**, 50 (1974).
Anonymous, Car Runs on Biogas, *Bull. Inst. of Engineers (India)*, **28(8)**, 42-43 (1974).
Anonymous, *Anaerobic Treatment Process and Methane Production*, Notes on Water Pollution No. 64, Water Poll. Lab., Herts, England, 1974.
Anonymous, Production of Bio-Fertilizer and Biogas from Agricultural Wastes, *Khadi Gramodyog*, **20(7)**, 6 (1974).
Anonymous, Gobar Gas for Diesel Unit, *Jagriti*, **20(2)**, 6 (1976).
Anonymous, The Ball Digester, *Biogas Newsletter (Nepal)*, **3**, 1-8 (1978).
Anonymous, New Model of Biogas Plant, *Agricultural Situation in India*, **33**, 823 (1979).
Anonymous, *Methane and Other Vendable Products from Animal Wastes*, Proc. Symp., Orlando, Florida, 1979.
Anonymous, Low Cost Biogas Digester, *Urja*, **9(1)**, 15-35 (1981).
Anonymous, Low Cost Biogas Stove, *Science for Villages*, **5(11)**, 7 (1982).
Anonymous, *Methanisation*, Proc. Seminar, Sophia – Antipolis, Valbonne, France, 1982.
Anonymous, Biogas Bag Digester, *Changing Villages*, **4(1)**, 7-20 (1982).
Anonymous, FORD Gets into Methane-fuelled Automobile Market, *Biomass Digest*, **4(2)**, 5 (1982)
Anonymous, Biogas Plants Made of Bamboo, *Invention Intelligence.*, **17**, 230-231 (1982).
Anonymous, *Anaerobic Digestion*, Proc. Third Inter. Symp., Dynatech, 99 Erie Street, Cambridge, MA, 1983.
Anonymous, Anaerobic Digestion of Rice Crop Residues, *Agricultural Wastes*, **17(4)**, 263-269 (1986).

Anonymous, *Compressed Biogas at the Christchurch Drianage Board*, Report No. 128, New Zealand Energy Research Dev. Committee, Auckland, New Zealand, 1986.

Anthony, W.B., *Animal Wastes Management*, Cornell Univ., Ithaca, NY, 1969.

ASAE, *Managing Livestock Wastes*, Proc. of the Third Inter. Symp. on Livestock Wastes, ASAE Pub. No. Proc. 275, St. Joseph, Michigan, 1975.

Ashare, E. and Wilson, E.H., *Analysis of Digester Design Concepts*, A Report, Dynatech R&D Co., Cambridge, MA, 1979.

Ashare, E., Analysis of Systems for Purification of Fuel Gas, in 'D.L. Wise (ed.), *Fuel Gas Production from Biomass*, Vol 2, CRC Press, Florida, 1981,' 1-22.

ASRCT, *Final Report on Pre-feasibility Study of the Biogas Technology: Application in Rural Areas of Thailand*, Applied Scientific Research Corporation of Thailand, Env. and Dev. Res. Dept., Bangkok, Apr., 1979.

Atkinson, B., *Biochemical Reactors*, Pitman, London.

Aubert, C.H., *et al.*, *Anaerobic Digestion of Pig Manure, Poultry Waste and Rabbit Waste on a Conventional Digester*, Proc. 2nd EC Conf. on Energy from Biomass, Applied Science, London, 599-603 (1982).

Aubert, C.H. and Bully, F., Anaerobic Digestion of Rabbit Wastes and Pig Manure Mixed with Rabbit Wastes in Various Experimental Conditions, *Agricultural Wastes*, **10(1)**, 1-13 (1983).

Aubert, C.H., *et al.*, Anaerobic Digestion of Poultry Waste: 1, Biogas Production and Pollution Decrease in Terms of Retention Time and Total Solids Content, *Process Biochemistry*, 30-37 (1983).

Aubert, C.H. and Bully, F., *Development of Installation for the Production of Biogas from Stock-farming Waste*, Proc. of the Information Symp under the EEC Programme on Recycling of Urban and Indus. Waste, Luxembourg, May 8-10, 318-322 (1984).

Baader, W., *et al.*, *Biogas in Theory and Practice*, KTBL, D-6100 Darmstadt 12, Bartningstrasse, 49, 1981.

BABA, *Landfill Gas*, Proc. Workshop, British Anaerobic and Biomass Assoc., Oxford, UK, 1983.

Babu, K.S., *et al.*, Improvement in Biomethanation of Mango Processing Wastes by Addition of Plant Derived Additives, *Biogas Forum*, **III** No **58**, 16-19 (1994).

Bacon, V.W., Digester Heating and Mixing Equipment, *Sewage Works Jour.*, **16(3)**, May, 534-539 (1944).

Baere, L. and Verstraete, W., High Rate Anaerobic Composting with Biogas Recovery, *Biocycle*, March 30-31, 1984.

Bahadur S., *et al.*, Effect of Nickel and Cadmium on Biogas Production, *Vijnana Parishad Anusandhan*, **24**, 383-388 (1981).

Baitha, D.L., Plastic Gas Holder for Gobar Gas Plant, *Science Reporter*, **19**, 440 (1982).

Balasubramanian, P.R. and Kasturibai, R., Recycling of Biogas Plant Effluent through Aquatic Plant (Lemna) Culture, *Bioresource Tech.*, **41**, 213-216 (1992).

Balasubramanian, P.R. and Kasturibai, R., Evaluation of Nutrient Recovery after Anaerobic Digestion of Cattle Dung in a Family Size Biogas Plant, *Biomass Bioenergy*, **3(5)**, 377-380 (1992).

Balasubramanian, P.R. and Kasturibai, R., Recycling of Cattle Dung, Biogas Plant Effluent and Water hyacinth in Vermiculture, *Bioversource Tech.*, **52(1)**, 85-87 (1995).

Balasubramanya, R.H., *et al.*, Production of Biogas from Willow-Dust by a Batch Fermentation Process, *Jour. of Textile Assoc.*, **42**, 145-149 (1981).

Bansal, M.L., Mittal, O.P., Sondhi, M.S. and Neelkantan, S., Biogas Production During Anaerobic Digestion of Livestock Excreta, *Indian Jour. of Dairy Sci.*, **30(4)**, 338-340 (1977).

Bansal N.K., *et al.*, A Study of a Greenhouse Concept on Conventional Biogas System for Enhancing Biogas Yield in Winter Months, *Inter. Jour. of Energy Research*, **9(2)**, 119-128 (1985).

Bargman, R.D., *et al.*, *Anaerobic Sludge Digestion*, Manual of Practice, No.16, Water Poll. Control Fed., Washington, DC, 1966.

Barker, H.A., Biological Formation of Methane, *Indus. Eng. Chem.*, **48**, 1438-42 (1956).

Barker, U.A., *Bacterial Fermentations*, John Wiley and Sons, NY, 1956.

Barlaz, M.A., *et al.*, Gas Production Parameters in Sanitary Landfill Simulators, *Waste Mgt. Res.*, **5**, 27-39 (1987).

Barlett, H.D., *et al.*, *Experiences from Operating a Full Size Anaerobic Digester*, ASAE Paper No.77-4035, Amer. Soc. of Agri. Engrs., St. Joseph, Michigan, 1977.

Barry, M. and Colleran, E., *Anaerobic Digestion of Silage Effluent Using an Upflow Fixed Bed Reactor*, A Report, Dept. of Microbiology, Univ. College, Galway, 1980.

Baugh, K. and McCarty, P.L., *Biological Conversion of Lignocellulose to Methane*, Quart. Progress Report on Heat Treatment of Organics for Increasing Anaerobic Biodegradability, Solar Energy Res. Inst., Colorado, Sep. 30, 1979.

Baumann, P.G., Digester Methane Utilisation can be Optimised, *Water Sewage Works*, 127, 44-45, 66-68 (1980).

Beba, A. and Atalay, F.S., Mathematical Models for Methane Production in Batch Fermenters, *Biomass*, **11(3)**, 173-184 (1987).

Bennet, A., Direct Combustion of Biogas for Water and Space Heating, in 'Savage, M. and Crocker, S. (eds.), *Using Biogas and Other Prodvcts of Anaerobic Digestion*, Proc. Symp., Edenbridge, IK, 1984,' p. 34-35.

BES, *Report on Utilisation of Waste Heat from Biogas Power Generation*, Biogas Experimental Station, Shanghai Municipality Inst. of Indus. Microbiology, Shanghai, China, Jul. 1980.

Bethel, J.S., *The Potential of Lignocellulosic Material for the Production of Chemicals*, Fuels and Energy, Div. of Policy Res. and Analysis, Nat. Sci. Foundation, Washington, DC, 1976.

Bhaduri, B., Gupta, V.K. and Verma, N., Possibility of Biogas from Community Latrines, *Khadi Gramodyog*, **27(3)**, 199-204 (1980).

Bhatt, J.V., *et al.*, Anaerobic Rotting of Coconut Husks, *Coir*, **19(2)**, 27-29 (1975).

Bhatt, M.S., *et al.*, *Design of Large Scale Biogas Plant Using City Wastes*, Proc. Solar Energy Conv., Bhopal, Feb. 10-12, 1985.

Bhattacharya, B.C., Economic Design of Gobar Gas Plant for Rural Communities, *Chemical Age of India*, **29(5)**, 301-303 (1978).

Bhattacharya, D., Activated Sludge Treatment of Dairy Wastewater: Determination of Kinetic Parameters for Continuous Stirred Tank Reactor, *Res. and Industry*, **39(1)**, 43-47 (1994).

Battacharya, S.C. and Khai, P.Q., Kinetics of Anaerobic Cow Dung Digestion, *Energy*, **12(6)**, 497-500 (1987).

Bhide, A.D., Energy from Solid Wastes. *Jour. of the Inst. of Public Health Engrs. (India)*, **3 & 4**, 14-18 (1976).

Bhosale, S.B., *et al.*, Biochemistry of Methane Formation from Carbohydrates, Mechanism Model and Energetics of Proposed Reactions, *Biovigyanam*, **6(2)**, 97-104 (1980).

Biswas, D.K., Towards Better Design of Biogas Plants for Fuel and Fertiliser, *Bhagirath*, **12**, 140-145 (1975).

Biswas, D.K., Biogas Plants for Environmental Improvement, *Khadi Gramodyog*, **23(1)**, Oct., 68-72 (1976).

Biswas, S.C., Fuel Gas, *Indian Farming*, **6**, 124 (1945).

Biswas, T.D., *et al.*, *Removal of Carbon Dioxide from Biogas*, Proc. Nat. Symp. on Biogas Technology, IARI, New Delhi, Nov. 29-30, 1977.

Biswas, T.D., *Cow Dung Gas Plant : Instructions, Installation, Operation and Maintenance*, A Report, IARI, New Delhi, 1979.

Block, D., *Recovery of Methane from Biogas*, Proc. 2nd EC Conf. on Energy from Biomass, Berlin, Sep. 20-23, 1982, Applied Science, London, 1982.

Boopathy, R., Effect of pH on Anaerobic Fermentation of Coffee Pulp, *Asian Env.*, **9(3)**, 23-27 (1987).

BORDA, *Report on the Inter. Biogas Workshop*, Bremen Overseas Res. and Dev. Assoc., Bremen, Germany, May 16-20, 1979.

BORDA, *Biogas Plants: Building Instructions*, Overseas Res. and Dev. Assoc., Bremen, Germany, 1979.

BORDA, Discussion on Fixed Dome Plants, Bremen Overseas Res. and Dev. Assoc. Bremen, Germany, *Biogas Forum*, **19**, 1-7 (1985).

Borja, R., *et al.*, Anaerobic Digestion of Block-Olive Wastewaters, *Bioresource Tech.* **45(1)**, 27-32 (1993).

Boshoff, W.H., Methane Gas Production by Batch and Continuous Fermentation Methods, *Tropical Sci.*, **5(3)**, 1963.

Boswell, A.M., Microbiology and Theory of Anaerobic Digestion, *Sewage Works Jour.*, **19(28)**, 1947

Bousefield, S., Hobson, P.N. and Summers, R., A Note on Anaerobic Digestion of Cattle and Poultry Wastes, *Agricultural Wastes*, **1**, 161-164 (1979).

Brazil, *Biogas: Information Guidebook and Basic Bibliography*, Ministry of Industry and Commerce, Secretariat of Indus. Tech. Brazil, 1983.

Brouard, F., *et al., Advances in Anaerobic Digestion of Aquatic Plants*, Proc. 2nd EC Conf. on Energy from Biomass, London, Applied Science, 334-338 (1982).

Brown, J. G., *Florida Water Hyacinths to Methane*, Symp. on Commercialisation of Solar and Conversion Tech., Miami Beach, Dec., 299-300 (1978).

Bruce, A.M., *et al., Anaerobic Digestion of Sewage Sludge and Organic Agricultural Wastes*, Elsevier Science Pub. Co. Inc., New York, 1986.

Brule, M.R., Design of Small Scale Biogas Plants, in 'S.S. Sofer and O.R. Zabrosky (ed.), *Biomass Conversion Processes for Energy and Fuels*, Plenum Press, NY, 1981,' 297-313.

Brune, D.E., *C:N Ratio and Anaerobic Digestion*, M.S. thesis, Univ. of Missouri, Columbia, MO, 1975.

Byrant, M.D., Microbial Methane Production: Theoretical Aspects, *Jour. Animal Sci.*, **48**, 193-201 (1979).

Byrant, M.P., *The Microbiology of Anaerobic Digestion and Methanogenesis with Special Reference to Sewage*, Proc. Seminar on Microbial Energy Conversion, Erich Goltze, Cottingen, Germany, 107-117 (1976).

Buhr, H.O. and Andrews, J.F., The Thermophilic Anaerobic Digestion Process, *Water Research*, **11(2)**, 129-143 (1977).

Buren, A.V., *A Chinese Biogas Manual*, Intermediate Tech. Pub., London., 1979.

Bushness, M.E. and Slater, J.H., *Mixed Culture Fermentations*, Academic Press, NY, 1981.

Bushwell, A.M., Important Considerations in Sludge Digestion: II-Microbiology and Theory of Anaerobic Digestion, *Sewage Works Jour.*, **11**, 28 (1947).

Bushwell, A.M., Microbiology and Theory of Anaerobic Digestion, *Sewage Works Jour.*, **11**, 28 (1947).

Bushwell, A.M. and Muller, H.P., Mechanism of Methane Fermentation, *Indus. Eng. Chemistry*, **44**, 550-552 (1952).

Bushwell, A.M., Fundamentals of Anaerobic Treatment of Organic Wastes, *Sewage and Industrial Wastes*, **29**, 717 (1957).

Bushwell, A.M. and Boruff, G.S., Mechanical Equipment for Continuous Fermentation of Fibrous Materials, *Indus. Eng. Chemistry*, **25**, 147-149 (1983).

Buvet, R., Fox, M.F. and Picken, D.J. (eds.), *Biomethane, Production and Uses*, Proc. Course, Dijon, France, 1984.

Cabafi, K. and Stojanovic, A., *Some Experiences with Designed and Built Biogas Processing and Utilization Systems*, Proc. of the Conf. on Industrial Systems in Agro Complex, Yugoslavia, 242-247 (1987).

Cail, R.G. and Barford, J.P., Thermophilic Semi-Continuous Anaerobic Digestion of Palm-Oil Mill Effluent, *Agricultural Wastes*, **13(4)**, 295-304 (1985).

Callander, I.J. and Barford, J.P., Recent Advances in Anaerobic Digestion Technology, *Process Biochemistry*, **18(4)**, 24-30 (1983).

Callander, I.J. and Barford, J.P., Improved Anaerobic Digestion of Pig Manure Using a Tower Fermenter, *Agricultural Wastes*, **11**, 1-24 (1984).

Calzada, B.F., *et al.*, Biogas Production from Coffee Pulp Juice: One and Two-Phase System, *Agricultural Wastes*, **9(3)**, 217-230 (1984).

Carty, M., Anaerobic Waste Treatment Fundamentals, Part 1, 2, 3 and 4, *Public Works*, Sep., Oct., Nov., Dec., 1964.

CAZRI, *World Bibliography of Biogas Technology and its Utilization*, Central Arid Zone Research Institute, Jodhpur, 1977.

CEMIG, *Energy from Biomass and Waste, Proc. Seminar*, Companhia Energetica de Minas Gerais, Belo Horizonte, Brazil, 1988.

Centre of Science for Villages, Low Cost Biogas Stove, *Science for Villages*, **5(11)**, 6-7 (1982).

Chakraborty, R.K., *Immobilised Enzymes and Enzymes Cultures in Biogas Generation*, M. Phil thesis , Indian School of Mines, Dhanbad, Jul., 1989.

Chanakya, H.N., *et al.*, Solid-phase Biogas Production with Garbage or Water-hyacinth, *Bioresource Tech.*, **46(3)**, 227-231 (1993).

Chanakya, H.N., *et al.*, Two-phase Fermentation of Whole Leaf Biomass to Biogas, *Biomass Bioenergy*, **5(5)**, 359-367 (1993).

Chandra, A., *et al.*, Hydrodynamical Modelling of a Biogas Burner, *Energy Conv. and Mgt.*, **32(4)**, 353-358 (1991).

Chandra, A., Tiwari, G.N. and Yadav, Y.P., Performance Evaluation of Biogas Burners, *Energy Conver. and Mgt.*, **32(4)**, 395-401 (1991).

Chang, A.C., Wastewater Recycling in Anaerobic Digestion of Beef Cattle Wastes, *Agricultural Wastes*, **7(1)**, 1-12 (1983).

Chartier, P., *Biomass for Energy, Environment, Agriculture and Industry*, Proc. Eighth European Community Conf., Vienna, Austria, Oct. 3-5, 1994, Elesevier Sci. Ltd. (Pergamon), London, 3 vols., 1995.

Chatterjee, S.N., *et al.*, Gobar Gas as on Alternative Source of Energy: A Case Study, *Khadi Gramodyog*, **37(12)**, 477-483 (1991).

Chauhan, N.D., Fuel from Marine Algal Biomass, *Changing Villages*, **6(5)**, 345-349 (1984).

Chavade, S., *Combined Anaerobic and Aerobic Biological Treatment of Piggery Waste*, M.Sc. thesis, Univ. of Newcastle, Upon Tyne, 1978.

Chavade, S., Anaerobic Filter for Biogas Production, *Regional Jour. of Energy, Heat and Mass Transfer*, **2(1)**, 31-44 (1980).

Chawla, O.P., *et al.*, Note on Stimulation of Anaerobic Fermentation of Dung in the Cow Dung Gas Plants, *Indian Jour. of Agri. Sci.*, **39(11)**, 1040-1043 (1969).

Chawla, O.P., Gobar Gas Plant: Its Efficiency and Manurial Returns, *Khadi Gramodyog*, **18(10)**, 547-549 (1972).

Chawla, O.P., Winter and the Cow Dung Gas Plant, *Indian Farming*, **23(8)**, 29-30 (1973).

Chawla, O.P., *Bioconversion of Agricultural Wastes: Recent Trends*, paper presented at the Nat. Symp. on Biogas Tech. and Usage, IARI, New Delhi, Nov 29-30, 1977.

Chawla, O.P., Utilization of Slurry as a Manure, *Khadi Gramodyog*, **25(5)**, Feb., 240-242 (1979).

Chawla, O.P., Physical and Chemical Properties of Biogas, *Khadi Gramodyog*, **26(7)**, Apr., 295-298 (1980).

Chawla, O.P., Night-Soil for Fuel and Fertilizer, *Indian Farming*, Mar., 1980.

Chawla, O.P., Advances in Biogas Technology, Pub. and Inf. Div., ICAR, New Delhi, 1986.

Chawla, S.K. and Ghosh, T.K., Thermodynamic Analysis of Hydrogen Production for Biogas for Phosphoric Acid Fuel Cell, *Inter. Jour. of Hydrogen Energy*, **17**, 405-412 (1992).

Chayovan, S., *et al.*, Biogas Production from Dairy Manure : The Effects of Temperature Perturbation, *Biological Wastes*, **25(1)**, 1-16 (1988).

Chen, L., Two-Step Fermentation Technology for Biogas Production, *Jour. Agri. Eng. Soc., Dec, No. 6*, **23** (1985).

Chen, Y.R. and Hasimoto, A.G., Kinetics of Methane Fermentation, *Biotech. and Bioeng. Symp.*, **8**, 269-282 (1978).

Chen, Y.R., Kinetic Analysis of Anaerobic Digestion of Pig Manure and Its Design Implications, *Agricultural Wastes*, **8**, 65-81 (1983).

Chengyong, X., *et al.*, Study on the Mixed Plug Flow Biogas Digester with Small Size and High Efficiency, *Biogas Forum (BORDA)*, 2 Special Issue on China, 26-29 (1991).

Chin, K.K. and Goh, T.N., *Bioconversion of Solar Energy : Methane Production Through Water-hyacinth*, Symp. on Energy from Biomass and Wastes, Washington, Aug, 215-228 (1978).

Chiranjivi, C., *Design Analysis of Small Scale Anaerobic Digesters in India*, Proc. Symp. on Clean Fuels from Biomass and Wastes, Washington, Aug., 449-464 (1978).

Chopra, N., Biogas Slurry Boosts Crop Yield : Economics of Slurry Application, *Indian Farming*, **41(4)**, 35-36 (1991).

Chou, T.C. and Weng, H.S., *Purification of Biogas, Proc. Inter. Symp. on Biogas*, Microalgae and Livestock Wastes, Taipei, Taiwan, Rep. of China, Sep 15-17, 1980.

Chowdhury, R.B.S. and Fulford, D.J., Batch and Semi-Continuous Anaerobic Digestion System, *Renewable Energy — An Inter Jour.*, **2(4 & 5)**, 391-400 (1992).

Chung, P.O., *Production and Use of Methane from Animal Wastes in Taiwan*, Proc. of the Inter. Biomass Energy Conf., Winnipeg, Canada, May, 1973.

Chynoweth, D.P., *Methane Production from Marine Biomass*, Proc. Inter. Symp. on Biogas, Microalgae and Livestock Wastes, Taipei, Taiwan, Rep. of China, Sep. 15-17, 1980.

CIBR, *Design and Construction of Large Scale Biogas Digesters in China*, Chengdu Inst. of Biogas Res., Chengdu, P.R. China, Jul., 1980.

Class, D.C., *Anaerobic Digestion for Methane Production : A Stutus Report*, Proc. Bioenergy World Congress and Exposition, Atlanta, GA, 1980, p. 30.

Coe, W.B. and Turk, M., *Processing Animal Wastes by Anaerobic Fermentation*, paper presented at the Inter. Symp. on Processing Agri. and Municipal Wastes, Westport, Connecticut, 1973.

Colber, P.J., *et al.*, *Heat Treatment of Organics for Increasing Anaerobic Biodegradability*, Civil Eng. Tech. Report No. 253, Stanford Univ., California, 1980.

Colleran, E., *Methane from Agricultural Wastes and from Energy Crops*, A Report, Dept. of Microbiology, Univ. College, Galway, Ireland, 1981.

Colleran, E., *et al.*, *Anaerobic Digestion of Agricultural Residues and of Silage Effluent*, Proc. European Community Contractors Meeting on Energy from Biomass, Brussels, May 5, 1982, G. Grazzi and W. Plaz (ed.), Reidel Pub., Dordrecht, 1982.

Colleran, E., *Application of the Anaerobic Filter to Biogas Production from Agricultural Wastes*, A Report, Commission of the European Communities, Luxembourg, Jan., 1984.

Converse, J.C. and Graves, R.E., *Facts on Methane Production from Animal Manure*, Fact Sheet, Univ. of Wisconsin, 1974.

Converse, J.C., *et al.*, *Performance of a Large Size Anaerobic Digester for Poultry Manure*, ASAE Paper 77-0457, Amer. Soc. of Agri. Eng., St. Joseph, Michigan, 1977.

Converse, J.C., Graves, R.E. and Evans, G.W., Anaerobic Degradation of Dairy Manure Under Mesophilic and Thermophilic Temperatures, *Trans. Amer. Soc. of Agri. Eng.*, **20(2)**, 336-340 (1977).

Coombs, J. and Meynell, P.J., Cleaning Biogas, *The Digest (BABA, UK)*, **10**, 5-9 (1983).

Cooney, C. and Wise, D., Thermophilic Anaerobic Digestion of Solid Wastes for Fuel Gas Production, *Biotech. Bioeng.*, **17**, 1119 (1975).

Coppinger, E., *et al.*, Operation of a Dairy Farm Digester Using Available Equipment and Technology, in 'D.L. Wise, *Fuel Gas Production from Biomass*, Vol I, CRC Press, Florida, 1981', 183-200.

CORT, Biogas Slurry Manures : Rural News and Views, *Special Issue of Changing Villages*, **10(1)**, 1991. (CORT : Consortium on Rural Technology), D-320, Laxmi Nagar, Delhi 110092.

CORT, *Alternative Feedstock for Biogas*, A Report of the Two Workshops held during Oct. 29-30, 1990 at Khori and Feb 21-22, 1991 in New Delhi, Consortium on Rural Technology, D-320, Laxmi Nagar, Delhi 110092.

Cramer, F.J., Curing a Sick Digester, *Jour. Water Poll. Control Fed.*, **37(9)**, 1317-1318 (1965).

Crook, C., A Chinese Biogas Manual, Intermediate Tech. Pub., London, 1979.

CRT, *Alternative Feedstock for Biogas*, Proc. of the two workshops organised by the Consortium on Rural Technology, 1992, New Delhi, containing 14 research papers on different feedstock. Papers included are:
- Production of Biogas from Pig Dung.
- Poultry Litter as an Alternative Feedstock for Biogas Production.
- Potential of Biogas Production from Fruit and Vegetable Processing Wastes in India.
- Use of Non-edible Oilseed Cakes for Biogas Production.
- Production of Biogas from Eucalyptus, *Cassia* and *Vilayati* Plants.
- Production of Biogas from Solid Cellulosic Wastes.
- Dry Anaerobic Fermentation of Cowdung : Subabul Leaves Mixture in Multiple Batchfed Digesters.
- Vanaspati Kachra Biogas Plant and its Operation.
- Garbage Gas Manure Plant.
- Sugarcane Filter Cake : A Supplementary Biogas Feedstock.
- Semicontinuous Biogas Plant for Water-hyacinth.

Cuzin, N., Methanogenic Fermentation of Cassava Peel Using a Pilot Plug Flow Digester, *Bioresource Tech.*, **41(3)**, 259-264 (1992).

Dagnall, S., UK Strategy for Centralised Anaerobic Digestion, *Bioresource Tech.*, **52(3)**, 275-280 (1995).

Dange, R.R., Application of Digestion Theory to Digester Control, *Jour. Water Poll. Control Fed.*, **40**, 2021-2032 (1968).

Dahiya, A.K. and Vasudevan, P., Biogas Plant Slurry as an Alternative to Chemical Fertilisers, *Biomass*, **9(10)**, 67-74 (1986).

Dandewade, L.N. and Joglekar, V.R., The Performance of a Night Soil-based Biogas Plant, *Biogas Forum (BORDA)*, **46**, 4-6 (1991).

Das, B., *et al.*, *Effect of Fermentation Temperature and Additives on Efficiency of Biogas Production*, Proc. Nat. Symp. on Biogas Tech. and Usage, IARI, New Delhi, Nov. 29-30, 1977.

Dass, R.S., *et al.*, Biogas as a Replacement Fuel to Diesel Engines, *Agri. Eng. Today*, **2(2)**, 14-19 (1978).

Dave, J.M., *et al.*, *Studies of the Design of Gobar Gas Plant for Rural Areas : Part 1*, All India Seminar on Gobar Gas Plant, KVIC, Mumbai, 1962.

De Baere, L., *et al.*, Anaerobic Fermentation of Refuse, *Resources Conservation*, **14**, 295-308 (1987).

Delgado, M., *et al.*, Methane Generation from Water-hyacinth Biomass, *Jour. of Env. Sci. & Health*, **A 27(2)**, 347-36 (1992).

Demant, D., *et al.*, Possible Applications of Bioslurry for the Purpose of Fertilisation, *Biogas Forum (BORDA)*, I special, 11-18 (1990).

Demant, D., Adaption of Appliances for Biogas Use, *Biogas Forum (BORDA)*, **40**, 13-16 (1990).

Demuynck, M., Nyns, E.J. and Palz, W., *Biogas Plants in Europe : A Practical Handbook*, Reidel Pub. Co., Dordrecht, the Netherland, 1984, p. 339.

Demuynck, M., *et al.*, *Agricultural Uses of Digested Effluents*, Proc. of the Int. Symp. under EEC Programme on Recycling of Urban and Indus. Waste, Luxembourg, May 8-10, 409-422 (1984).

De Poli, F., *et al.*, Membrane Cleaning of Biogas for Injection to Pipelines, in 'Palz, W. *et al.*, *Energy from Biomass*, CEC-DG XII Commission Euro. Communities – Directorate General XII, 1985,' p. 625-629.

Desai, B.P., Combustible Gas From Cattle Dung, *Poona Agri. College Magazine*, **42**, 74-82 (1951).

Desai, S.V. and Biswas, S.C., Manure and Gas Production by Anaerobic Fermentation of Organic Wastes, *Indian Farming*, **6(2)**, 67-71 (1945).

Desai, S.V., Combustible Gas from Cow Dung, *Indian Farming*, **6(2)**, 140-143 (1945).

Deshpande, M.V., Recovery of Vitamin B_{12} from Gobar Gas Plant Effluent Slurry, *Indian Jour. of Microbiology*, **18(4)**, Oct.-Dec., 255-256 (1978).

Deshpande, P., *et al.*, Use of Water-hyacinth as an Additive in Biogas Production, *Current Science*, **48(11)**, Jun., 490-492 (1979).

Deshpande, V., *et al.*, *Utilisation of Cellulosic Waste for Energy Production*, Proc. of the Conf. on Energy From Biomass and Wastes, Univ. of Peradeniya, Sri Lanka, Nov. 26-30, 1979.

Devkota, G.P., Report on Utilization of Waste Heat from Biogas Power Generation, *Biogas Forum*, No. **15**, 3-4 (1983).

Devlesavev, N.D., *et al.*, *Anaerobic Digestion of Cellulosic Wastes*, Proc. 5th Canadian Bioenergy R & D Seminar, Canada, March 26-28, 1984.

Dhone, E., *Gas Storage and Utilisation*, Proc. First Inter. Symp. on Anaerobic Digestion, Cardiff, UK, 1979.

Dhone, E., *Generation of Current from Biogas*, Proc. Ist EC Conf. on Energy from Biomass, Brighton, Nov., 1980, Applied Science, London, 1981, p. 915-921.

Dhussa, A. and Ibrahim, M., *Biogas Sanyantra : Model I Nirman Pustika*, Planning Res. and Action Inst., Lucknow, 1980.

Dhussa, A., *et al.*, *Construction Details of Janata Biogas Plant*, A Manual, Gobar Gas Res. Station, Ajitmal, 1981.

Diaz, L.F., *Energy Recovery Through Biogasification of Municipal Solid Wastes and Utilisation of Thermal Wastes From An Urban Energy Agro-Waste Complex*, Ph.D. thesis, Dept. of Mech. Eng., University of California, Berkeley, 1976.

DNES, *Guide to Biogas Plants*, Department of Non-Conventional Energy Sources, New Delhi, 1985.

Dongbiao, L. and Changshu, M., The Biogas Digester with a Movable Blocking Board Treating the Mixture of Manure and Straw, *Biogas Forum (Jour. BORDA)*, **2nd** Special Issue, 17-19 (1992).

Doshi, D.S., Use of Fibreglass for Gasholders in Biogas Plant, *Jour. Biowaste Treatment*, **1(10)**, 25-27 (1981).

Durand, J.H., *et al.*, Modelling, Stimulating and Optimising the Anaerobic Digestion of Swine Manure, *Biological Wastes*, **24(1)**, 1-15 (1988).

ECOTOPE GROUP, *Process Feasibility Study: The Anaerobic Digestion of Dairy Cow Manure at the State Reformatory Honour Farm*, Manroe, Washington, ECOTOPE GROUP, Seattle, WA, 1975.

Edie, D.L., *Instruction Manual for Installation of Waste Digesters*, Indus. Tech. Res. Inst., Taiwan.

El-Shim, B.A., *et al.*, Biogas Generation for Food Processing Wastes, *Resources, Conserv. & Recycling*, **6(4)**, 315-327 (1992).

ESCAP, *Guidebook on Biogas Development*, United Nations, New York, 1980.

ESCAP, *Updated Guidebook on Biogas Development*, United Nations, New York, 1985, p. 178.

Eusuf, M., *et al.*, Design and Economics of Floating Cover Biogas Plant, Part IV: Construction and Performance of a Cheaper Design, *Bangladesh Jour. of Scientific and Indus. Res.*, **21(1-4)**, Jan.-Oct., 167-175 (1986).

Fair, G.M. and Moore, E.W., Time and Rate of Sludge Digestion and Their Variation with Temperature, *Sewage Works Jour.*, **6**, 3-13 (1934).

Fairbank, W.C., Fuel from Livestock Wastes, *Agri. Eng.*, **H9**, 20-21 (1974).

FAO, *China: Azolla Propagation and Small Scale Biogas Technology*, Soils Bull. No.41, FAO, Rome, 1978.

Felizardo, I.S., Improved Multipurpose Biogas Apparatus, *NSTA Tech. Jour.*, **10(3)**, 81-83 (1985).

Fernandez, N. and Forster, C.F., A Study of the Operation of Mesophilic Anaerobic Filters Treating a Synthetic Coffee Waste, *Bioresource Tech*, **45(3)**, 223 (1993).

Ferranti, M.P., *et al.* (eds.), *Anaerobic Digestion : Results of Resrearch and Demonstration Projects*, Proc. Workshop, Villeneuve d'Asc, France, March, 1986.

Finlay, J.H., *Operation and Maintenance of Gobar Gas Plants*, Report prepared for the United Mission Eco. Dev. Agency, United Mission to Nepal, Butwal, Nepal, 1976.

Fischer, A.J. and Greene, R.A., Plant Scale Tests on Thermophilic Digestion, *Sewage Works Jour.*, **17(4)**, 718-729 (1945).

Fischer, J.R., *et al.*, Anaerobic Digestion of Wastes, in 'Jewell, W.J., (ed.), *Energy Agriculture and Waste Management*, Ann Arbor, Sci. Pub., Ann Arbnor, MI, 1975,' 307-316.

Fischer, J.R., *et al.*, *Design and Operation of a Farm Size Anaerobic Digester for Swine*, paper No.77-4052, paper presented at the Annual Meeting of Amer. Soc. of Agri. Eng., St. Joseph, Michigan, 1977.

Fischer, J.R., *et al.*, Producing Methane Gas from Swine Manure in a Pilot Size Digester, ASAE-Paper MC 77-604, Amer. Soc. of Agri. Eng., St. Joseph, Michigan, 1977.

Fischer, J.R., *et al.*, *Biological and Chemical Fluctuations During Anaerobic Digestion of Swine Manure*, ASAE Paper No.78-4011, paper presented at the Amer. Soc. of Agri. Eng., St. Joseph, MI, 1978.

Fischer, J.R., Anaerobic Digestion Manure from Swine Fed on Various Diets, *Agricultural Wastes*, **3(3)**, 201-214 (1981).

Fletcher, P., *Landfill Gas Extraction — Vertical Wells or Horizontal Trenches*, Proc. Inst. Mech. Engg. I., Mech. E. Conf. 1, 37-46 (1988).

Fogh, S. and Petersen, J.H., *Studies on Anaerobic Conversion of Carbohydrates and Performance of Danish Biogas Plants*, Inst. of Plant Physiology, Univ. of Copenhagen, Copenhagen, 1979.

Fong, W.Y., *Methane Production from High Rate Anaerobic Digestion of Hog and Dairy Cattle Manure*, M.Sc. thesis, Univ. of Manitoba, Winnipeg, Manitoba, Canada.

Foth, M., *Biogas Bibliography on Your Personal Computer*, GmbH, Schiffbauerweg, 4, 2800 Bremen 21, Germany.

Fretlöh, G., Stratification of Slurry Inside the Digester, *Biogas Forum (BORDA)*, **II** No. **45**, 9-10 (1991).

FRI, *Biological Treatment of Industrial Wastes*, Fermentation Res. Inst., Agency of Indus. Sci. and Tech., Japan, 1974.

Friedberg, J.L., Mathane and Propane as Auto Fuels, *Alternative Sources of Energy*, No. **23**, Dec. 12-18, 1976.

Fry, J.L., *Power and Electric Light from Pig Manure*, Farm & Country (London), Apr., 1960.

Fry, J.L., *Methane Power Plants*, DA Knox, USA, 1975.

Fry, J.L., Farmer Turns Pig Manure into Horsepower, *Farmer's Weekly* (S. Africa), Feb. 22, 1961.

Fry, J.L., Methane Gas Power for This Farm is Fuelled by 900 Pigs, *Farmer's Weekly*, Nov. 3, 1961.

Fry, J.L., *Practical Building of Methane Power Plants for Rural Independence*, Standard Printing, Santa Barbara, California, 1974.

Fulford, D.J., Nepali Version of the Chinese Design of Biogas Plant, *Biogas Newsletter*, Nepal, No. **10**, Autumn, 6-7 (1980).

Gaddy, J.L., Park, E.L. and Rapp, E.B., Kinetics and Economics of Anaerobic Digestion of Animal Wastes, *Jour. Water, Air and Soil Poll.*, **3**, 161-169 (1974).

Gadge, S.R. and Madansure, V.N., Development of an Efficient Low Cost Biogas Burners, *New Agriculturist*, **2(1)**, 27-30 (1991).

Gamal-El-Din, H., *et al.*, Microbiology Control of Biogas Plants, *Routine Operation Assistance*, **34(5)**, 261-262 (1983) (in German).

Garcia, J.A., *Water-hyacinth as an Unconventional Energy Resource*, A Report, H. SKAT, San Salvador, 1981, p. 29.

Garde, R.V. and Godbole, S.H., Studies on Anaerobic Digestion of Distillery Wastes, *Chemical Age of India*, **33**, 439-442 (1982).

Gardner, N., *et al.*, Harnessing Landfill Gas from Refuse-Dumping Sites in the U.K., *Applied Energy*, **29(3)**, 201-215 (1988).

Garg, H.P., Family Size Biogas Plant, *Invention Intelligence*, **12(1)**, 149 (1977).

Garg, H.P., Designing a Suitable Biogas Plant for India, *Appropriate Technology*, **7(2)**, 29-31 (1980).

Garg, M.K., *Pilot Project on Gobar Gas*, A Report, PRAD, Lucknow, 1962.

Garry, M.M., *Health Aspects and the Potential for Transmission of Human Diseases through the Biogas Unit*, Inter Dev. Res. Center, Ottawa, Canada.

GATE, *Survey of Biogas Plants*, German Appropriate Technology Exhange, Postfach 5180, D-6236, Eschborn 1, Fed. Rep. of Germany (erstwhile), 1978.

GATE, Biogas From Toilet Wastes, *Biogas Forum (BORDA)*, I No **48**, German Appropriate Technology Exchange, 6237 Eschborn, Germany, 1992.

Gawade, B.R., *Kamadhenu* (New Model of Clay), paper presented at the Panel Meeting on Biogas Tech., Bordoli, Nov. 5-6, 1981.

Georgacakis, D. and Dalis, D., Controlled Anaerobic Digestion of Settled Olive-Oil Wastewater, *Bioresourse Tech.*, **46(3)**, 221 (1993).

Ghanem, K.M., Methane Production from Beet Pulp, *Resources, Conserv. and Recycling*, **6(3)**, 267 (1992).

Ghate, P.B., Biogas: A Pilot Project to Investigate a Decentralized Energy System, Proc. Inter. Seminar on Energy, Jan. 4, 1979, Hyderabad, John Wiley and Sons, NY, 344-369 (1980). Also published in *Eco. Pol. Weekly*, **14(27)**, 1132-1136 (1979).

Ghosh, A.B., *Research and Development of Cow Dung Gas Plant*, Proc. Symp. on Radio Isotopes, Fertilizers and Cow Dung Gas Plant, New Delhi, 1961, p. 364-369.

Ghosh, A.B., et al., Use of Digested Cowdung Slurry in Improving the Soil Structure, *Science Culture*, **34**, 307-308 (1978).

Ghosh, S. and Klass, D.L., *Feasibility Analysis of the Biological Gasification of Solid Wastes*, Ins. of Gas Tech., Chicago, Dec., 1972.

Ghosh S. and Pohland, F.G., Kinetics of Substrate Assimilation and Production Formation in Anaerobic Digestion, *Jour. Water Poll. Control Fed.*, **46**, 748-759 (1974).

Gosh, S. and Klass, D.L., *Two Phase Anaerobic Digestion, Symp. on Clean Fuels form Biomass*, Inst. of Gas Tech., Chicago, IL, 1977.

Ghosh, S., *Kinetics of Acid-Phase Fermentation in Anaerobic Digestion,* Proc. Third Symp. on Biotechnology in Energy Production and Conservation, Inst. of Gas Tech., Chicago, **11**, 301-313 (1981).

Ghosh, S., *et al.*, Methane Production from Industrial Wastes by Two-phase Anaerobic Digestion, *Water Res.*, **19**, 1083-1088 (1985).

Ghosh, T.K., Utilisation of Wastes, *Calcutta Science and Culture*, **25(3)**, 1958.

Ghosh, T.K., *et al., Biochemical Engineering Feasibility of Biogas Production as an Appropriate Technology*, paper presented at the 2nd National Conv. on Env. Eng., New Delhi, Feb., 1976.

Ghosh, T.K., *et al., Increased Methane Production in Biogas*, Biochem. Engg. Res, Center, IIT, New Delhi, 1979.

Gichohi, P.M., Potential of Biogas as an Engine Fuel, *Biogas Forum*, **53**, 19-23 (1993).

Gnanamani, A. and Kasturibai, R., Influence of Biodigested Slurry on Rice-Gram Cultivation, *Bioresource Tech.*, **41(3)**, 217-221 (1993).

Goleuke, C.G., Oswald, W.J. and Gotaas, H.B., Anaerobic Digestion of Algae, *Applied Microbiology*, **5**, 47-55 (1957).

Goleuke, C.G., Temperature Effects on Anaerobic Digestion of Raw Sewage Sludge, *Sewage and Indus. Wastes*, **30**, Oct., 1225 (1958).

Goleuke, C.G. and Oswald, W.J., Biological Conversion of Light Energy to the Chemical Energy of Methane, *Applied Microbiology*, **7**, 219-227 (1959).

Goleuke, C.G. and Oswald, W.J., Power and Solar Energy via. Algae-Produced Methane, *Solar Energy*, **7(3)**, 86-92 (1963).

Goleuke, C.G., *Bioconversion Energy Studies at the University of California* (Berkeley), Proc. of the Conf. on Bioconversion Energy Res., Univ. of Massachussetes, Boston, 1973.

Goleuke, C.G., *Biological Reclamation of Sold Wastes*, Rodale Press, Emmaus, PA, 1977.

Gooding, B., Power from Pigs, *New Zealand Farmer*, Apr., **5**, 4-15 (1976).

Gopalakrishnan, K.V. and Murthy, B.S., Energy for IC Engines from Wastes and Biomass, *Regional Jour. of Energy, Heat and Mass Transfer*, **1**, 265-279 (1979).

Gopalakrishnan, N.K. and Mazumdar, A., *Biogas Programmes Directory*, TERI, Mumbai, 1980, p. 236.

Gosset, J.M., *Heat Treatment of Refuse for Increasing Anaerobic Degradability*, Ph.D thesis, Stanford Univ., California, 1976.

Goswami, K.P., *Development of a 'Kachara' Gas Plant for Biogas and Manure*, Proc. Nat. Symp. on Biogas Tech. and Usage, IARI, New Delhi, Nov. 29-30, 1977.

Goswami, K.P., A *Continuous Fermentation 'Kachara' Gas Plant for Producing Biogas and Manure from Straw-like Wastes*, Proc. FAO Seminar on the Dev. of Complementary Use of Mineral Fertiliser and Organic Materials in India, Vigyan Bhavan, New Delhi, Sep. 14-19, 1978.

Goswami, K.P., *Batch Fermentation Studies on Paddy Straw for Biogas Production*, Proc. Nat. Symp. on Recycling of Residues of Agri. and Industry, Punjab Agri. Univ., Ludhiana, Mar. 28-29, 1980.

Goswami, K.P., A New Batch Scale Unit for Continuous Fermentation of Straw-like Plant Wastes to Produce Biogas and Manure, *Agricultural Wastes*, 1981.

Goswami, K.P., Cooking Efficiency of Biogas as Affected by the Gas Supply Pressure, *Urja*, 263-268 (1983).

Gramms, L.C., Polkowski, L.B. and Witzel, S.A., Anaerobic Digestion of Farm Animal Waste (Dairy Bull., Swine and Poultry), *Trans. Amer. Soc. of Agri. Engrs.*, **14**, 7-11 and 13 (1971).

GRCDA. *Landfill Gas*, Proc. Sixth Symp., Governmental Refuse Collection and Disposal Association, Industry Hills, California, 1983.

Grewal, N.S. and Sagar, B., *Possibilities of Biogas as an Energy Supplement During Peak Hours*, Dept. of Civil Eng., Punjab Agri. Univ., Ludhiana , 1975.

Grewal, N.S. and Sagar, B., Efficient Layout for a Gobar Gas Plant: Waste Disposal Structures, *Progressive Farming,* **12(11)**, Jul 12-13, 1976.

Grewal, N.S. and Singh, M., *Maintenance of Gobar Gas Plants, Dept. of Civil Eng.*, Punjab Agri. Univ., Ludhiana, 1976.

Grewal, N.S., *Higher Biogas During Winter*, Proc. Nat. Seminar on Biogas Tech., Ludhiana, Jul. 9-11, 77-82 (1981).

Grewal, N.S. and Duggal, K.N., *Drying Beds for Biogas Plants*, Proc. Nat. Seminar on Biogas Tech., Ludhiana, Jul. 9-11, 72-76 (1981).

Grewal, N.S. and Sagar, B., Suitability of Soil Stabilized Waste Digestion Pit, *Jour. Inst. of Engrs.*, Env. Eng. Div., **62**, 92-94 (1982).

Grewal, N.S., *et al.*, Gas Distribution Pyne for Biogas, *Progressive Farming*, Feb., 1985.

Gron, G., *Engineering Design of Digesters for the Biogas Research and Development Project in Denmark*, Proc. Inter. Symp. on Anaerobic D, Cardiff, Wales, Sep 17, 1979, Applied Science Pub., Essex, 377-393 (1980).

Guang-qian, C., *et al.*, *A Survey of Rural Biogas Fermentatative Technology in China,* Proc. of the Second European Community Conf. on Biomass, Berlin, Sep. 20, 1982, Applied Science Pub., Essex, 1072-1084 (1983).

Guha, B.R., *et al.*, Fuel Gas and Compost Manure from Water-hyacinth, *Khadi Gramodyog,* **22(11)**, 483-486 (1976).

Gunaseelan, N.V., Parthenium as an Additive with Cattle Manure in Biogas Production, *Biological Wastes*, **21(3)**, 195-202 (1987).

Guo, M. and Ma, Y., Studies on the Decomposition Rate of Raw Materials in Biogas Fermentation, *Microbiology Jour.* (Chinese), **8(4)**, 155-157 (1981).

Gupta, C.L., Biogas Centered Domestic Waste Recycling System, *Urja*, 235-236 (1983).

Gupta, S.C., *Energy Through Solar Aided Biogas Systems*, Proc. Inter. Solar Energy Congress, New Delhi, Jan., 791-796 (1978).

Gyan, S., *Cooking Gas and Manure from Community Latrines for Rural Areas and Slums,* Env. Eng. Div., Annual Meeting of the Inst. of Engrs. (India), 1977.

Haberl, R., *et al.*, *Biogas from Molasses and Raw-Juice Slops*, Proc. of Biogas Status Seminar, Graz, Austria, May 6, 1982. DBV Verlag, Graz, 183-199 (1982).

Habig, C. and Rythev, B.H., Methane Production from the Anaerobic Digestion of Some Marine Macrophytes, *Resource and Conservation*, **8(3)**, 271-279 (1983).

Habig, C., *et al.*, Nitrogen Recycling and Methane Production Using *Gracilaris Tikahiea:* A Closed System Approach, *Resources and Conservation*, **10(4)**, 303-313 (1984).

Hadjitofi, M., *The Kinetics of Gas Production in the Anaerobic Digestion of Cellulose Wastes*, Ph.D thesis, Univ. of London, 1976.

Halderson, J., *Dynamic Response of an Anaerobic Digester with Dairy Cow Manure Substrate*, Ph.D thesis, Purdue Univ., Lafayette, Indiana, 1972.

Hall, E.R., *Automated Monitoring of Methane Fermentation Factors*, 5th Canadian Bioenergy R&D Seminar, Canada, Mar. 26-28, Elsevier, London, 1984.

Hamad, M.A., *et al.*, Evaluation of the Performance of Two Rural Biogas Units of Indian and Chinese Design, *Energy in Agri.*, **1(3)**, Feb., 235-250 (1983).

Hanaki, K., Characterisation of Accumulated Biomass in Anaerobic Filter Treating Various Types of Substrates, *Bioresource Tech.*, **47(3)**, 275-282 (1994).

Hanseen, J.F., et al., Anaerobic Digestion of *Laminaria Spp* and *Ascophyllum Nodosum* and Application of End Products, *Biomass*, **14(1)**, 1-13 (1987).

Hanson, G., Methane Production from Marine Green Microalgae, *Resources and Conservation*, **83**, 185-194 (1983).

Hao, P.L.C., *Design and Development of Red Mud Plastic Digester*, Proc. Inter. Symp. on Biogas, Microalgae, and Livestock Wastes, Taipei, Taiwan, Rep. of China, Sep. 15-17, 1980.

Harper, J.H. and Seckler, D.W., *Engineering and Economic Overview of Alternative Livestock Waste Utilisation Techniques*, paper presented at the Inter. Symp. on Livestock Wastes, Urbana, Illinois, Apr. 21-24, 1975.

Hashimoto, A.G., *et al.*, *Anaerobic Fermentation of Crop Residues*, US Animal Mean Animal Res. Center, Clay Center, NE, 1977.

Hashimoto, A.G., *et al.*, Anaerobic Fermentation of Agricultural Residue, in 'M.L. Shuler (ed.), *Utilization and Recycle of Agricultural Wastes and Residues*, CRC Press, Boca Raton, Florida,' 136-196 (1980).

Hashimoto, A.G., Themophillic and Mesophillic Anaerobic Fermentation of Swine Manure, *Agricultural Wastes*, **6(3)**, 175-191 (1983).

Hashimoto, A.G. and Robinson, S.A., Pilot Stage Operation and Economic Assessment of Two Stage Straw Manure Fermentation System, *Resources and Conservation*, **12(1)**, 29-45 (1985).

Hashimoto, A.G., Ammonia Inhibition of Methanogenesis from Cattle Wastes, *Agricultural Wastes*, **17(4)**, 241-261 (1986).

Hassan, A.E., *et al.*, Energy Recovery and Fuel Production from Poultry Wastes, in 'W.J. Jewell (ed.), *Energy, Agriculture and Waste Management,* Ann Arbnor Sci. Pub. Ann Arbor,' 289-303 (1975).

Hassan, A.E., *Characterisation of Methane Production from Poultry Wastes*, Proc. Third Inter. Symp. on Livestock Wastes, Univ. of Illinois, Urbana-Champaign, 21-24, Apr., 1975, ASAE, St. Joseph, Michigan, 244-247, 251 (1975).

Hattingh, W.H.J. and Toercin, D.F., Anaerobic Digestion : 1 – the Microbiology of Anaerobic Digestion, *Water Research,* **3**, 385 (1969).

Hawkes, D.L., Horton, R. and Stafford, D.A., Application of Anaerobic Digestion to Produce Methane Gas and Fertiliser from Farm Wastes, *Process Biochemistry,* **12(2)**, 32-36 (1976).

Hawkes, D.L. and Horton, R., Anaerobic Digester Design Fundamentals: Part I, *Process Biochemistry*, **14(9)**, 10-12 (1979).

Hawkes, D.L., *Factors Affecting Net Energy Production from Anaerobic Digesters*, Proc. 1st Inter. Symp. on Anaerobic Digestion, AD 79, Cardiff, Sep., 1979.

Hawkes, D.L., *Computers Aided Design of Anaerobic Digesters for Energy Production*, Proc. 1st EC Conf. on Energy from Biomass Brighton, London, Applied Science Pub., 429-434 (1981).

Hawkes, F.R., *et al.*, Mesophillic Anaerobic Digestion of Cattle Slurry after Passage through a Mechanical Separator: Factors Affecting Gas Yield, *Agricultural Wastes*, 10, 241-250 (1980).

Hayes, T.D., *et al.*, *Methane Generation from Small Scale Dairy Farms*, paper presented at the Seminar on Biogas and Alcohol Fuels, Chicago, Illinois, Oct. 25, 1979.

Hayes, T.D., *et al.*, *Anaerobic Digestion of Cattle Manure*, First Inter. Symp. on Anaerobic Digestion, Cardiff, 1979.

Hayes, T.D., *Methane Generation from Small Scale Farms*, Proc. of the Seminar on Biomass Energy for City, Farm and Industry, Emmaus, JG Press, 88-117 (1980).

Hein, M.E., *Anaerobic Digestion of Beef Manure and Corn Straw*, paper presented at the Winter Meeting of the Amer. Soc. of Agri. Engrs., Chicago, Dec. 15-18, 1975.

Held, W. and Konig. A., *Biogas Driven Engines*, Proc. 2nd EC Conf. on Energy from Biomass, Berlin, Sep. 20-23, 1982, Applied Science Pub., London, 511-516 (1982).

Henrich, R.A., Advances in Biogas to Fuel Conversion, *Biocycle*, **24(40)**, 28-31 (1983).

Hernandez, E.P.S., Effects of Pyrite Addition on Anaerobic Digestion, *Bioresource Tech.*, **47(2)**, 189-190 (1994).

Hills, D.J. and Schroeder, E.D., Temperature Effects on the Rate of Methane Fermentation, *Water Sewage Works*, **6**, Jul., 46 (1969).

Hills, D.J. and Dykstra, R.S., Anaerobic Digestion of Cannery Tomato Solid Wastes, *Jour. of Env. Eng. Div.*, ASCE, **106,** 257 (1980).

Hills, D.J., Methane Gas Production from Dairy Manure at High Solids Concentration, *Trans. of the ASAE*, **23**, 122-126 (1980).

Hills, D.J. and Roberts, D.W., Anaerobic Digestion of Dairy Manure and Field Crop Residues, *Agricultural Wastes*, **3**, 179-189 (1981).

Hill, D.T., Modelling Techniques and Computer Simulation of Agricultural Waste Treatment Processes, *Agricultural Wastes*, **2(2)**, 135-156 (1980).

Hill, D.T., Design Parameters and Operating Characteristics of Animal Waste Anaerobic Digestion Systems: Swine and Poultry, *Agricultural Wastes*, **5**, 157-178 (1983).

Hill, D.T., Simplified Monod Kinetics of Methane Fermentation of Animal Wastes, *Agricultural Wastes*, **5**, 1-16 (1983).

Hill, D.T., Design Parameters and Operating Characteristics of Animal Waste Anaerobic Digestion Systems: Beef Cattle, *Agricultural Wastes*, **5(4)**, 205-218 (1983).

Hilpert, R., Agricultural Feed Additives and Disinfectants as Inhibitory Factors in Anaerobic Digestion, *Agricultural Wastes*, **10(2)**, 103-116 (1984).

Hitte, S.J., Anaerobic Digestion of Solid Wastes and Sewage Sludge into Methane, *Compost Science*, Jan.-Feb., 26-30 (1976).

Hobson, P.N., The Anaerobic Digestion of Waste from an Intensive Pig Unit, *Water Research*, **7**, 437-449 (1973).

Hobson, P.N., The Bacteriology of Anaerobic Sewage Digestion, *Process Biochemistry*, **8**, 19-25 (1973).

Hobson, P.N., The Bacterial Population of Piggery Waste Anaerobic Digesters, *Water Research*, **8**, 507-516 (1974).

Hobson, P.N., Bouefield, S. and Summers, R., Anaerobic Digestion of Organic Matter, *CRC Critical Reviews in Env. Control*, June, 131-191 (1974).

Hobson, P.N., Robertson, A.M. and Mills, P.J., Anaerobic Digestion of Agricultural Wastes, *ARC Review*, **1**, 82-85 (1975).

Hobson, P.N. and Shaw, B.G., Inhibition of Methane Production by *Methanobacterium Formicicum*, *Water Research*, **10**, 849-859 (1976).

Hobson, P.N., Production of Methane from Wastes and Crops, in '*Processes for Chemicals from Renewable Raw Materials*, Inst. of Chem. Engrs., 1979', 1-8.

Hobson, P.N., Biogas Production from Agricultural Wastes, in '*Energy from Biomass*, Watt Committee on Energy, Report No. 5, London', 37-44.

Hobson, P.N., Straw as Feedstock for Anaerobic Digesters in 'E. Crossbard (ed.), *Straw Decay and its Effect on Disposal and Utilisation*, John Wiley, 1979,' 217-224.

Hobson, P.N., Bousfield, S., Summers, R. and Mills, P.J., Anaerobic Digestion of Farm Animal Waste, in 'J.C.. Hawkins (ed.), *Engineering Problems with Effluents from Livestock*, Commission of the European Communities, 1979,' 492-506.

Hobson, P.N., *et al.*, The Digestion of Piggery and Poultry Wastes, in 'Stafford, D. (ed.) *Anaerobic Digestion*, Applied Science Pub., 1980,' 237-253.

Hobson, P.N. and McDonald, I., Methane Production from Acids in Piggery Waste Digesters, *Jour. Chem. Tech. Biotech.*, **30**, 405-408 (1980).

Hobson, P.N., Bousfield, S. and Summers, R., *Methane Production from Agricultural and Domestic Waste*, Applied Science Pub., Apr, 1981.

Honda, K., Studies Concerning the Methane Fermentation Treatment and the Utilisation of Livestock Excrement, *Bull. of the Kawagawa Livestock Experimental Station*, **71**, 15-38 (1981) (In Japanese).

Honda, K., Methane Fermentation and Gasification Technology of Livestock Excreta, *Animal Husbandary*, **36(6)**, 765-771 (1982) (In Japanese).

Horton, R. and Hawkes, D.L., Anaerobic Digester Design Fundamentals: Part II, *Process Biochemistry*, **14(9)**, 12-16 (1979).

House, D., *Biogas Handbook*, Peace Press, California, 1981.

Huang, S.Y., *et al.*, *Economic Evaluation of Fuel Gas Production by Anaerobic Digestion of Swine Waste*, Proc. Inter. Symp. on Biogas, Microalgae and Livestock Wastes, Taipei, Rep. of China, Sep. 15-17 (1980).

Huang, S.Y., *et al.*, *Improvement of Digester Performance Based on Kinetic Studies on Swine Waste*, Proc. Inter. Symp. on Biogas, Microalgae and Livestock Wastes, Taipei, Rep. of China. Sep. 15-17, 1980.

IARI, *Cow Dung Gas Plant: Instruction for Installation*, Operation and Maintenance, IARI, New Delhi, 1979.

Ichikawa, S. and Kanai, M., Studies on Methane Fermentation added with Carbon Dioxide, *Jour. of Japan Sewage Works Assoc.*, **19(216)**, 62-67 (1982).

Idnani, M.A., Some Improvements in the Biogas Plant, *Indian Farming*, **12(11)**, 24-25 (1963).

Idnani, M.A. and Acharya, C.N., *Biogas Plants, their Installation, Operation, Maintenance and Use*, ICAR, New Delhi, 1963.

Idnani, M.A. and Chawla, O.P., Biogas Plants Slurry is Now Easier to Dispose off, *Indian Farming*, 6-13 (1964).

Idnani, M.A. and Chawla, O.P., Improve the Performance of Your Biogas Plant in Winter, *Indian Farming*, **19**, 37-38 (1969).

Ifeadi, C.N. and Brown, J.B., Technologies Suitable for Recovery of Energy from Livestock Manure, in 'Jewell, W.J. (ed.), *Energy, Agri. and Waste Management*, Ann Arbor Sci. Pub., MI, 1975'.

Imhoff, K.. Digester Gas for Automobiles, *Sewage Works Jour.*, **18(1)**, 17-25 (1946).

Imhoff, K. and Keefer, C., Sludge Gas as a Fuel for Motor Vehicles, *Water Sewage Works*, **99**, 284 (1952).

Inglett, G.E. (ed.), *Symp. on Processing Ag. ultural and Municipal Wastes*, Avi. Pub. Co., Westport, CT, 1973.

IOB, *Electricity Generation with Biogas*, Institute of Biogas, Chinese Acad. of Agri. Sci., P.R. China, Jul., 1959.

Isaacson, R., *Methane from Community Wastes*, Elsevier Applied Sci, London, 1992.

ISF, *Utilisation of Biogas Fermentation Residue-Sludge and Effluent*, A Report, Inst. of Soil and Fertilizer, Sichuan, P.R. China, Jul., 1980.

Islam, M.N., Production of Biogas from Different Types of Wastes, paper presented at the Seminar on '*Feed and Energy from Unconventional Sources*, Dacca, Jul., 1976.'

Islam, M.N., *Design, Construction and Operation of a Biogas Plants Suitable for the Rural Area of Bangladesh*, Report Submitted to the Univ. Grants Commission, Dacca, 1980.

Iwai, S., *et al.*, *Experimental Studies on High-rate Digestion of Night-Soil*, Inter. Conf. on Water Poll. Res., London, 1962.

Jagdeesh, A. and Vaiderbhi, A., Biogas Research in India, in 'T.N. Veziroglu (ed.), *Alternative Energy Sources V : Part-D — Biomass, Hydrocarbon, Hydrogen*, Elsevier, Amsterdam, 1983,' 149-161.

Jagdeesh, K.S. and Geeta, C.S., Supplementing Soyabean Waste for Augmenting Biogas Production, *Urja*, **28(6)**, 63-64 (1990).

Jain, M.C., *Conversion of Petrol Engine to Run on Gobar Gas*, Proc. Nat. Symp. on Biogas Tech. and Usage, IARI, New Delhi, Nov. 29-30, 1977.

Jain, M.C., *Low Cost Burners for Biogas*, Proc. Nat. Symp. on Biogas Tech. and Usage, IARI, New Delhi, Nov. 29-30, 1977.

Jain, M.C., *Small Service Holder-cum-Pressure Regulator for Drumless Biogas Plant*, Proc. Nat. Seminar on Biogas Tech., Ludhiana, Jul 9, 1981, S.K. Vyas and N.S. Grewal (eds.), USG Pub. and Distributors, Ludhiana, 103-107, 1981.

Jain, M.C., and Kumar, S., Semi-Continuous Laboratory Digester for Biogas Studies, *Indian Research*, Sep., 203-205 (1986).

Jain, M.K., *et al.*, Anaerobic Digestion of Cattle and Sheep Wastes, *Agricultural Wastes*, **3(1)**, 65-73 (1981).

Jain, M.K., *et al.*, Cattle Wastes Anaerobic Digestion in a Semi-Continuous and Batch Digester, *Agricultural Wastes*, **7(4)**, 251-259 (1983).

Jain, S.K., *et al.*, Production of Biogas from Aquatic Biomass : A Comparison with Terrestrial Biomass, *Research and Industry*, **35(2)**, 104-107 (1990).

Jain, S.K., *et al.*, Production of Biogas from *Azolla Pinnate R. Br* and *Lemma Minor L.* : Effect of Heavy Metal Contamination, *Bioresource Tech.*, **41(3)**, 273-277 (1992).

Jain, S.K., *et al.*, Modelling of Hydrolysis Controlled Anaerobic Digestion, *Jour. of Chem. Tech. Bio. Tech.*, **53(4)**, 337-344 (1992).

Jain, S.P., Gobar Gas : A Study of the Working of Plants in Andhra Pradesh, Haryana and Karnataka, *Rural Dev. Digest*, 231-274 (1980).

Jayachandran, S. and Kasturibai, R., *Influence of Different Organic Wastes on Gobar Gas Production*, paper presented at the 19th Annual Conf. of Microbiologists in India, Baroda.

Jeffrey, E.A., Blackman, W.C. and Ricketts, R.L., *Aerobic and Anaerobic Digestion Characteristics of Livestock Wastes*, Engg. Series Bull. No. 57, Univ. of Missouri, Columbia, Missouri, 1964.

Jeffrey, E.A., *et al.*, Treatment of Livestock Wastes : A Laboratory Study, *Trans. Amer. Soc. of Agri. Engrs.*, 113-117 (1965).

Jewell, W.J., *et al.*, *Methane Generation from Agricultural Wastes: Review of Concept and Future Applications*, Paper No. NA 74-107, presented at the Northeast Regional Meeting of ASAE, Morgantown, West Virginia, Aug. 18-21, 1974.

Jewell, W.J., and Morris, G.R., *et al.*, The Economic and Technical Feasibility of Methane Generation from Agricultural Wastes, in 'P.J. Catania (ed.), *Food, Fuel and Fertiliser*, Canadian Plains Res. Centre, Regina, Saskatchewan, Canada, 1974,' 132-164.

Jewell, W.J., *Energy from Agricultural Waste : Methane Generation*, Agri. Eng. Extension Bull. No. 397, New York State College of Agri. and Life Sci., Cornell Univ., Ithaca, NY, 1974.

Jewell, W.J., and Morris, G.R., Technological Suitability for the Recovery of Energy from Livestock Manures, In 'Jewell, W.J. (ed.), *Energy, Agri. and Waste Management*, Ann Arbor Sci. Pub., Ann Arbor, MI, 1975.'

Jewell, W.J., *et al.*, *Bioconversion of Agricultural Wastes for Pollution Control and Energy Conservation*, Final Report, New York State College of Agri. and Life Sci., Cornell Univ., Ithaca, NY, Sep., 1976.

Jewell, W.J., *Small Farm Methane Generation Potential*, paper presented at the Second Annual Symp. on Fuels from Biomass, Rensselaer Polytechnic Inst., Troy, NY, Jun. 20-22, 1978.

Jewell, W.J., Anaerobic Fermentation of Agricultural Residues into Biogas, *New York's Food and Life Sciences*, **11**, 1978.

Jewell, W.J., and Switzenbaum, M.S., Agricultural Wastes, Jour. Water Poll. Control Fed., *Literature Review*, **50(6)**, 1245-1258 (1987).

Jewell, W.J., *et al.*, *Anaerobic Fermentation of Agricultural Residue : Potential for Improvement and Implementation*, Final Report HCP/T2981-07, UC-61, prepared for US Dept. of Energy, Dept. of Agri. Eng., Cornell Univ., Ithaca, NY, 1978.

Jewell, W.J., and Morris, J.W., Agricultural Wastes, *Jour. Water Poll. Control Fed. Literature Review*, **51(6)**, 1360-1384 (1979).

Jewell, W.J., *et al.*, *Design Manual for Anaerobic Digesters*, Dept. of Agri Eng., Cornell Univ., Ithaca, NY, 1978.

Jewell, W.J., *et al.*, *Low Cost Methane Generation on Small Farms*, paper presented at the Third Annual Symp. on Biomass Energy Systems, Solar Energy Res. Inst., Golden, Colorado, 1980.

Jewell, W.J., *et al.*, *Low Cost Approach to Methane Generation, Storage, and Utilisation from Crop and Animal Residues*, paper presented at the Symp. on Biotech. in Energy Production, Amer. Inst. of Chem. Eng., Chicago, Illinois, Nov. 17, 1980.

Jewell, W.J., *et al.*, *Dry Anaerobic Methane Fermentation*, Proc. of the Second Seminar on Biomass Energy for City, Farm, and Industry, Biogas and Alcohol Fuels Production , Vol 2, Chicago, IL, Mar. 13, 1981.

Jewell, W.J., *Cogeneration of Electricity and Heat from Biomass*, Sixth Quarter Program Report for Period from 16.12.82 to 15.3.83, Dept. of Agri. Eng., Cornell Univ., Ithaca, NY, 1983.

Jeybal, A. and Thanu, N.M., Biogas Generation from Glue Industry Wastes, *Urja*, **29(6)**, 67-69 (1991).

Jianxiang, Z., *et al.*, Upflow Biogas Fermentor with Floating Gasholder, *Biogas Forum (BORDA)*, **2** Special Issue on China, 39-42 (1991).

Jaishi, S. and Reng, Z., A Study on Applying Biogas Fertiliser to Present Plant Diseases and Insect Pests, *Biogas Forum (BORDA)*, **2** Special Issue on China, 49-52 (1991).

Joshi, D.S., Use of Fibre Glass for Gasholder in Biogas Plant, *Jour. of Biowaste Treatment*, **1(11)**, 25-27 (1981).

Joshy, D., Low Cost Biogas Digester, *Biogas Newsletter (Nepal)*, **1**, Summer, 2-3 (1978).

Kader, J., *et al.*, *Tapioca Wastes as a Source of Energy: Preliminary Work on Biogas Production*, Energy '80 Seminar, Kualalumpur, Aug. 21-23, 1980.

Kakkar, V.K. and Sharma, P.B., Effect of Utilisation of Cattle Dung for Producing Gobar Gas Economically in Winter Month, *Jour. of Res. (Haryana Agri. Univ.)* **15**, 112-119 (1978).

Kakkar, V.K., *A Simple and Economical Model of Biogas Plant*, Proc. Nat. Seminar on Biogas Tech., Ludhiana, Jul. 9-11, 128-135 (1981).

Kalia, A.K., *Findings of the Drumless Biogas Plant*, Proc. Nat. Solar Energy Conv., Bangalore, Jan. 22-24, 1982, IISc, Bangalore, 1982.

Kalia, A.K., *Bioconversion of Lantana Canara to Biogas*, Proc. Nat. Solar Energy Conv., Baroda, Dec. 15-18, 183-184 (1983).

Kalia, A.K. and Kanwar, S.S., Filtration pit for Dewatering Biogas Plant Slurry in Hills, *Biogas Forum (BORDA)*, **I** No **56**, 19-24 (1994).

Kalia, A.K., Development and Evaluation of Fixed Dome Plug Flow Anaerobic Digester, *Biomass*, **16**, 225-235 (1988).

Kalia, V.C., *et al.*, Biomethanation of Plant Materials, *Bioresource Tech.*, **41(3)**, 209-212 (1992).

Kamata, S. and Uchida, K., *Studies on Livestock Excreta Disposal by Methane Fermentation: Optimum Load & Comparison Between Mesophilic & Thermophilic Fermentation*, Dept. of Animal Hygiene, Nippon Veterinary & Zootechnical College, Japan.

Kanakasabapathy, T., Gobar Gas Holder Built of Ferrocement, *Indian Concrete Jour.*, **49(12)**, 355-359 (1975).

Kang, H., Ultimate Anaerobic Biodegradability of Some Agro-Industrial Waste, *Bioresource Tech.*, **43(2)**, 107-111 (1993).

Kanotra, S. and Mathur, R.S., Biodegradation of Paddy Straw with Cellulolytic Fungi and its Application on Wheat Crop, *Bioresource Tech.*, **47(2)**, 185-188 (1994).

Kanwar, S.S. and Kalia, A.K., Anaerobic Fermentation of Sheep Dropping for Biogas Production, *World Jour. of Microbiology and Biotechnology*, **9**, 174-175 (1993).

Kanwar, S.S., Nayyar, H. and Walia, D.P., Influence of Biogas Slurry on Germination and Early Seedling Growth of Bread Wheat, *Biogas Forum (BORDA)*, **54**, 10-12 (1993).

Kanwar, S.S. and Guleri, R.L., Influence of Recycled Slurry Filtrate on Biogas Production in Anaerobic Digestion of Cattle Dung, *Biogas Forum*, **II** No **57**, 1994.

Kanwar, S.S. and Guleri, R.L., Effect of Recycling of Digested Slurry on Biogas Production, *Biogas Forum*, **IV** No **59**, 12-13 (1994).

Kanwar, S.S. and Guleri, R.L., Biogas Production from Mixtures of Poultry Litter and Cattle Dung with an Acclimatised Inoculum, *Biogas Forum*, **I** No **60**, 21-23 (1995).

Karbue, I., *et al.*, Methane Production from Waste Waters by Immobilized Methogenic Bacteria, *Biotechnology and Bioeng.*, **22(4)**, Apr., 847-857 (1980).

Karim, A., Microbiological Decomposition of Water-hyacinth, *Jour. of Soil Science*, **66**, 401-416 (1968).

Kashimoto, A.G., Methane from Swine Manure : Effect of Temperature and Influent Substrate Concentration on Kinetic Parameters (K), *Agricultural Wastes*, **9**, 299-308 (1984).

Kasturibai, R., *et al.*, *Influence of Different Organic Wastes on the Intermediate Products of Fermentation, Gobar Gas Production and Manurial Value*, paper presented at the Inter. Seminar on Management of Env., Mumbai, Feb., 1980.

Kasturibai, R., and Tirumalasundari, T., *Anaerobic Fermentation of Biomass : A Study on Two Phase Digestion Process : Part I*, paper presented at Biogas Production Tech. Seminar, T.N. Agri. Univ., Coimbatore, May 13-14, 1981.

Kasturibai, R., *et al.*, *Assessment of Various Available Organic Wastes for Bio-energy Using Single-phase and Two-phase Anaerobic Digestion*, paper presented at the 6th Inter. Conf. of Women Engrs. and Scientists, Mumbai, Sep. 7-12, 1981.

Kasturirangan, K., *et al.*, *Use of Gobar Gas as Diesel Engine Fuel*, Proc. 3rd Nat. Conf. on IC Engines and Combustion, Univ. of Roorkee, Dec., 177-188 (1976).

Kaveti, S., *Biogas Burners : Design and Parametric Studies*, M. Tech. thesis, Centre for Energy Studies, IIT, Delhi, Dec., 1991.

Kate, T., Low Cost Biogas Plants in Clay, *Moving Technology*, **4(5)**, 13-16 (1989).

Kate, T., Utilisation of Anaerobically Digested Biogas Effluent Slurry as Sludge, *Changing Villages*, **10(1)**, 9-17 (1991).

Kaur, K. and Sehgal, G.K., Effect of Biogas Slurry on Ovarian Maturation of Common Corp, *Biomass Tech.*, **44(1)**, 89-91 (1993).

Kawahara, A.A. and Williams, P.M., An Application of Dempster-Shafer Theory to the Assessment of Biogas Technology, *Energy*, 205-214 (1992).

Ke-Yun, D. (ed.), *Anaerobic Digestion-1985*, Proc. 4th Intes. Symp., Guangzhou, PR China 1985.

Khan, R.R. and Gupta, R.C., Wealth and Waste : Water Hyacinth for Benefit of Man, *Yojana*, **21(1)**, 23-24 (1977).

Khandelwal, K.C., *Influence of pH on Gobar Gas Production*, paper presented at the 15th Annual Conf. of Assoc. of Microbiologists of India, Bangalore, Dec., 1974.

Khandelwal, K.C., Mechanism of Methane Production and Design Criteria, *Agri. Eng. Today*, **1(9)**, 17-18 (1977).

Khandelwal, K.C., *Technical Feasibility of Dome-Shaped Biogas Plant*, Proc. Nat. Symp. on Biogas Tech. and Usage, IARI, New Delhi, Nov. 29-30, 1977.

Khandelwal, K.C., Dome-Shaped Biogas Plant : Waste Recycling, Methane Gas Production and Equipment, *Compost Science*, **19(2)**, 22-23 (1978).

Khandelwal, K.C., This Biogas Plant is for you, *Intensive Agri.*, **16(12)**, 28-30 (1979).

Khandelwal, K.C., *Janata Model Gobar Gas Plant*, AFPRO, New Delhi, 1979.

Khandelwal, K.C., Biogas System and Production of Manure, in '*Compost Technology*, Project Field Document No.13, 183-190 (1980).'

Khandelwal, K.C. and Mahdi, S.S., *Biogas Technology : A Practical Handbook*, Tata McGraw-Hill Pub. Co. Ltd., New Delhi, 1986.

Khandelwal, K.C. and Maithani, P.C., *Guide to Family Type Biogas Plants*, DNES, New Delhi, 1989.

Khandeparkar, V.G., Biogas from Willow-Dust for Textile Mills, *Indian Cloth Manufac. Fed. Jour.*, Dec. 1-5, 1981.

Khandeparkar, V.G., *Production of Biogas from Willow-Dust : Present and Future*, Proc. of the 20th Tech. Conf. on Energy Conservation and Management in Textile Industry, BTRA, Mumbai, Feb. 3-4, 123-130 (1984).

Khanna, S.K., Gobar Gas Plant, *Science Reporter*, **12(4)**, 186-187 (1975).

Kiely, P.V., *Anaerobic Digestion of Slurry from Cattle-fed High Roughage Diets*, Proc. of the Inf. Symp. under EEC Programme on Recycling of Urban and Indus. Waste, Luxembourg, May 8-10, 1984, Elsevier Applied Science, 251-258 (1984).

Kimber, M., *Anaerobic Filter Treatment For the Soluble Fraction of a Pig Waste*, M.Sc. thesis, Univ. of Newcastle, Upon Tyne, 1976.

Kishor, J., Goyal, I.C., Sawhney, R.L., Singh, S.P., Sodha, M.S. and Dayal, M., Solar Assisted Biogas Plants III : Energy Balance of Fixed Dome Biogas Plants and Enhancement of Production in Winters, *Inter. Jour. of Energy Res.*, **12(4)**, 711-737 (1988).

Kishore, V.V.N., Raman, R. and Rangarao, V.V., *Fixed Dome Biogas Plants : A Design, Construction and Operation Manual*, TERI, New Delhi, June, 1987.

Kivaisi, A.K., *et al.*, Conversion of Cereal Residues into Biogas in a Rumen-Derived Process, *World Jour. of Microbiology and Biotech.*, **8**, 428-433 (1992).

Kiwitt, W., The Flexible Biogas Plant FLXI, *Biogas Forum (BORDA)*, **III (54)**, 23-26 (1993).

Klass, D.L., Ghosh, S. and Conard, J.R., *Conversion of Grass to Fuel for Captive Use*, paper presented at the Symp. on Clean Fuels from Biomass, Sewage, Urban Refuse and Agri. Wastes, Orlando, Florida, 27-30 (1976).

Klass, D.L. and Ghosh, S., *Anaerobic Digestion of Macrocystic Pyrifera Under Mesophilic Condition*, Symp. Proc. on Clean Fuels from Biomass Wastes, Orlando, Inst. Gas Tech., Chicago Illinois, 1977.

Klass, D.L. and Ghosh, S., *Methane Production by Anaerobic Digestion of Water-hyacinth*, paper presented at the Amer. Chem. Soc. Meeting, Las Vegas, Aug., 1980.

Klass, D.L. (ed.), *Energy from Biomass and Wastes X*, Proc. Symp., Washington DC, Elsevier Appl. Sci. Pub., London, UK, 1987.

Klass, R., *Model as an Instrument for Determining the Optimal Layout of Biogas Plants with Regard to Technical and Economical Aspects*, Proc. 2nd EC Conf. on Energy from Biomass, Sep. 20-23, 644-647 (1982).

Klass, R., High Rate Plants for Anaerobic Treatment of Wastewater and Production of Biogas, *Biogas Forum (BORDA),* **I** No **44**, 4-14 (1991).

Klein, S.A., Anaerobic Digestion of Solid Wastes, *Compost Science*, **13(1)**, 6-11 (1972).

Köberte, E., *et al.*, De-sulphurisation of Biogas, *Biogas Forum (BORDA)*, **II** No **49**, 18-22 (1992).

Kon, M.T. and Chiu, T.H., *Testing Methane as a Fuel for Gasoline Engines*, Proc. Inter. Symp. on Biogas, Microalgae and Livestock Wastes, Taipei, Taiwan, Rep. of China, Sep. 15-17, 1980.

Konstandt, H.G., *Engineering, Operation and Economics of Methane Gas Production*, Proc. Seminar on Microbial Energy Conversion, Erich Goltze, Gottingen, Germany, 379-398 (1976).

Kositrana, N., *Anaerobic Digestion Study of Distillery Waste*, Master's thesis, AIT, Bangkok, 1981.

Kothandaraman, C.P., *Prospects of Biogas and Producer Gas as Alternate Fuels for Engines*, Proc. of the Workshop on 'Alternate Fuels for Engines, BHU, Varanasi, Mar. 3-4, 1979.'

Kotze, J.P., *et al.*, A Biological and Chemical Study of Several Anaerobic Digestion, *Water Research*, **2**, 195-213 (1968).

Kotze, J.P., *et al.*, Anaerobic Digestion-II: The Characterisation and Control of Anaerobic Digestion, *Water Research*, **3(7)**, 459-493 (1969).

Keratidadpanya, P., *Design Parameters for Biogas Production in Tropical Rural Villages*, M.E. thesis, AIT, Bangkok, Thailand, 1978.

Krebs, H.A., The Impact of Stable Disinfectants and Veterinary Medicament on the Production of Biogas in Livestock Husbandry, *Biogas Forum (BORDA)*, **II** No **45**, 11-13 (1991).

Krishna, C.V., *Kalinga Biogas Plant: An Economic Design of Fixed Dome Plant*, paper presented at the 17th Annual Convention of the Indian Soc. of Agri. Engrs., Karnal, Feb. 26-28, 1981.

Krishna, C.V. and Kar, B.K., *Gyatri Fibreglass Reinforced Plastic Dome Biogas Plant*, Proc. Nat. Solar Energy Convention, Baroda, Dec. 15-18, 175-176 (1983).

Krishna, C.V. and Kar, B.K., *Bajwa : KVIC Model Biogas Plant*, Proc. Nat. Solar Energy Convention, Baroda, Dec. 15-18, 181-182 (1983).

Krishna, K., Installation of Gobar Gas Plants, *Kurukshetra*, **22(1)**, 9 (1974).

Krishnan, J., Biogas from Water-hyacinth, *Economic Times*, Calcutta, Sep. 23, p. 4 (1985).

Krishnappa, A.M., *et al.*, Gobar Gas Plant Sludge is Superior to FYM, *Current Science*, **8(3)**, 41-42 (1979).

Krishnaswamy, K.V., *et al.*, Reinforced Concrete Biogas Plant, *Indian Jour. of Env. Health*, **23(1)**, 41-52 (1981).

Korekar, E.J., *et al.*, *Methane Production from Animal Wastes II*, Paper No. 76-208, Dept. of Agri. Eng., Univ. of Manitoba, Winnipeg, Manitoba, 1976.

Kulkarni, M.K., *Kirloskar Gobar Gas Engine*, Workshop Proc. on Biogas Utilisation on Engines, IIT, Dehradun, Apr. 19-20, 1978, 19-20

Kulkarni, M.K., Kirloskar Dual-Fuel Biogas Engines, *Khadi Gramodyog*, **27(3)**, 188-189 (1980).

Kumar, A., *et al.*, Solar Assisted Biogas Plants II: Energy Balance of Floating Drum Type Biogas Plants, *Inter. Jour. of Energy Research*, **12(2)**, 253-273 (1988).

Kumar, A., *Biogas from Biomass Waste*, M. Tech. thesis, Centre for Energy Studies, IIT, Delhi, May. 1994.

Kumar, K., Installation of Gobar Gas Plants, *Kurushetra*, **22(15)**, 9-10 (1974).

Kumar, L., A 600-Rupees Janata Biogas Plant, *Science for Villages*, **65,** 8-9 (1983).

Kumar, R. and Sharma, M., The Gas Plants: A Comparative Study, *Gram Praudyogiki*, **1(1)**, 28-37 (1981) (In Hindi).

Kumar, S. and Biswas, T.D., *Anaerobic Digestion of Livestock Wastes for the Production of Methane : A Batch Study*, Proc. Nat. Seminar on Biogas Tech., Ludhiana, Jul. 9-11, 20-25 (1981).

Kumar, S. and Jain, M.C., Dry Anaerobic Fermentation of Cowdung-Water Hyacinth Mixture in Multiple Batch Fed Digester, *Research and Industry*, **33(2)**, 162-166 (1988).

Kuroda, N. and Skakibara, T., Analysis of Anaerobic Digestion Treatment of Animal Manure, *Regional Jour. of Energy, Heat and Mass Transfer*, **5(3)**, 195-214 (1983).

Kurup, G.R., Chawla, S.K. and Mathur, A., Biogas Fuel Cell-based Cogeneration in the Pulp and Paper Industry, in 'A Kumar (ed.), *Proc. Conf. Indian Pulp and Paper Tech. Assoc.*, Saharanpur, 1992', 157.

Kurup, G.R., Continuous Bio-methanation of Solid Wastes: Technical Constraints and Their Solution, , in 'N.K. Bansal (ed.), *Proc. Nat. Solar Energy Conv.*, Dec. 28-30, 1992, Tata McGraw Hill Pub. Co. Ltd., New Delhi, 1993,' p. 377-383.

Kurup, G.R., Joseph, B. and Parameswaran, M.P., Innovative Dry Anaerobic Digestion Process for Municipal Solid Wastes of Trichur Town, Kerala, in 'N.K. Bansal (ed.), *Proc. Nat. Solar Energy Conv.*, Dec. 28-30, 1992, Tata McGraw Hill Pub. Co. Ltd., New Delhi, 1993.'

KVIC, *Gobar Gas Appliances*, Directorate of Gobar Gas Scheme, Mumbai.

KVIC, *Recent Developments in Gobar Gas Technology*, Mumbai, 1978.

Lakshmanan, A.R., *et al.*, Biogas Generation from Sugarcane Filterscale, Laboratory Scale and Pilot Plant Studies, *Urja*, **27(2)**, 25-28 (1990).

Lakshmanan, A.R., Kuppuswamy, G. and Jeyabal, A., Sugarcane Pressmud Charged Community Biogas Plant, *Changing Villages*, **11(1)**, 1-5 (1992).

Lam, J. and van Nes, W.J., Enforcement of Quality Standards upon Biogas Plants in Nepal, *Biogas Forum*, **II** No **57**, 4-14 (1994).

Landers, T., *Development of Farmer Built Methane Digesters*, Proc. of the 1980 Bioenergy World Congress and Exposition, Atlanta, 138-140 (1980).

Lane, A.G., Pilot-Scale Anaerobic Digester for Fruit and Vegetable Waste Solids, *Process and Chem. Eng.*, **32(6)**, 20-23 (1979).

Lane, A.G., Anaerobic Digestion of Spent Coffee Grounds, *Biomass*, **5**, 247-268 (1983).

Lapp, H.M., *et al.*, *Start-up of Pilot Scale Swine Manure Digesters for Methane Production*, Proc. of the Third Inter. Symp. on Livestock Wastes, ASEA Pub. No. Proc-2751, St. Joseph, Michigan, 234-237.

Lapp, H.M., Methane Production for Agri. By-products: An Overview, In 'Catania, P.J. (ed.), *Uses of Agri. Wastes : Food, Fuel and Fertilizer*, Univ. of Ragina, Saskatchewan, Canada,' p. 5-11

Laura, C. and Idnani, M.A., Note on the Use of Anaerobic Fermentation of Dung in the Cow Dung Gas Plant, *Indian Jour. of Food & Agri. Sci.*, **39(10)**, 1040-1048 (1969).

Laura, R.D. and Idnani, M.A., Increased Production of Biogas from Cow Dung by Adding Other Agricultural Waste Materials, *Jour. of Food & Agri. Sci.*, **11**, 165 (1971).

Lavagno, E., Proposals for the Analytical Modelling of a Digester, Proc. 1st EC Conf. on Energy from Biomas, Brighton, London, *Applied Science*, 423-428 (1981).

Lavagno, E., An Analytical Model to Study the Performance of an Anaerobic Digester, *Agricultural Wastes*, **5(1)**, 37-50 (1983).

Lawrence, A.W. and McCarty, P.L., *Kinetics of Methane Fermentation*, Tech. Report, Dept. of Civil Eng., Stanford Univ., Standford, 1967.

Lawrie, J., *Methane : Its Production and Uses*, Chapman and Hall, London, 1940.

Leake, H.M. and Howard, L.R., *Methane Gas from Farm Wastes*, Pub. No.9, Albert Howard Foundation of Organic Husbandry, UK, 1952.

Leese, T.M., Converting Seaweed to Biogas, *Hydrocarbon Processing*, **55(4)**, 86-89 (1976)

Lemanna, C. and Mallette, M.F., *Basic Bacteriology*, Williams and Wilkins Co., Baltimore, 1965.

Li, S., Three Aquatic Plants : Good Raw Material for Biogas Production, *Agri. Eng. Sci.*, Apr., No. **2**, 27 (1984) (In Chinese).

Liao, P.H., *et al.*, Effect of Liquid-Solids Separation on Biogas Production from Dairy Manure, *Energy in Agriculture*, **3(1)**, 61-69 (1984).

Libin, W. and Maegaard, P., Computerisation in Anaerobic Digestion Technology, *Biogas Forum (BORDA)*, **4(55)**, 13-17 (1993).

Libman, J.C., Energy from Refuse by Bio-conversion, Fermentation and Residue Disposal Processes, *Resources, Recovery and Conservation*, **1(3)**, 295-314 (1976).

Lindley, J.A. and Roehl, L., *Feasibility of a Methane Production System for a Dairy Farm in a Cold Climate*, Proc. Inter. Symp. on Biogas, Microalgae, and Livestock Wastes, Taipei, Taiwan, Rep. of China, Sep. 15-17, 1980.

Lingaiha, V. and Rajasekran, P., Biodigestion of Cowdung and Organic Wastes Mixed with Oil Cake in Relation to Energy, *Agricultural Wastes*, **17(3)**, 161-173 (1986).

Lingappa, B.T. and Lingappa, Y., Family Size Methane Generators, *Solar Age*, **4**, 16-18 (1979).

Lo, K.V., Effect of Solids-Separation Pre-treatment on Biogas Production from Dairy Manure, *Agricultural Wastes*, **8**, 155-165 (1983).

Lo, K.V., *et al.*, *An Integrated Dairy Manure Treatment and Biogas Production System*, Proc. 5th Canadian Bioenergy R & D Seminar, Canada, Mar. 26-28, 1984, S. Hasnain (ed.), Elsevier, London, 305-309 (1984).

Lo, K.V., *et al.*, Anaerobic Treatment of Swine Waste Waters Using Hybrid UASB Reactors, *Bioresource Tech.*, **47(2)**, 153-157 (1994).

Loehr, R.C., Alternatives for the Treatment and Disposal of Animal Wastes, *Jour. Water Poll. Control Fed.*, **43**, 668 (1971).

Lohmann, J., Biogas Processing to Natural Gas Quality : An Alternative to Conventional Usage, *Water Sci. Tech.*, **16(12)**, 399-410 (1983).

Loll, U., Engineering Operation and Economics of Biodigestion, in 'Schlegel, H.G. and Barnea, J., *Microbial Energy Conversion*, Verlag E., Goltze, K.G., Gottingen, (1976),' 5361-5378.

Luck, H., *Separation of Biogas and Recovery of CO_2 for Cooling Purposes*, Proc. Inter. Symp. on Biogas, Microalgae and Livestock Wastes, Taipei, Sep. 15-17, 1980.

Madanjwar, D.B. and Mithal, B.M., Effects of Surfactants on Anaerobic Digestion of Cattle Dung, *Jour. of Microbiology*, **27(1)**, 84-87 (1987).

Mah, R.A., *Microbiology of Methanogensis*, A Report, Div. of Env. and Nutritional Sciences, Univ. of California, LA, Calif., 1980.

Mahajan, B.A. and Singh, M.K., *Biogas as an Alternative Source of Energy for Rural Development*, Proc. Nat. Solar Energy Conv., Baroda, Dec. 15-18, 1983.

Maheshwari, R.C. and Vasudevan, P., Biogas Production : Chemical and Biochemical Aspects, *Chemical Weekly*, **27**, 104-106 (1982).

Malik, R.K., *et al.*, Brackish Water for Biogas Production, *Urja*, Oct., 284 (1983).

Malik, R.K., *et al.*, Economic Use of Water in Biogas Production, *Urja*, **15(3)**, 180 (1984).

Malik, R.K., *et al.*, Optimisation of Biogas Production from Cattle Waste, *Urja*, **15(6)**, 429 (1984).

Malik, R.K., *et al.*, Effect of Inorganic Nitrogen Supplementation on Biogas Production, *Biological Wastes*, **21(2)**, 139-142 (1987).

Malik, R.K. and Taruo, P., Effect of Predigestion and Effluent Slurry Recycling on Biogas Production, *Indian Jour. of Microbiology*, **35(3)**, 295-209 (1995)

Malik, R., *Energy Recovery from Textile Mill Wastes Using Anaerobic Film Fixed Bed Reactor System*, M.E. thesis (Env. Engg.), MS Univ., Baroda, 1994.

Maly, J. and Fadrus, H., Influence of Temperature on Anaerobic Digestion, *Jour. Water Poll. Control Fed.*, **43**, 641 (1971).

Manocha, A., *Anaerobic Digestion of Plant Residues into Biogas*, M.Sc. thesis, Kurukshetra Univ., Kurukshetra, 1977.

Manocha, A. and Neelakantan, S., *Anaerobic Digestion of Plant Residues into Biogas*, Proc. Nat. Symp. on Biogas Tech., IARI, New Delhi, Nov. 29-30, 1977.

Mantell, C.L., *Solid Wastes : Origin, Collection, Processing and Disposal*, John Wiley & Sons, NY, 1975.

Marchaim, U. and Krause, C., Propionic to Acetic Acid Ratios in Overloaded Anaerobic Digestion, *Bioresource Tech.*, **43(3)**, 195-203 (1993).

Maramba, F.D., *Biogas and Waste Recyling : Philippines Experience*, Maya Farms Division, Liberty Flour Mills, Metro, Manila, Philippines, 1978.

Maramba, F.D., *Biogas System at Maya Farm*, Proc. of the Bio Energy World Congress and Exposition, Atlanta, Apr., 459-461 (1980).

Martin, J.H. and Gene, D., *Energy Recovery from Manure Using Plug Flow Digester*, Sheaffer and Roland, Inc., 1978.

Martin-Leake, H. and Howard, L.E., *Methane Gas From Farm Manure*, Albert Howard Foundation of Organic Husbandry, Pub. No. 9, London, 1952.

Massey, V., *Utilisation of Biogas in Running Engines for Chopping Hay*, Proc. Nat. Symp. on Biogas Technology and Usage, IARI, New Delhi, Nov. 29-30, 1977.

Mata-Alverez, J., A Simple Device to Measure Biogas Production in Laboratory Scale Digester, *Biotechnology Letters*, **8**, Oct., 719-720 (1986).

Mathew, B.J., *Study of Biogas Generation Technology : Problems and Some Suggested Imporvements,* M. Tech. thesis, Centre for Energy Studies, IIT, Delhi, Nov., 1994.

Mathur, H.B., *et al.*, *Performance and Emission Characteristics of a Biogas Operated Compression Ignition Engine*, Proc. Nat. Conf. on Fuels from Crops, Melbourne, Sep. 25-29, 47.1-47.7, 1981.

Mathur, H.B., *et al.*, *Biogas as an Alternate Fuel for Combustion Engines*, Proc. Nat Solar Energy Conv, New Delhi, Dec. 17-19, 1982.

Mawson, A.J., Bioconversions for Whey Utilisation and Waste Abatement, *Bioresource Tech.*, **47(3)**, 195-203 (1994).

Mcabe, J. and Bekenfelder, W.W., *Biological Treatment of Sewage and Industrial Wastes*, Reinhold Pub. Corp., NY, 1958.

Mc Caffrey, P.M. (ed.), *Methane and Other Valuable Products from Animal Wastes*, Symp. Proc., Coordinating Council on the Restoration of the Kissimmee River Valley and Taylor Creek-Nubbin Slough Basin, Tallahassee, FLA, Jun. 26-29, 1979.

Mc Carty, P.L., *et al.*, *Microbiology of Anaerobic Digestion*, MIT Press, Cambridge, MA, 1962.

Mc Carty, P.L., Anaerobic Waste Treatment Fundamentals — Part 3 & 4, *Public Works*, **11(11)**, 91-99 (1964).

Mc Carty, P.L., *et al.*, *Heat Treatment of Biomass for Increasing Biodegradability*, Proc. Third Annual Biomass Systems Conf., Solar Energy Res. Inst., Golden, Colorado, 411 (1979).

Mc Carty, P.L., *Pretreatment of Lignocellulosic Materials for Increasing Anaerobic Biodegradability to Methane*, paper presented at the Technical Workshop on Fuels from Biomass, Antigua, Guatemala, Aug. 4, 1986.

Mc Garry, M.G., *Health Aspects and the Potential for Transmission of Human Diseases through the Biogas Plant*, Inter. Dev. Res. Center, Ottawa, Canada, 1976.

Mc Garry, M.G. and Stanforth, J., *Compost, Fertilizer and Biogas Production from Human and Farm Wastes in the People's Republic of China*, Report No. IDRC-T58e, Inter. Dev. Res. Center, Ottawa, Canada, 1978.

Mc Geer, P. and Durbin, E., *Methane : Fuel for Future*, Plenum Press, NY, 1982.

Meenaghan, G.F., *et al.*, *Gas Production from Beef Cattle Wastes*, ASAE Paper No. 70-907, Amer. Soc. of Agri. Engrs., St. Joseph, Michigan, 1970.

Mehta, B.J., Mody, K.H. and Chauhan, V.D., Effect of Salinity on Biogas Production, *Jour. for Entrepreneurs and Technologists*, **33(1)**, 16-18 (1988).

Mehta, R.K., *et al.*, A Model for Recycle of Feed Nutrients in a Crop, Animal and Biogas Systems in India, *Agricultural Systems*, **21(3)**, 159-169 (1986).

Meier, U., Biogas from Animal Dung and Other Organic Matter : The Case of Third World Countries, *African Env.*, **5 (4)** and **6(1-2)**, 109-122 (1987).

Menon, M.R.F., *Full Scale Treatment of Distillery Waste by Anaerobic Method Using Methane-Forming Bacteria*, paper presented at the All India Seminar on Distillery Waste Disposal Methods, Sangli, 1974.

Menon, T.C.K., *Biogas from Weeds and Plant Litter*, paper presented at the Seminar on Selected Crops of Work in Progress on Adaptive Technology, Ullar, Jul., 1977.

Merwe, M.V.D. and Britz, T.J., Anaerobic Digestion of Bakers Yeast Factory Effluent Using an Anaerobic Filter and a Hybrid Digester, *Bioresource Tech.*, **44(1)**, 71-76 (1993).

Metcalf, E., *Dung Digestion*, Dadar Sewage Works, Dadar, Mumbai, 1954.

Meynell, P.J., *Methane : Planning a Digester*, Prism Press, Dorset, London, 1976.

Mi, Z. and Libin, W., *Proc. Inter. Symp. on Biogas Dev. and Environmental Protection*, Chengdu Biogas Research Institute, Chengdu, P.R. China, Oct. 4-14, 1994.

Mill, D.T., Design Parameters and Operatory Characteristics of Animal Waste Anaerobic Digestion System : Dairy Cattle, *Agricultural Wastes*, **5(4)**, 219-230 (1983).

Mills, P.J., Minimisation of Energy Input Requirements of an Anaerobic Digester, *Agricultural Wastes*, **1(1)**, Feb., 57-66 (1979).

Mills P.J., Hobson, P.N. and Boufield, S., *Anaerobic Digestion of Farm Wastes*, Proc. Annual Res. Meeting, Inst. of Chem. Engrs., U.K., Apr., 1979.

Mishra, S., Digestion of Human Faecal Matter, *Sewage Works Journal*, **7(5)**, 798-809 (1935).

Mishra, U.P., *Biochemical Studies on the Production of Combustible Gas and Manure from Bullock Dung and other Organic Material*, M.Sc. thesis, IARI, 1954.

Misra, S.K. and Tiwari, R., *Brief Report on Dome Shaped Biogas Plant, Workshop Proc. on Biogas Utilisation in Engines*, IIP, Dehradun, Apr 19-20, 1978, 30-33.

Mital, K.M., *A Survey of Properties of Biogas*, Proc. of the 6th Annual Conf. of the Indian Society of Mech. Eng., Delhi College of Engg., New Delhi, Dec. 22-23, 1985.

Mital, K.M., Scope of Industrial Engineering in Promoting CNG as an Alternative Vehicle Fuel, *Udyog Pragati (NITIE Jour.)*, **15 (3&4)**, 3-9 (1991).

Mital, K.M., Biogas Systems, Ch. 8, in 'Mital, K.M., *Non-Conventional Energy Systems : Principles, Progress and Prospects*, Wheeler Publishing, New Delhi, 1996.

Mital, K.M., *Biogas systems : Policies, Progress and Prospects*, New Age Inter. Ltd., New Delhi, 1996.

Mitchell, J.C.V., Review of Current Research and Development in Biogas Production, *Process Biochemistry*, Oct., 32-33 (1975).

MNES, Biogas : A Rural Energy Source, 14 CGO Complex, Lodhi Road, New Delhi 110003, Oct., 1995.

Mody, K.H., *et al.*, Alginate Industry Waste : A Source of Biogas, *Research and Industry*, **35(4)**, 248-250 (1990).

Mohan, S. and Gopalan, M., A Study on the Use of Biogas from Cowdung for the Storage Inset Control, *Bioresource Technology*, **39**, 229-232 (1992). Also summarised in *Biogas Forum*, **II** No **57**, 15-16 (1994).

Mohanrao, G.J., Fundamentals of Anaerobic Digestion, *Env. Health*, **5**, 169-188 (1963).

Moorhead, K.K. and Nordstedt, R.A., Batch Anaerobic Digestion of Water Hyacinth : Effects of Particle Size, Plant Nitrogen Content and Inoculum Volume, *Bioresource Tech.*, **44(1)**, 71-76 (1993).

Morris, G.R., *et al.*, Alternative Animal Waste Anaerobic Fermentation Design and Their Costs, in 'Jewell, W.J. (ed.), *Energy, Agriculture and Waste Management*, Ann Arbor Sci. Pub., Ann Arbor, MI, 1975,' p. 317-335.

Morris, G.R., Jewell, W.J. and Loehr, R.C., *Anaerobic Fermentation of Animal Wastes : A Kinetic Design Evaluation*, Proc. 32nd Ind. Waste Conf., Purdue Univ., Lafayette, Indiana, 689-700 (1977).

Mother Earth News, The Marvellous Chicken Powered Motor Car, Issue No. **10**, Jul. 14-19, 1971.

Mulcock, A.P., Fuel from Crops, *New Zealand Energy Jour.*, Apr., 56-58 (1976).

Muraleedharan, K., A Simple Model of Gobar Gas Plant, *Invention Intelligence*, **12**, 304 (1977).

Murthy, G.R.R., Guruswamy, T. and Venkataramiah, K., Using Treated Dung Slurry of Biogas Plant as Manure and Fuel, *Invention & Intelligence*, **26(4)**, 181-182 (1991).

Myles, R.M., *Construction of 2 cum (70 cft) Demonstration-cum-Training Model Janata Biogas Plants in the Union Territory of Delhi*, AFPRO, New Delhi, 1980.

Myles, R.M. and Dhussa, A., *Construction Manual on Janata Biogas Plants*, Course material for the FAO/UNDP Training Programme on Low Cost Biogas Tech., New Delhi, Dec. 28-Jan. 14, 1982, p. 47.

Myles, R.M. and Dhussa, A., *Comparison of Performances of Janata and Deenbandhu Biogas Plants*, Proc. of the Nat. Solar Energy Convention on these Energy Options for the 90's, Tata McGraw-Hill Pub. Co. Ltd., New Delhi, 590-595 (1988).

Myles, R.M. and Kumar, T.S., Biogas Pressure Regulator, *Biogas Forum (BORDA)*, **I** No. **56**, 25-26 (1994).

Myles, R.M., *A Practical Guide to Janata Biogas Plant Technology*, AFPRO, New Delhi, 1986.

Nagori, G.P., *et al.*, *Development of Low Cost Biogas Plants*, Proc. Nat. Solar Energy Conv., Baroda, Dec. 15-18, 191-192 (1983).

Nagori, G.P., and Patel, B.D., Performance of a 5 m^3 Biogas Plants Using Banana Stems, *SESI Jour.*, **2(1-2)**, 1-11 (1988).

Nair, K.V.K., Biogas Generation Using Microalgae and Macrophytes, *Indian Jour. of Env. Health*, **24(4)**, 277-284 (1982).

NAL, Utilisation of Sludge Gas for Operation of Gas Turbine Engines, *NAL Newsletter*, July, 1978, p. 67.

Namasivayam, C. and Yamuna, R.T., Waste Biogas Residual Slurry as an Adsorbent for the Removal of *Pb (II)* from Aqueous Solution and Radiator Manufacturing Industry Wastewater, *Bioresource Tech.*, **52(2)**, 125-131 (1995).

Namasivayam, C. and Yamuna, R.T., Adsorption of Direct *Red 12B* by Biogas Residual Slurry : Equilibrium and Rate Processes, *Env. Pollution*, 1995.

Naramba, F.D., *et al., Small Biogas Plants : Its Construction, Operation and Use,* A Report, Chamber of Commerce of Philippines, Manila, 1977.

Narasimhamurthy, G.S.R. and Purushothaman, A., *A Bioconversion System to Generate Biogas : A Substitute for LPG,* Proc. Nat. Seminar on Adoption of Renewable Sources of Energy, Ahmedabad, Sep. 20-23, 1982.

Narayanaswamy, V.P., *et al., Optimum Design of Biogas Plants,* paper presented at the Nat. Sem. on Tech. for Agri. Dev., Inst. of Engrs. (India), Chandigarh, Apr. 9-11, 1977.

Narayanaswamy, V.P., *et al., Analysis, Design and Construction of Ferrocement Biogas Holder,* paper presented at the Symp. on Shell Structures, Univ. of Oulu, Finland, Jun. 9-14, 1980.

Narayanaswamy, V.P., Sankar, K., Sekran, P.M.C. and Lalitha, K., Biomethanation of *Leucaena Leucocephala* : A Potential Biomass Substrate, Fuel, **65(8)**, 1129-1133 (1986).

NAS, *Methane Generation from Human, Animal and Agricultural Wastes,* Nat. Acad. of Sci., 2101 Constitution Avenue, Washington, DC, 1977.

Nayyar, H., Kanwar, S.S. and Walia, D.P., Effect of Biogas Slurry as Seed Dressing Agent on Seed Performance of Maize and Rice, *Biogas Forum (Jour. BORDA),* **4(55)**, 10-20 (1993).

Nazir, M., Biogas Plants Construction Technology for Rural Areas, *Bioresource Technology,* **35(3)**, 283-289 (1991).

Neelakantan, S., Sondhi, M.S. and Mittal, C.P., Effect of Temperature, Innoculation and Agitation on Biogas Production from Cattle Dung, *Indian Jour. of Dairy Sci.,* **29**, 226 (1976).

Neelakantan, S., *Microbial Digestion of Organic Matter, Production of Biogas and its Utilisation,* paper presented at the 63rd Session of the Indian Sci. Congress Association (Section of Agri. Sci.), Waltair, 1976.

Neelakantan, S., *Microbiology and Chemistry of Biogas Production,* paper presented at the Summer Inst. on Waste Mgt. by Recycling, 238-416 (1976).

Neelakantan, S., *Bioconversion of Solar Energy into Fuel Gas,* paper presented at the Nat. Symp. on Biogas Tech. and Usage, IARI, New Delhi, Nov. 29-30, 1977.

Neelakantan, S., *Biodegradation of Livestock Excreta and Plant Materials by Anaerobic Digestion,* paper presented at the 18th Annual Conf. of Assoc. of Microbiolgists of India, Madurai, Dec. 21-23, 1977.

Neelakantan, S., Anaerobic Fermentation of Plant Materials into Acids and Biogas, *Current Science,* **47(50)**, 147-151 (1978).

Neelakantan, S., (ed.), *Janata Biogas Technology and Fodder Production,* Seminar-cum-Workshop Proc., Nat. Dairy Res. Inst., Karnal, Oct. 3-10, 1980.

Neelakantan, S., *Performance of Janata Biogas Plants and Biogas Utilisation Appliances,* Seminar-cum-Workshop Proc., Nat. Dairy Res. Inst., Karnal, Oct. 3-10, 1980.

NEERI, Digestion of Night-soil and Cow Dung, *Tech. Digest,* No. **6**, Nagpur, 1970.

Nekrasov, B.G., Biogas Cleaning from Impurities, *Biogas Forum* (*BORDA*), IV No **47**, 12-17 (1991).

Neupane, B.D., *et al.,* Development of a Low Cost Mixing Device, *Biogas Forum (BORDA),* **IV** No **51**, 22-25 (1992).

Neyeloff, S. and Gunker, W.W., Methane-Carbon Dioxide Mixture in an IC Engine, in 'W.J. Jewell (ed.), *Energy, Agriculture and Waste Management,* Ann Arbor Sci. Pub., Ann Arbor, MI, 1975,' 304-408.

Norris, A., Scrubbing Sewage Gas, *Water and Sewage Works*, **90,** 61 (1943).

NERAES, *Methane Technology for Agriculture*, Proc. Conf., North-East Regional Agri. Engg. Service, Cornell Univ., Ithaca, New York, 1981.

NSI, *Research Studies on Biogas Production from Bagasse*, National Sugar Institute, Kanpur, 1970.

NTIS, *Converting Waste to Energy*, Proc. Workshop, Savannah, Georgia (USA), 1982.

Nussberger, F.E., Determination of Hydrogen Sulphide in Digester Gas, *Sewage Works Jour.*, **20(30)**, May, 499-501 (1948).

Obias, E., *Biogas System at the Maya Farm*, Internal Report, Philippines Council of Agri. and Resources Res., Manila, 1970.

Obias, E., *Waste Utilization at Maya Farms*, paper presented at the ESCAP Workshop on Biogas Tech. and Utilization, Manila, 1975.

Odeyemi, O. and Adewuuoni, A.A., *Relative Biogas Generation from Five Animal Manures in Nigeria*, Proc. of a Forum on Energy Self-reliance, Observation, Production and Consumption held at Regina, Saskatchewan, Canada, Aug. 23-29, 1982, F.A. Curtis (ed.), Solar Energy Soc. of Canada, Winnipeg, Manitoba, 1982.

ORD, Design *Manual for Methane Gas Producing Facilities*, Office of Rural Dev., Suweon, Rep. of Korea, 1976.

ORD, Utilization *Effects of Methane Gas as a Rural Fuel Source*, Office of Rural Dev., Suweon, Rep. of Korea, 1976.

Oswald, W.J., and Golueke, C.G., Biological Transformation of Solar Energy, *Advances in Applied Microbiology*, **2**, 223-262 (1960).

Oswald, W.J., *Integrated Systems for Community Waste Management*, Proc. Inter. Symp. on Biogas, Microalgae and Livestock Wastes, Taipei, Taiwan, Rep. of China, Sep. 15-17, 1980.

Owsianowski, R.P., *et al.*, *Biogas Models in the World*, Proc. Inter. Sym. on Biogas, Microaalgae and Livestock Wastes, Taipei, Taiwan, Rep. of China, Sep. 15-17, 1980.

Pai, P.C. and Abbey , D.K., *Biogas Production from Sewage : A Preliminary Study*, Report No. NAL TM-PR 000/1-78, Nat. Aero. Lab., Bangalore, Mar., 1978.

Pal, P.C. and Kalra, R., *Low Cost Clay Energy Appliances for Rural Areas in India : A Clay Biogas Plant and a Clay Solar Cooker*, Proc. Nat. Solar Energy Conv., Bangalore, Jan. 22-24, 1982, p. 9.028-9.031.

Pale, D.L. and Freeman, C., *Methane Production by Anaerobic Fermentation : An Annotated Bibliography*, Intermediate Tech. Group, London, 1977.

Pande, B.R., *et al.*, A Simple Method of Transporting Biomass-Biogas in Rural Areas, in 'T.N. Veziroglu (ed.), *Alternative Energy Sources V : Part-D — Biomass, Hydrocarbon, Hydrogen*, Elsevier, Amsterdam, 1983,' 119-123.

Pande, D.R. and Kate, T., Laboratory Table Model for Biogas Generation, in 'T.N. Veziroglu (ed.), *Alternative Energy Sources V : Part-D — Biomass, Hydrocarbon, Hydrogen*, Elsevier, Amsterdam, 1983,' 145-149.

Pandey, A., *Solid State Fermentation*, Wiley Eastern Ltd., New Delhi, 1995.

Pandey, P.N., Description of Janata Biogas Plant, *Changing Villages*, **6(4)**, 233-240 (1984).

Pandya, J.D., *et al.*, A Phosphoric Acid Fuel Cell Coupled with Biogas, *Energy (Oxford)*, **13(4)**, 383-388 (1988).

Pangotra, P.N., Techno-Economic Aspects of Operating Diesel Engines on Biogas, *Agri. Eng. Today*, **1(9)**, 11-16 (1977).

Parikh, M., *Garbage Gas Plant*, paper presented at the All India Seminar on Tech. for Rural Dev., Allahabad Polytechnic, Allahabad, Jul. 2-3, 1977. Also published in Khadi Gramodyog, **24**, 310-317 (1978).

Parikh, M., *Gobar Gas Plant Without Iron Gasholder*, Proc. Res. Planning Workshop on Energy for Rural Dev., Ahmedabad, Dec., 1979.

Parikh, M., *Biogas Manure Plant*, Proc. Nat. Seminar on Biogas Tech., Ludhiana, Jul. 9-11, 114-120 (1980).

Parikh, M. and Kalia, M., Garbage Gas Plant, *Khadi Gramodyog*, **24(6)**, 310-317 (1978).

Parikh, P.P. and Jain, C.K., Gobar Gas as Supplementary Fuel for Diesel Engines, *Urja*, **4(1)**, Apr. 27-29, 1977.

Parikh, P.P., *Problem and Prospects of Utilisation of Biogas Plants in IC Engines*, paper presented at the Workshop on Biogas Utilisation in IC Engines, IIP, Dehradun, Apr. 19-20, 1979, 60-67.

Parikh, R., Garbage Gas Manure Plant, *Changing Villages*, **8(4)**, 184-189 (1989).

Patankar, G.L., *Recent Developments in Gobar Gas Technology*, A Report, Gobar Gas Dev. Center, KVIC, Mumbai, 1977.

Patankar, G.L., *Utilisatin of Biogas in IC Engines*, Proc. Workshop on Biogas Utilisation in Engines, IIP, Dehradun, Apr. 19-20, 1978, 53-59.

Patankar, G.L., Plastics in Setting-up of a Gobar Gas Plant, *Popular Plastics*, **23(4)**, 55-59 (1978).

Patankar, G.L., Corrosion in Gobar Gas Plants, *Khadi Gramodyog*, **24**, 541-545 (1978).

Patankar, G.L., *Gobar Gas As Alternate Fuel for IC Engines*, Proc. Workshop on '*Alternate Fuels for IC Engines*', BHU, Varanasi, Mar. 3-4, 1979.

Patankar, G.L., Substitute Material for Gasholder, Khadi Gramodyog, 25(10), Jul., 1979.

Patel, H.C. and Narasimhan, M.V., Small IC Engine for Gobar Gas Utilisation, *Khadi Gramodyog*, Jun., 598-600 (1966).

Patel, J.H., *et al.*, *Biogas Plant for Banana Stems*, Proc. Nat. Solar Energy Conv., Baroda, Dec. 15-18, 185-186 (1983).

Patel, J.J., Digestion of Organic Matter for Production of Methane Gas, *Poona Agri. College Magazine*, **42(33)**, 150-159 (1951).

Patel, J.J., On Constructing a Gobar Gas Plant, *Khadi Gramodyog*, 1963.

Patel, J.J., *Installation of Biogas Plants*, A Report, KVIC, Mumbai, 1964.

Patel, J.J., and Patankar, G.L., Burner Design for Gobar Gas, *Khadi Gramodyog*, March, 406-413 (1966).

Patel, J.J., *Gobar Gas Plant Fuel and Manure from Organic Wastes*, 'Proc. FAI/FAO Seminar on Optimizing Agri. Production Under Limited Availability of Fertilisers, New Delhi, Dec. 13-14, 1974,' 473-485.

Patel, J.J., Research and Development for Gobar Gas, *Commerce*, **131 (3358)**, 30-36 (1975).

Patelunas, G.M. and Regan, R.W., Biological Energy Recovery Using Dairy Cow Waste, *Jour. of Env. Eng. Div.*, Amer. Soc. of Civil. Engrs., **103**, 851-861 (1977).

Pathak, B.N., *et al.*, Anaerobic Digestion of Night-Soil : Seasonal Variation in the Volatile Matter Reduction and Gas Production, *Env. Health*, **7**, 202-207 (1963).

Pathak, B.N., *et al.*, Effect of Gas Recirculation in a Pilot Scale Cowdung Digester, *Env. Health*, **7**, 208-212 (1965).

Pathak, B.N. and Rajagopal, G., Effect of Volatile Acid Accumulation in Dung Digestion, *Env. Health*, **8(13)**, 194-199 (1966).

Pathak, B.N., *Gobar Gas Plant*, Internal Report, NEERI, Nagpur, 1966.

Pathak, P.P., *et al., Biogas Generation from Cotton Dust from Textile Mills*, Proc. of the Nat. Seminar on Biogas Tech., Punjab Agri. Univ., Ludhiana, Jul., 1981. Also pub. in *Env. Health*, **24(4)**, 298-302 (1982).

Pathak, S.N. and Pandey, P.N., Biogas Production from Sheep Wastes and with Cattle Dung, *Rural Technology*, **9(2)**, 5-7 (1992).

Patra, S.K. and Maheshwari, R.C., *Biogas Work at the Central Institute of Agricultural Engineering*, paper presented at the Panel Meeting on Biogas Tech. organised by the AFPRO at the Agri. Tools Res. Centre, Bardoli, Gujarat, Nov. 5-6 1981.

Paul, T.M., *Waste Management, Pollution Control, Compost and Biogas Production, and Intensive Farming Through Liquid Compost Plant*, Proc. Inter. Symp. on Biogas, Microalgae, and Livestock Waste, Taipei, Taiwan, Rep. of China, Sep. 15-17, 1980.

Pawar, S.H. and Torane, J.V., *Anaerobic Biogasification*, Proc. of the Nat. Seminar on Solar Energy and Rural Dev., Kolahapur, 69-82 (1987).

Pawar, V.P., Construction of a Gobar Gas Plant, *Jour. of Inst. of Military Engrs.*, **28(3)**, 19-20 (1976).

Persson, S.P.E., *Progress Report from the First Year of Operation of the Pennsylvania State Anaerobic Digester*, Meeting of the North Atlantic Region of the ASAE at Rutgers Univ., New Jersey, 1976.

Persson, S.P.E., *et al., Agricultural Anaerobic Digesters : Design and Operation*, Bull. 827, Pennsylvania State Univ., Nov., 1979.

Pfeffer, J.T., Increased Loadings on Digester with Digested Solids, *Jour. Water Poll. Control Fed.*, **40**, 1020 (1968).

Pfeffer, J.T., Temperature Effects on Anaerobic Fermentation of Domestic Refuse, *Biotech. Bioeng.*, **16**, 771-787 (1974).

Pfeffer, J.T. and Liebman, J.C., *Biological Conversion of Organic Refuse to Methane*, Annual Report NSF/RANN/Sc/G1-39/91/75/2, Dept. of Civil Eng., Univ. of Illinois, Urbana Champaign, and Report No. UILU-ENG 75-2019, 1975.

Pfeffer, J.T. and Khan, K.A., Microbial Production of Methane from Municipal Refuse, *Biotech. Bioeng.*, **18**, 1179-1191 (1976).

Phadnis, B.A., *Cow Dung Gas Plant*, Symp. on Radioisotopes, Fertilisers and Cow Dung Gas Plant, New Delhi, 398-399 (1961).

Picken, D.J. and Hassaan, H.A., A Method for Estimating Overhaul Life f Internal Combustion Engines including Engines Operating on Biogas and Methane, *Jour. Agri. Engg. Res.*, **28**, 139-147 (1983).

Pillai, K.R., *et al.*, Water-hyacinths : Its Prospects as Energy Source, *Energy Management*, **3(4)**, 315-318 (1979).

Poels, J., *et al.*, Performance, Operation and Benefits of Anaerobic Digestion of Piggery Wastes, *Agricultural Wastes*, **8**, 233-249 (1983).

Poels, J., *et al.*, Effects of Disinfectants and Antibiotics on the Anaerobic Digestion of Piggery Waste, *Agricultural Wastes*, **9(4)**, 239-247 (1984).

Po, C., *Production and Use of Methane from Animal Wastes in Taiwan*, Proc. Biomass Energy Conf., Biomass Energy Inst., Winnipeg, Manitoba, 1973.

Po, C., *et al.*, Small Methane Generator for Waste Disposal, Proc. of the Third Inter. Symp. On '*Managing Livestock Wastes*, Urbana, II, 1975,' 238-240.

Pohland, F.C., *Anaerobic Biological Treatment Process*, Adv. in Chem. Series 105, Amer. Chem. Soc., Washington, DC, 1971.

Polprasert, C., Low Cost Biogas Digester, *Appropriate Technology*, **6(3)**, Nov. 22-24, 1979.

Polprasert, C., *et al.*, A Ferrocement Digester : Biogas and Biomass Production, *Jour. of Ferrocement*, **12(1)**, 26-34 (1982).

Prakash, M.S.S. and Srinivasan, M.N., *Biogas-fired Melting Furnances*, ASTRA Annual Report, IISc., Bangalore, 1976.

Prasad, C.M.V., On the Prospects of Gobar Gas as Engine Fuel, Proc. of the Workshop on '*Anaerobic Fuels for Engines*', BHU, Varanasi, Mar. 3-4, 1979.

Prasad, C.R., *Studies on Production of Biogas and Changes in the Chemical Constituents of Organic Material During Anaerobic Fermentation*, Ph.D. thesis, IARI, New Delhi, 1964.

Prasad, C.R., Popularising Biogas Production, *Khadi Gramodyog*, **12**, 661-663 (1966).

Prasad, C.R., *et al.*, Studies on Biogas Production-I : Methane Fermentation of Some Carbohydrates, *Jour. Nat. Acd. of Sci. of India*, **38A (1 & 2)**, 179-184 (1968).

Prasad, C.R., *et al.*, Methane Formation of Some Organic Materials Mixed with Cowdung, *Indian Jour. of Agri. Sci.*, 360-365 (1970).

Prasad, C.R., *et al.*, Changes in Biochemical Constituent of Some Organic Waste Materials Under Anaerobic Methane Fermentation, *Indian Jour. of Agri. Sci.*, **40(10)**, 921-924 (1970).

Prasad, C.R. and Sathyanarayana, S.R.C., *Studies in Biogas Technology, Part III : Thermal Analysis of Biogas Plants*, Proc. Indian Acad. of Sci., Vol C2, Part 3, Sep., 377-386 (1979).

Pretorius, W.A., Anaerobic Digestion-III: Kinetics of Anaerobic Fermentation, *Water Research*, **3**, 545-558 (1969).

Proujon, C., The Biobattery : Electricity from Bacteria, *Science World*, Apr. 3, 11-13, 1965.

Pundir, B.P. and Singhal, S., Biogas Utilisation in Engines : An Appraisal, Proc. Workshop on '*Biogas Utilisation in Engines*,' IIP, Dehradun, April 19-20, 1978.

Pundir, B.P. and Zvonow, V.A., *Methane as an Engine Fuel*, paper presented at the Workshop on Alternate Fuels, BHU, Varanasi, Mar. 4-5, 1979.

Qiang, C.G., Overview of Biodigester Development in China, *Biogas Forum (BORDA)*, **I (8)**, 7-19 (1992).

Rajabapaiah, K., *et al.*, *Studies in Biogas Technology, Part I : Performance of a Conventional Biogas Plant*, Proc. Indian Acad. of Sci., Vol C2, Sep., 357-363 (1979).

Rajakumar, J. and Subramanian, V., A Modified Digester to Improve Biogas Production, *Yojna*, **36(11)**, 23-24 (1992).

Rajasekaran, P., *et al.*, Influence of Temperature on Microbial Numbers and Biogas Production of Some Anaerobically Digested Wastes, *Agricultural Wastes*, **17(2)**, 83-89 (1986).

Rajasekaran, P., Palaniswamy, A. and Swaminathan, K.R., Bio-Processing of Canteen Wastes in Relation to Energy, *Madras Agri. Jour.*, **73(8)**, Aug., 421-424 (1986).

Rajpakse, S., Simple Stirrer for a Biogas Plant, *Biogas Newsletter Nepal*, No. **11**, 1980.

Rajpal, J.K., Sondhi, M.S. and Neelakantan, S., *Microbial Decomposition of Municipal Waste into Manure and Biogas*, Proc. Nat. Symp. on Recyling of Residues of Agriculture and Industry, Ludhiana, Mar. 28-29, 1980, 325-332.

Rajput, R.K., A Treatise on Janata Model Gobar Gas Plant, *Changing Villages*, **6(4)**, 241-266 (1984).

Raju, N.R., *et al.*, Influence of Trace Elements on Biogas Production from Mango Processing Waste in 1.5 m^3 KVIC Digesters, *Biotechnology Letters*, **13(6)**, 461-464 (1991).

Ramakrishnan, J., *Bibliography on Anaerobic Digestion*, East-West Resources Systems Inst., Honolulu, 1980.

Raman, N.V., *et al.*, *Ferrocement Gasholder for Biogas Plants*, Proc. of the Inter Conf. on Materials of Construction for Developing Countries, AIT, Bangkok, Aug., 749-764 (1978). Also published in *Jour. of Ferrocement*, **9(2)**, Apr., 93-104 (1979).

Raman, P., *et al.*, Methods and Problems of Evaluating the Performance of Biogas Plants Under Field Conditions, in 'N.K. Bansal (ed.), *Proc. Nat., Solar Energy Conv.*, Dec., 28-30, 1992, Tata McGraw-Hill Pub. Co. Ltd., New Delhi, 1993'.

Ramana, P.V., State of Research and Development in Biogas, *Biogas Technology*, **4(5)**, 2-5 (1989).

Ramkumar, K.S., Banana Gas, *The Indian Express*, New Delhi, Jul. 1, 1979.

Ramprasad, T.N.C., *et al.*, *Integrated System of Treatment and Disposal of Night-Soil with Emphasis on By-product Utilisation*, Proc. of the Seminar on Utilisation of Farm Waste for Rural Indus. Growth, Bangalore, 1975.

Ramprasad, T.N.C., Shanta, S. and Rao, G.J.M., *Anaerobic Digestion of Night-Soil : An Effective Treatment and Disposal Method – A Pilot Plant Study*, Proc. of the Seminar on Solid Wastes Disposal organised by the Hyderabad Center of the Institution of Engineers, Hyderabad, 1976.

Ramyar, M.M., *Methane Production from Human and Animal Manure*, Master's thesis, AIT, Bangkok, 1980.

Ranade, D.R., *et al.*, Use of Dry Dung Cakes for Biogas Production, *Khadi Gramodyog*, **25(11)**, Aug., 502 (1979).

Ranade, D.R., *et al.*, Methanogenic Organisms from Fermenting Slurry of the Gobar Gas Plant, *Current Science*, **49(10)**, 395-397 (1980).

Rajan, K.K., *et al.*, *Use of Gobar Gas as Diesel Engine Fuel*, paper presented at 3rd Nat. Conf. on IC Engines and Combustion, Univ. of Roorkee, Roorkee, Dec. 10-12, 1976.

Rankin, R.S., Scum Control in Digesters, *Sewage Works Jour.*, **19(3)**, May, 405-414 (1947).

Rankin, R.S., Digester Capacity Requirements, *Sewage Works Jour.*, **20(3)**, May, 478-489 (1948).

Rao, B., *Design of Community Biogas Plant for Generating Electricity for Running Pumps*, Street Lighting and House Lighting, Proc. Nat. Seminar on Biogas Tech., Ludhiana, Jul. 9-11, 90-96 (1981).

Rao, B., *Design of Community Biogas Plant for Cooking and Lighting from Cowdung*, Proc. Nat. Seminar on Biogas Tech., Ludhiana, Jul. 9-11, 97-102 (1981).

Rao, E.C.K., Generation of Power from Biogas, in '*Radio Isotopes, Fertilizers and Cowdung Gas Plant*, ICAR, New Delhi, 1961,' 381-390.

Rao, J.K., R & D in Biogas Technology, *The Hindu*, Hyderabad, Sep. 4, 1984.

Rao, N.N., Gobar Gas and Other Fuels : A Comparative Study — Part I, *Kurukshetra*, **25(7)**, 4-7 (1977).

Rao, N.N., Gobar Gas and Other Fuels : A Comparative Study — Part II, *Kurukshetra*, **25(8)**, 8-10 (1977).

Rao, P.P., Shivraj, d. and Seenaya, G., Succession of Mircobial Population in Cow Dung and Poultry Litter Waste Digesters During Methanogenesis, *Indian Jour. of Microbiology*, **33**, 185-189 (1993).

Rao, S.P. and Subramaniam, T.V., Experimental Biogas Digestion, *Khadi Gramodyog*, **32(7)**, 346-351 (1986).

Ravindra, G., *Utilisation of Industrial and Farm Wastes in the Production of Biogas and Manures*, Ph.D. thesis, Univ. of Agri. Sci., Bangalore, 1980.

Ravishanker, *et al.*, Hydrogen Sulphide Removal from Anaerobic digester Gas, *Agri. Wastes*, **11**, 167-179 (1984).

Rebillat, J.M., *Methane Production and Recovery in Domestic Waste Landfills : The French Research and Experimental Programme*, Proc. of the 2nd EC Conf. on Energy from Biomass, Sep., 595-598 (1982).

Reddy, A.K.N., *et al.*, Studies in Biogas Technology – Part IV : A Novel Biogas Plant Incorporating a Solar Water heater and Solar Still, *Proc. Indian Acad. of Sci.*, Vol C2, Part 3, Sep., 169-175 (1979).

Remedios, E.C. and Boyd, W., Imperial Valley Biogas Project, in 'D.L. Wise (ed.), *Fuel Gas Production From Biomass*, CRC Press, Florida, 1981,' 233-242.

Rintala, J.A. and Puhakha, J.A., Anaerobic Treatment of Pulp and Paper Mill Waste Management : A Review, *Bioresource Tech.*, **47(1)**, 1-8 (1994).

Rinzema, A., *Anaerobic Treatment of Waste Water with High Concentration of Lipids or Sulphate*, Ph.D. thesis, Agricultural Univ. of Wageningen, The Netherlands, 1988.

Robin, G. and Hinterberger, J., Biogas Unit Slurry as a Fertiliser : The Experience of Dominica, *Biogas Forum (BORDA)*, **III** No **42**, 9-11 (1990).

ROPDP, *Experimental Research on Excreta Disposal by Biogas Plants*, Preliminary Report, Research Office for Parasitic Disease Prevention, Province of Szechuan, P.R. China, 1973.

Rosenberg, G., Power from Biological Sources, *Farm Mechanisation*, Nov., 425-428 (1951).

Rosenberg, G., Methane Production from Farm Waste as a Source of Tractor Fuel, *Agriculture (London)*, **58**, 487-494 (1952).

Ruchen, C., and Zhiping, X., *Digesters for Developing Countries : Water Pressure Digesters*, Guangzhen Institute of Energy Conversion, Guangzhen, 1979.

Ruchen, C., and Zhiping, X., *Design, Construction, Operation of Digesters in China*, Proc. Bio-Energy World Congress and Exposition, Atlanta, Apr., 435-438 (1980).

Ruchen, C., *Building Rural Digesters*, Proc. 2nd Inter. Symp. on Anaerobic Digestion, Travenunde, Germany, Sep. 6-11, 293-314 (1981).

Ruchen, C., *et al.*, Biogas Power Station in Foshan, China, in 'D.L. Wise (ed.), *Fuel Gas Production from Biomass*, CRC Press, Florida, 1981,' 225-232.

Safley, L.M. and Westerman, P.W., Performance of a Low Temperature Lagoon Digester, *Bioresource Technology*, **41**, 167-175 (1992).

Safley, L.M. and Westerman, P.W., Low-Temperature Digestion of Dairy and Swine Manure, *Bioresource Tech.*, **47(2)**, 165-171 (1992).

Sairam, M., Singh, L. and Alam, S.I., Effects of Sulphate and Nitrate on Anaerobic Degradation of Night-Soil, *Bioresource Tech.*, **45(3)**, 229 (1993).

Samaga, B.S. and Mahadevan, K., *Biogas Technology : Some Aspects of Separation of Biogas into Carbon Dioxide and Methane*, Proc. Symp. on Appropriate Technology for Rural Dev., Oct. 18-20, 103-106 (1978).

Samaga, B.S., *Some Studies on Power Generation from Biogas Plants*, Proc. Nat. Seminar on Biogas Technology, Ludhiana, July 9-11, 141-149 (1981).

Sanchez, E.P., *et al.*, Effect of the Organic Volumetric Loading Rate on Soluble COD Removal in Down-Flow Anaerobic Fixed-Bed Reactors, *Bioresource Tech.*, **47(2)**, 173-176 (1994).

Sanra, R., *Studies on the Utilisation of Waste Materials for Production of Biogas and Manure*, M.Sc. thesis, IARI, New Delhi, 1965.

Sans, C., *et al.*, Acidogenic Fermentation of Organic Urban Wastes in Plug Flow Reactor Under Thermophilic condition, *Bioresource Tech.*, **54(2)**, 1995.

Sarda, R. and Joseph, R., Studies on Factors Influencing Methane Production Processing from Tomato Waste, *Bioresource Tech.*, **47(1)**, 55-57 (1994).

Sarkar, A.N., Research and Development Work in Biogas Technology, *Jour. of Scientific and Industrial Research*, **41**, 279-291 (1982).

Sasse, L., Fixed dome Versus Floating Drum Plant, *Biogas Forum*, **9**, 4-5 (1982).

Sasse, L., Gas Pressure, Volume, Temperature, *Biogas Forum*, **13**, 4-5 (1983).

Sasse L., Levels of a Gas Plant, *Biogas Forum*, **13**, 1-3 (1983).

Sasse, L., Family Size Biogas Plants, *Biogas Forum*, **14**, 1-15 (1983).

Sasse, L., Storage Capacity and Gas Pressure, *Biogas Forum*, **15**, 5-8 (1983).

Sasse. L., Discussion on Fixed Dome Plants, *Biogas Forum*, **19**, 1-7 (1985).

Sasse, L., Bamboo Cement Biogas Holder, *Biogas Forum (BORDA)*, **IV** No. **43**, 4-8 (1990).

Sasse, L., Kellner, C. and Kimaro, A., *Improved Biogas Unit for Developing Countries : CARMATEC Biogas Unit*, GATE, Eschborn, Germany, 1991.

Sasse, L., Gas Tight Plaster for Fixed Dome Plants in India, *Biogas Forum*, **IV** No. **59**, 9-11 (1994).

Sathianathan, M.A., *Biogas : Achievements and Challgenges*, Lakshmi Book Store, New Delhi, 1975.

Sathianathan, M.A., *Research Project on Biogas*, Centre of Science for Villages, Wardha, 1979.

Sathianathan, M.A., *Discussion Paper on Research Planning for Biogas Technology*, Proc. Planning Workshop on Energy for Rural Dev., Ahmedabad, Dec., 1979.

Sathianathan, M.A., Research in Biogas Technology and Utilisation (Some Recent Trends in India) — Part I : The Process, *Science for Villages*, **5(4)**, 3-4 (1982).

Sathianathan, M.A., Research in Biogas Technology and Utilisation (Some Recent Trends in India) — Part II : Technology and Product Use, *Science for Villages*, **5(5)**, 5-8 (1982).

Satyanarayan, S., *et al.*, Anaerobic Treatment of Human Wastes, *Ind. Jour. of Env. Protection*, **7(1)**, 1987.

Saubolle, B.R., Gas in Bags, *Biogas Newsletter*, Nepal, No. **8**, Spring, 5 (1980).

Saubolle, B.R. and Bachmann, A., *Fuel Gas from Cowdung*, Sahayogi Press, Kathmandu, Sep., 1983.

Severy, C.W. and Curzan, D.C., Methane Recovery from Chicken Manure Digestion, *Jour. Water Poll. Control Fed.*, **44(12)**, Dec., 2349-2354 (1972).

Sawyer, C.N., *Biological Treatment of Sewage and Industrial Wastes*, Reinhold Pub. Co., NY, 1958.

Schell, W.J. and Houston, C.D., Use of Membranes for Biogas Treatment, *Energy Progress*, **3**, 96-100 (1983).

Schellenbach, S., Designing a High Quality System to Produce Biogas from Animal Manures, in 'D.L. Wise (ed.), *Fuel Gas Production from Biomass*, Vol I, CRC Press, Florida, 1981,' 245-256.

Schmidell. W., *et al.*, Anaerobic Digestion of Municipal Wastes, *Water Science Technology*, **18(12)**, 163-175 (1986).

Schmidt, F. and Eggersgluess, W., Gas from Agriculture Wastes, *Gas Jour.*, **27(9)**, 2861 (1954).

Schultz, P., Heat Supply for Biogas Plants, Part 2, Heating System with Exhaust Gases, *Bodenkultur*, **37**, 375-383 (1986) (In German).

Schwartz, R.D., Microbial Production of Methane on a Commercial Scale, *Amer. Soc. for Microbiology News*, **39**, 1973.

Sen, A., *et al.*, Production of Combustible Gas from Cane Molasses, *Science and Culture*, **11**, 562-564 (1954).

Sen, B.P. and Bhaskaran, T.R., Anaerobic Digestion of Liquid Molasses Distillery Waste, *Jour. Water Poll. Control Fed.*, Oct., 1972.

Sen, H.K., Water Hyacinth as a Source of Power, *Trans. Second World Power Conf.*, Berlin, **6**, 221-237 (1930).

Sen, H.K. and Chatterjee, H.N., Gasification of Water Hyacinth, *Indus. Chem. Soc.*, **8**, 1-6 (1931).

Shadduck, G.S. and Moore, J.A., *Anaerobic Digestion of Livestock Wastes to Produce Methane*, Dept. of Agri. Eng., Univ. of Minnesota, MN, 1976.

Shah, I.H., *Design and Construction of Biogas Plant : Site Selection, Digester and Gasholder*, paper presented at the ESCAP Expert Group Meeting on Biogas Dev., Bangkok, Jun., 1978.

Shah, I.H., *Use of Products of Biogas Plant*, paper presented at the ESCAP Expert Group Meeting on Biogas Dev., Bangkok, June, 1978.

Shah, I.H., *Biogas as a Fuel for Power Production on Agricultural Farms*, paper presented at the ESCAP Expert Group Meeting on Biogas Dev., Bangkok, June, 1978.

Shah, I.H., *Operation and Maintenance of Biogas Plant*, paper presented at the ESCAP Expert Group Meeting on Biogas Dev., Bangkok, June, 1978.

Shah, S.H., Develop a Balloon for Methane Gas, *Invention Intelligence,* Feb., **148,** 1977.

Shanta, S., *Survival of Helminthic Ova During Anaerobic Digestion of Night-Soil*, Research Report, NEERI, Nagpur, 1970.

Shanta, S. and Mohanrao, G.J., *Anaerobic Digestion of Night-Soil and Cow Dung Mixtures in Reducing Ammonia Toxicity*, Proc. of the Conf. of Soc. of Biological Chemists (India), Bangalore, 33, 46-47 (1974).

Sharma, A.K. and Panwar, J.S., Effect of Temperature on Biogas Production in a Water-hyacinth Digester, *Urja*, **22(1)**, 13-15 (1985).

Sharma, B., Precautions to Have and Maintain a Gobar Gas Plant, *Gram Praudyogiki*, **22(2)**, 89-91 (1982).

Sharma, M.P., Biogas Generation from Biomass Residues : An Overview, Rural Tech. *Jour.*, **8(3)**, 1-11 (1991).

Sharma, P.C. and Gopalakrishnan, V.S., *Ferrocement Biogas Holder,* A Report, Inter. Ferrocement Information Centre, AIT, Bangkok, 1980.

Sharma, P.C. and Narayanaswamy, V.P., *Use of Ferrocement as Alternate Construction Materials for Biogas Plants*, paper presented at the All India Seminar on Renewable

Sources of Energy, Institution of Engineers, Allahabad Sub-Centre, Oct. 20-21, 1982.

Sharma, P.K. and Mbise, H.A., *Biocrude and Biomethanation from Croton Bonplandiarium*, A Working Paper, Centre for Energy Studies, IIT, Delhi, 1988.

Sharma, S.K., Saini, J.S., Mishra, I.M. and Sharma, M.P., *Mirabilis* Leaves : A Potential Source of Methane, *Biomass*, **13(1)**, 13-24 (1987).

Sharma, S.K., *et al.*, Biogas Generation: Effect of Mixing Effluent Slurry with Fresh Cattle Dung, *Energy Management*, **11(3)**, 197-202 (1987).

Sharma, S.K., *Biogas from Ipomoea Fistulasa Leaves*, Annual Report, SBI Chair, AHEC, University of Roorkee, Roorkee, 1987.

Sharma, S.K., Starting a Biogas Plant, *Urja*, **23(1)**, 31-34 (1988).

Sharma, S.K., *et al.*, Biogasification of Banana Peeling and *Ipomea Fistulasa* Leaves, *Khadi Gramodyog*, Apr., 324-332 (1988).

Sharma, S.K., *et al.*, Biogas from Biomass, *Indian Jour. of Rural Tech.*, **1(1)**, 1-17 (1989).

Sharma, S.K., *et al.*, Generating Biogas from Agricultural Residues, *Rural Tech. Jour.*, **7(2)**, 17-22 (1990).

Sharma, S.P., *et al.*, *Small Capacity Biogas Plants : A Survey of Current Indian Design*, paper presented at Symp. on Renewable Sources of Energy, Nat. Acad. of Sci., Allahabad Univ., Oct. 24-26, 1980.

Sharma, U.P., and Paul, T.M., Studies on Biogas Sources, *Current Science*, **45(10)**, 376-377 (1976).

Sharma, U.P., *Energy from Wastes*, Summer Institute on Waste on Waste Management by Recycling, Mumbai, June, 176-182 (1976).

Shaw, B.G., *A Practical and Bacteriological Study of the Anaerobic Digestion of Piggery Waste*, Ph.D thesis, Aberdeen, 1971

Shelat, R.N. and Karia, G.L., Performance Studies on Biogas Plant Models, in 'T.N. Vezrigolu (ed.), *Alternative Energy Sources V : Part–D — Biomass, Hydrocarbon, Hydrogen*, Elsevier, Amsterdam, 1983,' 93-103.

Sheng, H.P., Energy Recovery from Municipal Solid Waste and Method of Comparing Refuse-Derived Fuels, *Resources, Recovery and Conservation*, **1(1)**, 85-93 (1976).

Shetty, K.S., *et al.*, *Studies on Use of Agricultural and Animal Wastes for Biogas Production*, paper presented at the Nat. Symp. on Recycling Residues of Agriculture and Industry, Punjab Agri. Univ., Ludhiana, March 28-29, 333-348 (1980).

Shian, Su-Te, *et al.*, Construction of Simple Biogas Digesters in the Province of Szechuan, China, *Agricultural Wastes*, **1(4)**, 247-258 (1979).

Shields, F.D. and Tragitt, G.N., *Methane Production from Anaerobic Fermentation of Beef Manure*, ASAE Paper No.78-4010, Amer. Soc. of Agri. Engrs., St. Joseph, MI, 1978.

Shih, J.C.H. and Huang, J.J.H., A Laboratory Study of Methane Production from Broiler-Chicken Litter, *Symp. on Biotechnology in Energy Production and Conservation*, John Wiley & Sons, Inc., NY, 1980.

Siddiqui, M.A., Economic Analysis of Biogas for Optimum Generation Temperatures of Four Vapour Absorption Systems, *Energy Conserv. Management*, **27**, 163-169 (1987).

Sievers, D.M., *et al.*, *Engineering and Economic Aspects of Farm Digesters*, ASEA Paper No.75-4541, Amer. Soc. of Agri. Engrs., St. Joseph, 1975.

Sievers, D.M. and Brune, D.E., Carbon/Nitrogen Ratio and Anaerobic Digestion of Swine Waste, *Trans. of Amer. Soc. of Agri. Engrs.*, **21**, 537 (1978).

Silverio, C.M., *et al.*, Utilisation of Food Waste Materials for Energy, Food and/or Animal Feed Production : Biogas from Dried Banana Peelings, *NSTA Tech. Jour.*, **8(1)**, 71-80 (1983).

Singh, A.K. and Pundir, B.P., Dual Fuel Operation of a Diesel Engine with Sewage Gas, *Agri. Eng. Today*, **2(2)**, 1978.

Singh, A.K and Pundir, B.P., *Dual Fuel Operation of a Diesel Engine with Sewage Gas — Part I*, Internal Report 4.981, IIP, Dehradun, March, 1978.

Singh, A.K and Pundir, B.P., *Dual Operation of a Diesel Engine with Sewage Gas*, Proc. of Workshop on Biogas Utilisation in Engines, IIP, Dehradun, Apr. 19-20 1978.

Singh, A.K. and Pundir, B.P., *Biogas Utilisation in Engines*, Proc. of the Workshop on Alternate Fuels for Engines, BHU, Varanasi, Mar. 3-4, 1979.

Singh, J.B., Modified Fixed Dome Type Biogas Plants, *Changing Villages*, **6(4)**, 267-273 (1984).

Singh, J.B., Myles, R. and Dhussa, A., *A Manual on Deenbandhu Biogas Plants*, Tata McGraw-Hill Pub. Co. Ltd., New Delhi, 1988.

Singh, J.B., *et al.*, A New Aspect of Corrosion in Biogas Plant, *Rural Technology*, **9(2)**, 8-11 (1992).

Singh, L., Maurya, M.S., Sairam, M. and Alam, S.I., Effects of Loading and Temperature on Biogas Production from Night-Soil, *Bioresource Tech.*, **45(1)**, 59 (1983).

Singh, R., *et al.*, Assimilation of Acetate and Production of Methane by Cattle Waste Slurry, *Current Science*, **49(10)**, 399-400 (1980).

Singh, R., Pre-Digestion to Improve Production of Biogas from Cattle Waste, *Agricultural Wastes*, **6(3)**, 167-174 (1983).

Singh, R., *et al.*, Biogas at Different Solids Concentration in Daily Fed Cattle Waste Digesters, *Agricultural Wastes*, **11(4)**, 253-257 (1984).

Singh, R. and Anand, R.C., Comparative Performances of Indian Small Solid State and Conventional Anaerobic Digesters, *Bioresource Tech.*, **47(3)**, 235-238 (1994).

Singh, R.B., Biogas Plant, A Report, Gobar Gas Res. Station Ajitmal. Etawah, 1971.

Singh, R.B., Gobar Gas Experiments in India, *Mother Earth News*, Nov. 12, 1971.

Singh, R.B., *Some Experiments with Biogas*, Gobar Gas Research Station, Ajitmal, Etawah, 1972.

Singh, R.B., Biogas Plant : Generating Methane from Organic Wastes, *Compost Science*, **13(1)**, Jan.-Feb., 20-25 (1972).

Singh, R.B., Building a Biogas Plant, *Compost Science*, **13(2)**, 1972.

Singh, R.B., *Technical Report on Biogas Plant*, Proc. Workshop on Biogas Tech. and Utilisation, New Delhi, Sep. 16-22, 1975.

Singh, R.B. and Gupta, K.K., Introduction of Solar Energy in Biogas Systems, *World Science News*, **15(4)**, 24-25 (1978).

Singh, R.B. and Singh, K.K., *Janata Biogas Plants : Preliminary Report on Design and Cost*, State Planning Institute, Lucknow, 1978.

Singh, R., Jain, M.K. and Tauro, P., *Effect of Addition of Rock Phosphate on Biogas Productionfrom Cattle Waste*, Proc. Nat. Symp. on Recycling Residues of Agriculture and Industry, Ludhiana, Mar. 28-29, 313-323 (1980).

Singh, R., Jain, M.K. and Tauro, P., Pressmud as Additive too Increase Biogas Production from Cattle Waste, *Current Science*, **50**, 640-643 (1981).

Singh, R.V., *Utilisation of Agricultural Wastes for Energy Production*, Ph.D. thesis, IIT, Kharagpur, 1979.

Singh, S.K., *et al.*, Effect of Temperature on Biogas Production for Water-hyacinth, *Rural Technology*, **8(3)**, 25-29 (1991).

Singh, T., Janata Biogas Plants, *Kheti Kissani*, A Monthly Newsletter of Kissan Vidyapeeth, Faizabad, A Special issue on Biogas Plants, Nov., 1978.

Slesser, M. and Lewis, C., *Biological Energy Resources*, E and FN Spon, UK, 1979.

Smith, M., Use of Heat Pump for Digester Heating, *Sewage and Indus. Wastes*, **25(12)**, 1369-1378 (1953).

Smith, R.J., *Anaerobic Digestion of Livestock Wastes and Prospect for Methane Production*, Proc. of the Conf. on Animal Waste Utilisation, Iowa State Univ., Ames, Iowa, 1973.

Sodha, M.S., *et al.*, Effect of PVC Greenhouse in Increasing Biogas Production in Cold Climatic Conditions, *Energy Conserv. Mgt.*, **27(1)**, 83-90 (1987).

Sodha, M.S., *et al.*, Solar Assisted Biogas Plant-II A : Experimental Validation of the Mathematical Models for Floating Drum Type Biogas Plants, *Inter. Jour. of Energy Res.*, **12(2)**, 369-375 (1988).

Solly, R.K., *Manual for Installation and Operation of Biogas Digesters*, Institute of Natural Resources, Univ. of South Pacific, P.O. Box 1168, Suva, Fiji.

Solly, R.K., *et al.*, *Variation of Biogas Production with Varying Feedstock in Simple Anaerobic Digesters*, Proc. Inter. Symp. on Biogas, Microalgae and Livestock Wastes, Taipei, Taiwan, Rep. of China, Sep. 15-17, 1980.

SPIIBD, *Construction of Marsh Gas Producing Tanks (Digesters in Simple Ways)*, A Report, Szechwan Provincial Inst. of Indus. Buildings Design, People's Rep. of China.

SPOBD, *How the Rongxian Country Distillery in Sichaun Exploits Biogas*, A Report, Sichuan Provincial Office for Biogas Dev., Sichuan, P.R. China, 1979.

SPOBD, *Biogas Technology and Utilisation*, Chengdu Seminar, Sichuan Provincial Office of Biogas Dev., Chengdu, Sichuan, P.R. China, 1979.

Sridhar, B.L., *et al.*, *Some Tests with Gobar Gas Diesel Fuel Engines*, Proc. of Fifth Nat. Conf. on IC Engines and Combustion, Warangal, 1978.

Srinivasan, H.R., *Design, Construction and Maintenance of Gobar Gas Plants, Summer Institute on Waste Management by Recyling*, Mumbai, Jun., 209-213 (1976).

Srinivasan, H.R., *Design Aspects of Biogas Plant*, paper presented, at the Expert Group Meeting on Biogas Development, ESCAP, Bangkok, Jun 20, 1978.

Srinivasan, H.R., Elements of Biogas Technology, *Khadi Gramodyog*, Dec., **17,** 176-181 (1980).

Sriramulu, N. and Bhargawa, B.N., Water-Hyacinth as a Source of Energy, *Energy Management*, **4(3)**, Jul.-Sep., 266-271 1980.

Schulz, W., Optimal Bioresource Fertilisation in West Germany, *Biogas Forum (BORDA)*, I Special, 7-8 (1990).

Stafford, D.A., *et al.*, *Energy from Biomass by Anaerobic Digestion*, Report No. 5, Watt Committee on Energy, London, 1979.

Stafford, D.A., *et al.*, *Methane Production From Waste Organic Matter*, CRC Press Inc., Boca Raton, USA, 1981.

Stafford, D.A., *Biogas Energy from Animal Wastes and Energy Crops*, Proc. 2nd EC Conf. on Energy from Biomass, Berlin, Sep 20-23, 1982. (Edited By A Strub, *et al.*, Applied Science Pub., London, 487-491).

Stahl, T., Energy Cogeneration with Biogas Fuel, Ph. D. thesis, Univ. of Missouri, Columbia, 1983.

Stewart, D.J., *et al.*, *Production of Methane from Energy Crops and Use as a Vehicle Fuel,* Proc. Inter. Gas Res. Conf., Los Angeles, CA, 1981, Govt. Institutes, Rockville, MD, 1982.

Subramanian, P.K., *et al.*, *Studies in Biogas Technology, Part II, Optimisation of Plant Dimensions*, Proc. Indian Acad. of Sci., Vol C2, Part 3, Sep., 365-375 (1979).

Subramanian, S.K., Biogas Plant : A New Design Approach, *Invention Intelligence*, Feb., 61-67 (1980).

Subramanian, T.V., Design of a Gobar Gas Plant, *Chemical Age of India*, **28(10)**, 805-807 (1977).

Subramanian, T.V., Gobar Gas Appliances, *Khadi Gramodyog*, **26(12)**, Sep., 532-535 (1980).

Subramanian, T.V., Gobar Gas Plant : Optimisation and Economic Feasibility, *Khadi Gramodyog*, **26(10)**, 450-458 (1980).

Subramanian, T.V., A Mathematical Model for Predicting Methanogenic Bacterial Activity in Methane Fermentation, *Indian Jour. of Env. Protection*, **10(6)**, 428-431 (1990).

Subramanyam, P.V.R., *et al.*, *Anaerobic Digestion : A Unit Operation for Waste Treatment and Biogas Production*, paper presented at the 18th Annual Conf. of the Assoc. of Microbiologists of India, Madurai, Dec. 21-26, 1977.

Suloway, M., *Clean Fuels from Biomass, Sewage Urban Refuse and Agricultural Wastes*, A Research Report, Inst. of Gas Tech., Chicago, 1976.

Summers, R. and Bousfield, S., Practical Aspects of Anaerobic Digestion, *Process Biochemistry*, **11(5)**, 3-6 (1976).

Summers, R. and Bousfield, S., Detailed Study of Piggery Waste Anaerobic Digestion, *Agricultural Wastes*, **2(1)**, 61-78 (1980).

Summers, R., Bousfield, S. and Hobson, P.N., Uses and Analysis of Digested Sludge, in 'Stafford, D. (ed.), *Anaerobic Digestion*, Applied Science Pub., 1980,' 409-414.

Sundaran, S. and Krithiga, S., Comparative Study of Stirred and Unstirred Biogas Reaction, *Jour. Inst. of Engineers*, **72(10)**, 18-20 (1991).

Swadas, B.P., *Use of Biogas as Cooking Fuel*, Proc. Nat. Solar Energy Conv., Baroda, Dec. 15-18, 189-190 (1983).

Swadas, B.P., *et al.*, *Domestic Biogas Plant Using Dried Water hyacinth*, Proc. Nat. Solar Energy Conv., Baroda, Dec. 15-18, 1983.

SWADI, *Collection of Simple Biogas Digester Designs*, South West Architectural Designing Inst., Sichuan, P.R. China, July, 1980

Swaminathan, K.R. and Palaniswamy, P.T., *Low Cost Biogas Burners*, A Report, College of Agri. Eng., Coimbatore, 1982.

Swift, W.R., Methane Digester Design, *Alternatives*, **1**, Autumn, 41-43 (1973).

Tabin, A.I., *Feasibility Study on the Marketing of Biogas Generators in Region XI : Valtab Drum Type*, Bureau of Animal Industry, Philippines, Sep. 15, 1985, 53.

Taiganides, E.P., *Animal Wastes*, Applied Science Pub. Ltd., Barking, Essex, England.

Taiganides, E.P., *et al.*, Anaerobic Digestion of Hog Wastes, *Jour. Agri. Eng. Res.*, **8**, 327-333 (1963).

Tanticharoen, M. and Chuntranuluck, S., *Biogas Production from Aquatic Weeds,* Proc. ISES Congress, Brigton, England, Aug. 23-28, 1981.

Tanticharoen, M., *Biomass Energy Through Microbial Processes*, Proc. US-ASEAN Seminar on Energy Technology : Biomass, Coal, Solar/Wind Energy Planning, Bandung, Indonesia, Jun. 7-18, 1982, p. 101.

Tentscher, W. and Chantsawang, S., Biomethanation of Animal Wastes with a Hybrid System of Digester-I : Power Production and System Design, *RERIC Inter. Energy Jour.*, **12(2)**, Dec., 9-34 (1990).

Tentscher, W. and Chantsawang, S., Biomethanation of Animal Wastes with a Hybrid System of Digester-II : Economic Consideration, *RERIC Inter. Energy Jour.*, **12(2)**, Dec., 35-52 (1990).

TERI, *Indian Biogas Resource Index*, Documentation Center, Mumbai, 1980.

TERI, *Selected Bibliography : Promotion of the Use of Biogas in Developing Countries*, Mumbai, 1981.

TERI, *Review of the Literature on Promotion of Biogas Systems*, Mumbai, Mar., 1982.

TERI, *Biogas Technology Handbook*, Mumbai, 1982.

TERI, *Biogas Technology : An Information Package*, New Delhi, 1985.

TERI, *Fixed-dome Biogas Plants*, New Delhi, 1995.

TERI, *Spherical Biogas Plant : A Manual of Extension Workers*, New Delhi, 1995.

TERI, *Spherical Biogas Plants : A Manual for Plant Owners*, New Delhi, 1995.

Theilen, U., Solar Heated Simple Biogas Plants for the Analine Area, *Biogas Forum (BORDA)*, **IV** No. **55**, 4-11 (1993).

Thirumalai, K. and Mukherjee, T.K., *Performance of a Solely Biogas Operated Engine and Pump Set*, Proc. Nat. Solar Energy Conv., Jan. 22-24, 1982, 9.039-9.043.

Thomas, E.B., Conversion of Solid Fuels to Low Gas, *Energy Sources*, **2(1)**, 11-31 (1975).

Ting-wei, L., *et al.*, Application of Low Pressure Soft Gas Storage Tank on Biogas Project, *Biogas Forum (BORDA)*, **2** Special issue on China, 10-14 (1991).

Tiwari, G.N. and Chandra, A., Solar-Assisted Biogas System : A New Approach, *Energy Conver. Management*, **26(2)**, 147-150 (1986).

Tiwari, G.N., *et al.*, A Simple Analysis of Conventional Biogas Plant, *Energy Conver Management*, **28(1)**, 1-4 (1988).

Tiwari, G.N., *et al.*, Design Criteria for an Active Biogas Plant, *Energy*, **17(10)**, 955-958 (1992).

Toerien, D.F. and Hattingh, W.H.J., Anaerobic Digestion-I : Microbiology of Anaerobic Digestion, *Water Research*, **3(6)**, 385-416 (1969).

Trivedi, R.C., *Studies on the Ammonia Nitrogen Behaviour in Anaerobic Digestion with Reference to the Digestion of Night-Soil and Other Organic Wastes*, Ph.D. thesis, Nagpur Univ., 1973.

Trivedi, R.C., *Production of Biogas through Anaerobic Fermentation of Organic Wastes*, IFTA-AFCT Regional Symp. on Alternative Energy Sources and Poll. Control, Trivandrum, Feb. 26-27, 1982.

Trujillo, D., *et al.*, Energy Recovery from Wastes : Anaerobic Digestion of Tomato Plant Mixed with Rabbit Wastes, *Bioresource Technology*, **45(2)**, 81-83 (1993).

Tucker, D., Electricity from Town Refuse, *Electronics and Power*, **22(1)**, 16-20 (1976).

Tumlos, E.T., *Biogas Production from Farming Wastes*, A Report, Bangkok, 1981.

Umesh, B.V. and Sheshadri, C.V., Jwala : The Low Cost Biogas Plant, *Invention Intelligence*, **19(2)**, Feb., 64-68 (1984).

Ushiyama, I. and Kondo, T., *Basic Study of Methane Fermentation*, Proc. Sixth Research Presentation Meeting of the Japan Solar Energy Soc., Tokyo, Dec. 8-9, 1980 (in Japanese).

UKIE, *Small Engines and their Fuels in Developing Countries,* Proc. Conf., Institute of Energy, Reading, UK, 1984.

USDOE, *Production and Utilisation of Methane from Anaerobic Sludge Digestion in US Wastewater Treatment Plants*, Report No DOE/CS/300-3, Pacific Northwest Laboratory, Richland, Washington, 1981.

van Nees, W.J., *et al.*, Slurry in the Gas Pipes of Biogas Plants in Nepal, *Biogas Forum (BORDA),* **II** No. **45**, 4-7 (1991).

van Nees, W.J., *et al.*, Gas Production from Cattle Manure in Nepal, *Biogas Forum (BORDA),* **II** No. **45**, 8-9 (1991).

van Velsen, A.F.M. and Lettinga, G., *Effect of Feed Composition on Digester Performance*, First Inter. Symp. on Anaerobic Digestion, Cardiff, 1979.

Varani, F.T. and Burford, J., *Design of a Large Scale Manure/Methane Facility*, A Report, Biogas of Colorado, Arvada, Colorado.

Varel, V.H., *et al.*, Thermophilic Methane Production from Cattle Waste, *Applied and Env. Microbiology*, **33**, 298-307 (1977).

Vasiliev, V.B., *et al.*, Simulation Model of Anaerobic Digestion of Organic Matter by a Micro-organism Consortium : Basic Equation, *Water Res.*, **20**, 633-643 (1993).

Vavilin, V.A., *et al.*, Simulation Model Methane as a Tool for Effective Biogas Production During Anaerobic Conversion of Complex Organic Matter, *Bioresource Technology*, **48(1)**, 1-8 (1994).

Vavilin, V.A., *et al.*, Modelling of Gas Pressure Effects on Anaerobic Digestion, *Bioresource Tech.*, **52(1)**, 25-32 (1995).

Vavilin, V.A., *et al.*, Modelling Ammonia and Hydrogen Sulphide Inhibition in Anaerobic Digestion, *Water Res.*, **29**, 827-835 (1995).

Veluswamy, M.A., Biogas Generation and Utilisation, *Steam and Fuel Users Jour.*, **34(2)**, Jul.-Sep., 13-18 (1984).

Viertel, C. and Miller, L., Models of Fixed Dome Plants, *Biogas Forum*, **II** No. **57**, 17-18 (1994).

Vijay, V.K., Pandey, S. and Jain, S., Performance Evaluation of Biogas Domestic Appliances on Different Pressure Conditions, in 'N.K. Bansal (ed.), *Proc. Nat. Solar Energy Conv.,* Dec., 28-30 1992, Tata McGraw Hill Pub. Co. Ltd., New Delhi, 1993,' 407-416.

Visser, A., Gao, Y. and Lettinga, G., Effects of pH on Methanogenesis and Sulphate Reduction in Thermophilic (85°C) UASB Reactors, *Bioresources Tech.*, **44**, 113-121 (1993).

VTA, *Three Cubic Meter Biogas Plant : A Construction Manual*. Volunteers in Technical Assistance, Rhode Island, 1980.

VTA, *Small Scale Biogas Production for Cooking*, Energy Fact Sheet, No. 6, Volunteers in Technical Assistance, Rhode Island, 1980.

Vyas, S.K. and Goswami, K.P., *Recycling of Biomass for Biogas and Manure*, paper presented at the Annual Conf. of Inst. of Engrs., Ludhiana. Sub-Centre, Ludhiana, Sep., 1980.

Vyas, S.K. and Grewal, N.S. (ed.), *Biogas Technology*, Proc. Nat. Seminar on Biogas Tech., Ludhiana, Jul. 9-11, 1981.

Vyas, S.K., An *Engineering Approach of Enriching Methane Contents*, Proc. Nat. Seminar on Biogas Tech., Ludhiana, Jul. 9-11, 83-89 (1981).

Wagenerk, K., *Algae-Derived Biogas : A Firewood Subsitute in Tropical Countries*, Applied Science, London, 1982, p. 604-608.

Wall, T.M.P., *Instrumentation and Control of an Engine Generator Set for Biogas*, Ph.D thesis, Iowa State Univ., 1985.

Wanjun, Z., Study on a New Design Biogas Digester, *Biogas Forum (Jour. BORDA)*, **2**nd Special Issue, 20-23 (1992).

Wasser. R., Reducing Water Contamination Caused by Coffee Industries with Anaerobic Digestion Technology, *Biogas Forum*, **II** No **61**, 8-13 (1995).

Waszkiewica, C., *et al.*, Survey of Biogas Production Methods, *Agricultural Mechanics*, **29(21)**, 7-10 (1980) (In Polish).

Wellinger, A., Anaerobic Digestion : A Review Comparison with Two Types of Aeration Systems for Manure Treatment and Energy Production on the Small Farm, *Agricultural Wastes*, **10(2)**, 117-133 (1984).

West, J., *Domboshawa Biogas Plant*, Commonwealth Science Council, London, 1979.

Wheatley, B.I., *Absorbers, Scrubbers and Strippers : An Introduction with Particular Reference to Biogas*, Proc. of BAB Workshop on Biogas Scrubbing, Reading, Jun. 13, 1980.

Wise, D.L., Fuel Gas Production of Agricultural Residues and Municipal Solid Wastes by Low Capital Cost Batch Digestion : A Review, *Resources Conserv.*, **13**, 83-102 (1987).

Wolverton, B.C., *et al.*, *Bioconversion of Water-hyacinth into Methane Gas*, Part I, NASA Tech. Memo. TM-X-72725, Jul., 1975.

Xianming, L., Anaerobic Digester for Stabilizing Domestic Sewage, *Biogas Forum (BORDA)*, **IV** No **(55)**, 4-9 (1993).

Xun, M., *et al.*, Energy Input and Output Analysis of Biogas Digestion, *Biogas Forum (BORDA)*, **II** No. **41**, 15-18 (1990).

Appendix II

DEFINITION AND NOTES

Acid Hydrolysis: It refers to decomposition or alteration of a chemical substance by water in the presence of acid.

Acidity: The measure of how many hydrogen ions a solution contains per unit volume; may be expressed in terms of pH.

Aerobic: It is a biological activity that takes place in presence of oxygen.

Alcohol: The family name of a group of organic chemical compounds composed of carbon, hydrogen and oxygen, a series of molecules that vary in chain length and are composed of a hydrocarbon plus a hydrogen group, $CH_3\text{-}(CH_2)_n\text{-}OH$. Some of the commonly used alcohols are methanol, ethanol, isopropyl alcohol and others.

Aldehydes: This covers a class of highly reactive organic chemical compounds obtained by controlled oxidation of primary alcohols and are characterised by the commcn group – CHO. These are mostly used for producing resins, dyes, and organic acids.

Ambient: It refers to the prevalent surrounding conditions usually expressed as functions of temperature, pressure and humidity.

Amino Acids: These are naturally occurring, nitrogen-containing building blocks of protein.

Amylaceous Feedstocks: These are generally items such as cereal grains and potatoes containing saccharides in the form of starches.

Amylase: These refer to class of enzymes that accelerate hydrolysis of starch and glycogen.

Anaerobic: It is a biological activity that takes place in oxygenless environment.

Anaerobic Digestion: A type of bacterial degradation of organic matter that occurs in absence oxygen and produces primarily methane and carbon dioxide.

Anhydrous: A compound that does not contain water either absorbed on its surface or as water of crystallization.

Atmospheric Pressure: It refers to pressure of surrounding air and atmosphere which keeps changing from day to day. It is taken as 14.7 psia.

Bacterial Spoilage: A phenomenon which occurs when bacterial contaminants take over the fermentation process in suppression of productive microorganisms.

Bagasse: A cellulosic residue left over following extraction of juice from sugar cane.

Barrel: A liquid measure equal to 42 American gallons or about 306 pounds of crude oil; one barrel equals 5.6 cubic feet or 0.159 cubic metres.

Basic Hydrolysis: Decomposition or alteration of a chemical substance by water in presence of an alkali.

Batch Fermentation: It refers to fermentation of a specific quantity of biodegradable material from start to finish in a single vessel.

Batch System: In this process plant is fed once in the beginning and no further material can be added or withdrawn during the cycle. After completion of digestion, digester is emptied and a new cycle started. To ensure constant gas production, several digesters are used and operated on a cyclical basis. Batch plants are especially suitable for treatment of straw type and fibrous material i.e. where previous homogenization is not possible or uneconomical.

Biogas: Gaseous product consisting of primarily methane and carbon dioxide obtained by anaerobic fermentation of organic materials. As methane is principal constituent of biogas, the term is also sometime loosely used as synonymous with methane. It is also known as *gobar* gas, sewer gas or marsh gas.

Biomass: It refers to organic matter such as trees, crops, animal wates and aquatic plants which is available on a renewable basis. It also includes wood , wood wastes and residues, municipal wastes, fruits and vegetable wastes, and several types of industrial wastes.

Biomass Fuel: It refers to any gaseous, liquid, or solid fuel derived trom of biomass.

BOD (Biochemical Oxygen Demand): It is a measure of organic water pollution potential examined through a biological procedure which attempts to simulate the natural processes of oxidation of organic matter.

BOD Test: It is used for monitoring the quality of domestic sewage, sewage effluents, river waters, industrial and biogas plant effluents. It is carried out by suitably diluting sample with aerated water and dividing the diluted sample between two glass-stoppered-bottles. The dissolved oxygen is determined immediately in one bottle, and in the second bottle after it has been incubated at a standard temperature of 20°C for 5 days; the period of 5 days is the time during which the rate of oxygen consumption would normally attain a minimum value under natural conditions though it may not be complete.

Boiling Point: It is the temperature at which transition from liquid to gaseous phase of a pure substance occurs at fixed pressure.

British Thermal Unit (BTu): A unit of heat energy defined as the amount of heat needed to raise the temperature of 1 pound of water by 1 °F under specified conditions of pressure and temperature (is equal to 252 calories, 778 foot-pound, 1055 joules, or 0.293 watt-hours).

Bulk Density: It refers to the mass (weight) of a material divided by the actual volume it displaces as a whole, expressed in lb/ft^3, kg/m^3 etc.

Calorie: It refers to the amount of heat needed to raise the temperature of one gram of water by one degree centigrade.

Carbohydrate: These are certain neutral chemical compounds made up of carbon, hydrogen and oxygen which include cellulose, starches and sugars. They have general chemical formula $C_x (H_2O)_y$.

Carbon-to-nitrogen Ratio (C : N): It is the ratio of carbon to nitrogen by weight in organic materials. Anaerobic bacteria produce maximum biogas when feed materials have a C/N ratio compatible with their metabolic need. If the ratio lies between 20 to 30, it is regarded as favourable.

Catalysis: Speeding effect caused by a small quantity of a substance (catalyst) on a chemical reaction in which catalyst remains unchanged.

Cellulase: It is an enzyme capable of decomposing cellulose into a simpler carbohydrates.

Cellulolytic: Property helpful for hydrolysis of cellulose by rendering it water-soluble.

Cellulose: Biological polymer which is composed of sugar molecules and provides the basic building material of plant fiber. Cellulose is basically the main polysaccharide in living plants, forms the skeletal structure of plant cells and can be hydrolysed to glucose. It has general chemical formula $(C_6H_{12}O_6)_n$.

Cellulosic Feedstocks: Materials such as wood, crop stalks etc. containing sugar units linked by bonds which are not easily rupturable.

Celsius (Centigrade): It is a temperature scale commonly used in all scientific work. At sea level water freezes at 0°C and boils at 100°C. Temperature in Celsius and Fahrenheit is given by the following relationship.

$$°C = \frac{5}{9} \times °(F - 32)$$

Centrifuge: A rotating device for separating liquids of different specific gravities or for separating suspended colloidal particles in particle-size fractions by centrifugal force.

Cetane Number: It is a measure of a fuel's ease of self-ignition; the higher the number the better the fuel has combustion characteristics.

COD (Chemical Oxygen Demand): It is measure of water pollution which determines oxygen consuming capacity of inorganic and organic matter present in water. This test gives a measure of the oxygen absorbed by a sample from boiling acidic potassium dichromate solution. Mercuric sulphate normally added to form an unoxidisable complex with any chloride present in the sample which would otherwise interfere. Excess dichromate is titrated with ferrous sulphate or ferrous ammonium sulphate solution, the amount of oxidisable organic matter being proportional to the potassium dichromate consumed.

Column: Vertical, cylindrical vessel containing a series of perforated plates or packed with materials through which vapours may pass, used to increase the degree of separation of liquid mixtures by distillation.

Compound: A chemical term denoting a specific combination of two or more distinct elements.

Concentration: It is given by ratio of mass or volume of solute present in a solution to the amount of solvent.

Continuous Fermentation: It is a steady state fermentation process which occurs without interruption due to a continuous feeding of plant every day and is accompanied by corresponding withdrawal of digested effluent at the same rate. In this process feed rates are adjusted to suit the desired retention time.

Continuous Plants: Plants in which feed materials are continuously added or withdrawn with volume of slurry kept at a constant level to ensure a constant gas supply. However, in such plants floating material creates a scum which hinders the gas flow. Continuous plants are specified according to the period of digestion and digester capacity corresponding to particular temperature and withdrawal rate.

C.P.: Candle Power which denotes light intensity.

Digester: Sealed tank or container in which the biological requirements of anaerobic digestion are controlled to achieve fermention and optimise biogas yield. Volume of digester is a function of plant capacity and retention time.

Digestion: Process by which complex organic molecules are broken down into simpler ones. In case of anaerobic fermentation this refers to the process by which anaerobic bacteria achieve this objective.

Dilution Rate: It is the reciprocal of the mean retention time of the flowing medium in the digester.

Effluent: It is the partially digested liquid manure or slurry which exists in the digester. During anaerobic fermentation, organic content of the influent slurry is reduced and stabilised making effluent as an almost odourless, homogenised liquid mass which does not attract rodents or flies. The biogas plant effluent is a resource almost all of the nitrogen that was originally present. The effluent is an excellent fertiliser as its nitrogen is more readily absorbed by plants than nitrogen in raw feed materials. Its addition thus improves biological, physical and chemical properties of soil.

Enzymes: The group of catalytic proteins that are produced by living micro-organisms. Enzymes facilitate decomposition of complex organic molecules into simpler ones even without themselves being altered or destroyed.

Evaporation: It is conversion of a liquid to vapour state by adding latent heat of vapourization. It generally refers to vapourization into the atmosphere.

Facultative (Anaerobe): A micro-organism that grows equally well under aerobic and anaerobic conditions.

Fahrenheit Scale: A temperature scale in which boiling point of water is 212° and its freezing point 32°.

Feedstock: The base feed material such as animal wastes which are fed to a plant to produce biogas through anaerobic fermentation. Feed concentration is expressed as percentage of dry matter or total solids in the feed.

Fermentation: Usually taken as synonym of digestion. Biological enzymatic transformation by which organic material is broken down into simpler constituents by micro-organisms such as yeasts; and generally accompanied by the evolution of gas.

Flame Lift Off: A flame which does not touch flame port of the stove. It can lift off so much that the flame goes out.

Flame Speed Factor: Measure of the speed at which a flame travels along a column of the gas.

Flash Point: It is the temperature at which a combustible mixture ignites when a flame is introduced.

Floculation: Aggregation of fine suspended particles to form floating clusters or dumps.

Gobar Gas: Alternative name for biogas where gas is formed from cow or buffalo dung (*Gobar* is the Hindi and Nepalese word used to denote dung).

Heat Exchanger: A device that transfers heat from one fluid (liquid or gas) to another or to the environment.

Heat Value: It denotes the amount of heat obtainable from a fuel. For a solid fuel it is expressed in kcal/kg or BTU/lb. Calorific value is the synonym of heat value.

Hydrolysis: Decomposition of a substance by chemically adding a water molecule to the unit at the point of bonding.

Influent: Raw material such as animal wastes or crop residues which are fed to produce biogas through anaerobic fermentation. The term is interchangably used for feed or feedstock.

Inoculum: Seed bacteria produced from a pure culture which is used to start a new culture. As it contains active micro-organisms, its addition to a substrate is helpful in starting and activating the fermentation.

Joule (J): The metric unit of energy with one BTU equal to 1055 joules. Since the joule is a very small unit, it is also expressed in kilojoules (KJ), a thousand joules, or megajoules (MJ).

Lignin: It is a polymeric, noncarbohydrate constituent of wood that functions as a binder and support for cellulose fibers. It is very much resistant to biodegradation.

Lipids: They refer to fatty materials.

Lipolytic: A property conducive for hydrolysis of lipid materials i.e. making them water-soluble.

Loading Rate: It refers to the rate at which amount of volatile solid is fed to a digester. The loading rate is usually expressed in weight of volatile solids per unit of digester volume per unit time. The volatile solids are the organic portion of the waste solids which are about 80 per cent of the raw waste (total solids). It is generally expressed as kg. VS/m^3. day.

Liquefaction: It refers to the change in the phase of a substance to liquid state. In the context of fermentation it refers to the conversion of water-insoluble carbohydrate to water-soluble carbohydrate.

Manure: It refers to animal faeces and urine, wasted feed, bedding, and anti-slip material obtainable from the barn and yard.

Marsh Gas: Alternative name for biogas as the gas is also formed in marshes.

Mesophilic: Bacteria which thrive in a temperature range around 95°F (35°C).

Methane: The simplest hydrocarbon consisting of one carbon atom and four hydrogen atoms (CH_4), a flammable, odourless gas. It is the largest constituent of both biogas and natural gas.

Methanogenic: A property conducive for producing methane.

Molecule: The chemical term for the smallest particle of matter which is chemically the same as the whole mass.

Municipal Solid Waste (MSW): It covers combined residential and commercial wastes generated within a municipal area and includes any materials, including food wastes which are discarded or rejected as spent, useless, or in excess and not wet enough to be free-flowing. These are also sometimes referred to as municipal refuse.

Net Energy Balance: It represents the amount of energy available from fuel when it is burnt, less the amount of energy it takes to produce the fuel.

Night-Soil: It refers to human faeces which are highly biodegradable and best-suited for anaerobic fermentation. It is however difficult to handle due to repelling stink involved.

Octane Number: It specifies a rating indicating the tendency to knock when a fuel is used in a standard internal combustion engine under standard conditions.

Pathogens: Organisms which can cause disease both in animals and human-beings e.g. typhoid. During anaerobic fermentation, most pathogens of night-soil are destroyed.

pH: It is a measure of acidity or alkalinity of a solution and describes free hydrogen ion concentration of a system; a solution of pH less than 7 is acidic; pH of 7 is neutral; pH over 7 is alkaline.

Phytotoxic: A material property poisonous to plants.

Polymer: A class of compounds made of molecules with long chains or cross-linked simple molecules.

Pounds Per Square Inch Absolute (PSIA): It indicates pressure at complete vacuum or 0 pressure.

Pounds Per Square Inch Absolute (PSIG): It is a measurement of pressure from above atmospheric pressure.

Protein: A class of high-molecular-weight polymer compounds comprising a variety of amino acids joined by peptide linkages.

Proteolytic: A property conducive for hydrolysis of proteins i.e. making them water-soluble.

Renewable Energy Sources: Energy resources that can be generated on renewable basis without depletion.

Retention Time: It represents the average time the slurry remains in digester. The volume of slurry in the digester remains constant; the incoming slurry displaces an equal amount of processed slurry from the digester each time it is loaded. Since volume is constant, fraction of digester's liquid volume replaced each day determines the retention time. For example, if slurry equalling one-tenth of the digester's liquid volume is added daily, digested slurry has an average retention time of 10 days. A very brief retention time does not allow the bacteria enough time to digest the organic matter, and a long retention time does not furnish enough fresh slurry to promote bacterial growth and consequent high gas output. Retention time can be estimated by dividing total weight of material in the system by weight removed per unit time such as hour, day, week etc.

Scrubbing: The process of cleaning a gas by removing chemical contaminants and particulates.

Scum: It is the waste material formed on top of the liquid slurry, and can be a severe problem in digesters receiving higher fatty or floatable material. Methane formation can be inhibited if the scum is allowed to accumulate and to set hard. Scum builds localised zone of acid accumulation which inhibits anaerobic fermentation.

Sludge: Pracipitated solid matter that gravitates at digester bottom.

Sludge Gas: Alternative name for biogas especially the one formed in sewerage.

Slurry: The liquid manure that flows out through plant outlet pipe. It is also called effluent and is primarily mixture of manure and water processed in the digester. Incidentally diluted input feed is called slurry.

Specific Gravity: It is the ratio of the mass of a solid or liquid to the mass of an equal volume of distilled water at 4°C.

Stoichiometric Ratio: It is the ratio of chemical substances necessary for a reaction to occur completely.

Substrate: It refers to material supplied for microbial action in a digester. It is used by the micro-organisms for its growth and metabolism ultimately leading to gas formation.

Supernatant: The liquid material which collects at the surface when a substrate is allowed to settle.

Thermophilic: Bacteria that are most active at temperature around 120°F (49°C - 60°C).

TKN: Total Kjeldahl Nitrogen, the amount of nitrogen obtained by Kjeldahl method of digesting organic material with sulfuric acid.

Total Solids: Weight of solid matter remaining after a sample is dried to constant weight at 102 ± 1 °C.

Toxins: Poisonous materials which inhibit anaerobic fermentation by killing micro-organism.

Volatile Acids: These are the intermediate materials produced in the digester by acid forming bacteria and used by methane-forming bacteria. These are essentially low molecular weight fatty acids.

Volatile Solids: Portion of solids volatilized at 550 ± 50°C; the difference between total solids content and ash remaining after ignition at 550 ± 50 °C. These are primarily the organic constituents of wastes.

Acknowledgement

Various definitions and notes given above have been framed or culled out from numerous sources comprising books, reports and papers published in various journals. Gratitude is conveyed to all the original authors who framed these definitions first. In particular, gratitude is conveyed to the following sources from which information was largely gathered for developing this glossary.

1. Barnett, A., Pyle, L. and Subramanian, S.K., *Biogas Technology in the Third World*, Inter. Dev. Res. Centre, Box 8500, Ottawa K16 3HP, Canada, 1978.
2. BORDA, *Biogas: Manual for the Realisation of Biogas Programmes*, GATE, P.P. Box 5180, 6236 Eschborn 1, Germany, 1980.
3. Chawla, O.P., *Advances in Biogas Technology*, Pub. and Inf. Div., ICAR, New Delhi, 1986.
4. Khandelwal, K.C., *Some Information about Biogas Technology and Developments in India*, A Report, Ministry of Agriculture, Krishi Bhawan, New Delhi, Sep., 1978.
5. Tarr, T.D., *Fuels from Farms: A Guide to Small-scale Ethanol Production*, SERI, Golden, Colorado, 1980, p. G1-G7.

Appendix III

IMPORTANT CONVERSION FACTORS

1. Weight

1 oz = 28.35 g

1 lb = 0.453592 kg = 0.009 cwt

1 kg = 2.20462 lb = 0.02 quintal

1 jin (Chinese) = 0.550 kg

1 cwt = 112 lb = 50.802 kg

1 metric tonne = 0.98421 long ton = 1.10231 short tons = 2204.6 lb

1 English or long ton = 1.01605 metric tonnes = 1.12 short tons

1 short ton = 0.892857 long ton = 0.907185 metric tonne = 2000 lb

2. Length

1 inch = 0.0254 m

1 ft = 0.333 yard = 12 inch = 0.305 m

1 yard = 3 ft = 36 inches = 0.914 m

1 m = 1.094 yards = 3.281 ft = 39.37 inches = 0.001 km

1 ft^2 = 0.093 m^2

1 square yard = 0.836 square meter

1 square meter = 1.196 square yard

1 mu (Chinese)[4] = 660 m^2 = 0.066 hectare

3. Volume

1 $inch^3$ = 16.387 cm^3

1 ft^3 = 28.317 litres = 7.4805 American gallons
= 6.2288 Imperial gallons = 0.17811 American barrel
= 0.028317 m^3

1 m^3 = 1000 litres = 264.17 American gallons
= 219.97 Imperial gallons = 6.2898 American barrels
= 35.315 ft^3

1 litre = 1000 cm^3 = 0.001 m^3 = 61.024 $inch^3$

= 0.26417 American gallon = 0.21997 Imperial gallon
= 0.035314 ft^3

1 kilolitre = 1000 litres = 6.2898 American barrels
1 hectolitre = 100 litres.

1 American gallon = 231 $inch^3$ = 0.133681 ft^3 = 0.0037854 m^3
= 0.0238095 American barrel
= 0.83268 Imperial gallon = 3.785 litres

1 Imperial gallon = 277.42 $inch^3$ = 0.160544 ft^3 = 0.004546 m^3
= 1.20094 American gallons = 0.028594 American barrel
= 4.55 litres

1 American barrel = 9702 $inch^3$ = 5.6146 ft^3 = 0.15899 m^3
= 158.99 litres = 42 American gallons
= 34.9726 Imperial gallons.

4. Miscellany

1 kcal = Amount of heat need to heat up 1 litre of water by 1°C
= 4187 Ws = 4187 J(1 Ws = 1 J = 1 Nm) = 3.968 BTU

1 kwh = performance of 1 kw for a period of 1 hour
= 860 kcal = 3.6×10^6 Ws = 3411 BTU = 1340 h.p. hour

1 kw = 1000 watts = 1.340 hp = 737 foot pounds per second

1 kgf/cm^2 = 0.0283 m^3 = 28.3 litres

1 m water column = 1 m Ws = 1 Mp/m^2 = 10 kp/cm^2 = 1 MN/m

1 atm = 1 kp/cm^2

1 lb/in^2 (psi) = 6.89 kN/m^2 = 0.0703 kgf/cm^2 = 27.7 in water gauge = 0.06805 atmospheres

1 kg/cm^2 = 14.22 lbs/m^2 = 0.9678 atm = 10,000 mm water gauge

1 mile = 1760 yards = 1.60934 km

1 international nautical mile = 6076.12 ft = 1.852 km

1 sq. km = 100 hectares = 0.386102 sq. miles

1 sq. mile = 640 acres = 258.999 hectares = 2.58999 km^2

1 acre (British) = 4047 m^2 = 4840 sq. yards = 0.404686 hectares

1 hectare = 10^4 m^2 = 2.47105 acres = 0.01 km^2

1 electron volt (eV) = $1.6021892 \times 10^{-19}$ J

1 cal = 4.18 J

1 kcal/min = 69.783 W = 9.351×10^{-2} h.p. = 3.9865 BTU

1 joule = 2.389×10^{-4} kcal = 2.389×10^{-7} kwh = 9.486×10^{-4} BTU

1 MJ = 238.8 kcal = 947.9 BTU

1 MJ/m^3 = 26.75 BTU/ft^3

1 BTU = 1.055 KJ = 0.252 kcal

1 BTU/ft^3 = 37.2 KJ/m^3 = 8.894 kcal/m^3

1 BTU/hour = 0.293 W

1 BTU/minute = 17.58 W = 2.356 × 10^{-2} h.p. = 0.252 kcal/min.

1 BTU/second = 1.06 kW

1 watt = 1.36 × 10^{-3} h.p. = 1.433 × 10^{-2} kcal/min = 5.688 × 10^{-2} BTU/min.

1 h.p. = 745.7 W = 10.68 kcal/min = 42.44 BTU/min

1 therm = 10^5 BTU = 25200 kcal = 105.5 MJ

1 tonne of crude oil = 7.5 barrels of crude oil

20,000 barrels of crude oil per day = 1 million tonne of oil per annum

1000 m^3 of natural gas = 1 tonne of oil equivalent = 35300 ft^3 of natural gas

1 tonne coal equivalent = 29.3 GJ (UN standard)

1 tonne oil equivalent = 42.6 GJ (UN standard)

1 ft^3 /lb = 0.0624 m^3/kg

1 m^3 /kg = 6.02 ft^3 /lb

Lumens = Candle power × 2

25 watt bulb ≈ 230 lumens

40 watt bulb ≈ 439 lumens

60 watt bulb ≈ 730 lumens

1 man hour = 250 kcal

1 woman hour = 200 kcal

1 child hour = 120 kcal

1 bullock hour = 2300 kcal

1 kg fuelwood = 3800 kcal

1 litre kerosene = 8980 kcal

1 barrel = 200 litres = 44 gallons

1 billion = 1000 million = 10^9

1 trillion = 1000 billion = 10^{12}

$$°F = °C \times 1.8 + 32°$$

$$°C = \frac{°F - 32}{1.8}$$

$$K = t + 273$$

Note :
1. **KJ is the short form of kilojoules which is equal to thousand joules.**
2. **MJ is the short form of megajoules which is equal to million joules.**
3. **GJ is the short form of gigajoules which is equal to 10^3 joules.**
4. **In Sichuan province of China mu is equal to 1.3 times that in other provinces, thus in this province 1 mu = 858 m^2 or 0.086 hectares.**

Author Index

Subject Index